Linear and Quasilinear Complex Equations of Hyperbolic and Mixed Type

Asian Mathematics Series
A Series edited by Chung-Chun Yang
Department of Mathematics, The Hong Kong University of Science and Technology, Hong Kong

Volume 1
Dynamics of transcendental functions
Xin-Hou Hua and Chung-Chun Yang

Volume 2
Approximate methods and numerical analysis for elliptic complex equations
Guo Chun Wen

Volume 3
Introduction to statistical methods in modern genetics
Mark C.K. Yang

Volume 4
Mathematical theory in periodic plane elasticity
Hai-Tao Cai and Jian-ke Lu

Volume 5
Gamma lines: On the geometry of real and complex functions
Grigor A. Barsegian

Volume 6
Linear and quasilinear complex equations of hyperbolic and mixed type
Guo Chun Wen

Linear and Quasilinear Complex Equations of Hyperbolic and Mixed Type

Guo Chun Wen

School of Mathematical Sciences,
Peking University, Beijing, China

CRC Press
Taylor & Francis Group

CRC Press
Taylor & Francis Group
6000 Broken Sound Parkway NW, Suite 300
Boca Raton, FL 33487-2742

First issued in paperback 2020

ISBN-13: 978-0-367-45480-7 (pbk)
ISBN-13: 978-0-415-26971-1 (hbk)

This book contains information obtained from authentic and highly regarded sources. Reasonable efforts have been made to publish reliable data and information, but the author and publisher cannot assume responsibility for the validity of all materials or the consequences of their use. The authors and publishers have attempted to trace the copyright holders of all material reproduced in this publication and apologize to copyright holders if permission to publish in this form has not been obtained. If any copyright material has not been acknowledged please write and let us know so we may rectify in any future reprint.

Trademark Notice: Product or corporate names may be trademarks or registered trademarks, and are used only for identification and explanation without intent to infringe.

Visit the Taylor & Francis Web site at
http://www.taylorandfrancis.com

and the CRC Press Web site at
http://www.crcpress.com

Every effort has been made to ensure that the advice and information in this book is true and accurate at the time of going to press. However, neither the publisher nor the authors can accept any legal responsibility or liability for any errors or omissions that may be made. In the case of drug administration, any medical procedure or the use of technical equipment mentioned within this book, you are strongly advised to consult the manufacturer's guidelines.

British Library Cataloguing in Publication Data
A catalogue record for this book is available from the British Library

Library of Congress Cataloging in Publication Data
A catalog record for this book has been requested

Contents

Second order quasilinear equations of mixed type   200

1   *Oblique derivative problems for second order
    quasilinear equations of mixed type 200*
2   *Oblique derivative problems for second order
    equations of mixed type in general domains 209*
3   *Discontinuous oblique derivative problems for second
    order quasilinear equations of mixed type 218*
4   *Oblique derivative problems for quasilinear equations
    of mixed type in multiply connected domains 227*

    References   240
    Index   250

Introduction to the Series

The *Asian Mathematics Series* provides a forum to promote and reflect timely mathematical research and development from the Asian region, and to provide suitable and pertinent reference on text books for researchers, academics and graduate students in Asian universities and research institutes, as well as in the West. With the growing strength of Asian economic, scientific and technological development, there is a need more than ever before for teaching and research materials written by leading Asian researchers, or those who have worked in or visited the Asian region, particularly tailored to meet the growing demands of students and researchers in that region. Many leading mathematicians in Asia were themselves trained in the West, and their experience with Western methods will make these books suitable not only for an Asian audience but also for the international mathematics community.

The *Asian Mathematics Series* is founded with the aim to present significant contributions from mathematicians, written with an Asian audience in mind, to the mathematics community. The series will cover all mathematical fields and their applications, with volumes contributed to by international experts who have taught or performed research in Asia. The material will be at graduate level or above. The book series will consist mainly of monographs and lecture notes but conference proceedings of meetings and workshops held in the Asian region will also be considered.

Preface

In this book, we mainly introduce first and second order complex equations of hyperbolic and mixed (elliptic-hyperbolic) type, in which various boundary value problems for first and second order linear and quasilinear complex equations of hyperbolic and mixed type are considered. In order to obtain the results on complex equations of mixed type, we need to first discuss some boundary value problems for elliptic and hyperbolic complex equations.

In Chapters I and II, the hyperbolic pseudoregular functions and quasi-hyperbolic mappings are introduced, which are corresponding to pseudoanalytic functions and quasiconformal mappings in the theory of elliptic complex equations. On the basis of hyperbolic notations, the hyperbolic systems of first order equations and hyperbolic equations of second order with some conditions can be reduced to complex forms. In addition, several boundary value problems, mainly the Riemann–Hilbert problem, oblique derivative problems for some hyperbolic complex equations of first and second order are discussed in detail.

In Chapter III, firstly the generalizations of the Keldych–Sedov formula for analytic functions are given. Moreover, discontinuous boundary value problems for nonlinear elliptic complex equations of first and second order are discussed. Besides some oblique derivative problems for degenerate elliptic equations of second order are also introduced.

In Chapter IV, we mainly consider the discontinuous boundary value problems for first order linear and quasilinear complex equations of mixed type, which include the discontinuous Dirichlet problem and discontinuous Riemann–Hilbert problem. In the meantime we give some a priori estimates of solutions for the above boundary value problems.

For the classical dynamical equation of mixed type due to S. A. Chaplygin [17], the first really deep results were published by F. Tricomi [77] 1). In Chapters V and VI, we consider oblique derivative boundary value problems for second order linear and quasilinear complex equations of mixed type by using a complex analytic method in a special domain and in general domains, which include the Dirichlet problem (Tricomi problem) as a special case. We mention that in the books [12] 1), 3), the author investigated the Dirichlet problem (Tricomi problem) for the simplest second order equation of mixed type, i.e. $u_{xx} + \mathrm{sgn} y\, u_{yy} = 0$ in

Preface

general domains by using the method of integral equations and a complicated functional relation. In the present book, we use the uniqueness and existence of solutions of discontinuous Riemann–Hilbert boundary value problem for elliptic complex equations and other methods to obtain the solvability result of oblique derivative problems for more general equations and domains, which includes the results in [12] 1), 3) as special cases.

Similarly to the book [86] 1), the considered complex equations and boundary conditions in this volume are rather general, and several methods are used. There are two characteristics of this book: one is that mixed complex equations are included in the quasilinear case, and boundary value conditions are almost considered in the general oblique derivative case, especially multiply connected domains are considered. Another one is that complex analytic methods are used to investigate various problems about complex equations of hyperbolic and mixed type. We mention that some free boundary problems in gas dynamics and some problem in elasticity can be handled by using the results stated in this book.

The great majority of the contents originates in investigations of the author and his cooperative colleagues, and many results are published here for the first time. After reading the book, it can be seen that many questions about complex equations of mixed type remain for further investigations.

The preparation of this book was supported by the National Natural Science Foundation of China. The author would like to thank Prof. H. Begehr, Prof. W. Tutschke and Mr Pi Wen Yang, because they proposed some beneficial improving opinions to the manuscript of this book.

Beijing   Guo Chun Wen
August, 2001   Peking University

CHAPTER I

HYPERBOLIC COMPLEX EQUATIONS OF FIRST ORDER

In this chapter, we first introduce hyperbolic numbers, hyperbolic regular functions and hyperbolic pseudoregular functions. Next we transform the linear and non-linear hyperbolic systems of first order equations into complex forms. Moreover we discuss boundary value problems for some hyperbolic complex equations of first order. Finally, we introduce the so-called hyperbolic mappings and quasihyperbolic mappings.

1 Hyperbolic Complex Functions and Hyperbolic Pseudoregular Functions

1.1 Hyperbolic numbers and hyperbolic regular functions

First of all, we introduce hyperbolic numbers and hyperbolic complex functions. The so-called hyperbolic number is $z = x + jy$, where x, y are two real numbers and j is called the hyperbolic unit such that $j^2 = 1$. Denote

$$e_1 = (1 + j)/2, \qquad e_2 = (1 - j)/2, \tag{1.1}$$

it is easy to see that

$$e_1 + e_2 = 1, \qquad e_k e_l = \begin{cases} e_k, & \text{if } k = l, \\ 0, & \text{if } k \neq l, \end{cases} \quad k, l = 1, 2, \tag{1.2}$$

and (e_1, e_2) will be called the hyperbolic element. Moreover, $w = f(z) = u(x, y) + jv(x, y)$ is called a hyperbolic complex function, where $u(x, y), v(x, y)$ are two real functions of two real variables x, y which are called the real part and imaginary part of $w = f(z)$ and denote $\operatorname{Re} w = u(z) = u(x, y)$, $\operatorname{Im} w = v(z) = v(x, y)$. Obviously,

$$z = x + jy = \mu e_1 + \nu e_2, \qquad w = f(z) = u + jv = \xi e_1 + \eta e_2, \tag{1.3}$$

in which

$$\mu = x + y, \qquad \nu = x - y, \qquad x = (\mu + \nu)/2, \qquad y = (\mu - \nu)/2,$$
$$\xi = u + v, \qquad \eta = u - v, \qquad u = (\xi + \eta)/2, \qquad v = (\xi - \eta)/2.$$

$\bar{z} = x - jy$ will be called the conjugate number of z. The absolute value of z is defined by $|z| = \sqrt{|x^2 - y^2|}$, and the hyperbolic model of z is defined by $\|z\| = \sqrt{x^2 + y^2}$. The operations of addition, subtraction and multiplication are the same with the real numbers, but $j^2 = 1$. There exists the divisor of zero and denote by $O = \{z \mid x^2 = y^2\}$ the set of divisors of zero and zero. It is clear that $z \in O$ if and only if $|z| = 0$, and z has an inversion

$$\frac{1}{z} = \frac{\bar{z}}{z\bar{z}} = \frac{1}{x+y}e_1 + \frac{1}{x-y}e_2 = \frac{1}{\mu}e_1 + \frac{1}{\nu}e_2,$$

if and only if $x + jy \notin O$, and if the hyperbolic numbers $z_1 = \mu_1 e_1 + \nu_1 e_2$, $z_2 = \mu_2 e_1 + \nu_2 e_2 \notin O$, then

$$\frac{z_1}{z_2} = (\mu_1 e_1 + \nu_1 e_2)\left(\frac{1}{\mu_2}e_1 + \frac{1}{\nu_2}e_2\right) = \frac{\mu_1}{\mu_2}e_1 + \frac{\nu_1}{\nu_2}e_2.$$

It is clear that $|z_1 z_2| = |z_1||z_2|$, but the triangle inequality is not true. As for the hyperbolic model of z, we have the triangle inequality: $\|z_1 + z_2\| \le \|z_1\| + \|z_2\|$, and $\|z_1 z_2\| \le \sqrt{2}\,\|z_1\|\|z_2\|$. In the following, the limits of the hyperbolic number are defined by the hyperbolic model. The derivatives of a hyperbolic complex function $w = f(z)$ with respect to z and $\bar{z}$ are defined by

$$w_z = (w_x + jw_y)/2, \quad w_{\bar{z}} = (w_x - jw_y)/2, \tag{1.4}$$

respectively, and then we have

$$w_{\bar{z}} = (w_x - jw_y)/2 = [(u_x - v_y) + j(v_x - u_y)]/2$$
$$= [(w_x - w_y)e_1 + (w_x + w_y)e_2]/2 = w_\nu e_1 + w_\mu e_2$$
$$= [\xi_\nu e_1 + \eta_\nu e_2]e_1 + [\xi_\mu e_1 + \eta_\mu e_2]e_2 = \xi_\nu e_1 + \eta_\mu e_2, \tag{1.5}$$
$$w_z = [(u_x + v_y) + j(v_x + u_y)]/2 = w_\mu e_1 + w_\nu e_2$$
$$= (\xi e_1 + \eta e_2)_\mu e_1 + (\xi e_1 + \eta e_2)_\nu e_2 = \xi_\mu e_1 + \eta_\nu e_2.$$

Let D be a domain in the (x, y)-plane. If $u(x, y), v(x, y)$ are continuously differentiable in D, then we say that the function $w = f(z)$ is continuously differentiable in D, and we have the following result.

Theorem 1.1 *Suppose that the hyperbolic complex function $w = f(z)$ is continuously differentiable. Then the following three conditions are equivalent.*

(1) $w_{\bar{z}} = 0$;   (1.6)

(2) $\xi_\nu = 0$, $\eta_\mu = 0$;   (1.7)

(3) $u_x = v_y$, $v_x = u_y$.   (1.8)

Proof From (1.5), it is easy to see that the conditions (1),(2) and (3) in Theorem 1.1 are equivalent.

The system of equations (1.8) is the simplest hyperbolic system of first order equations, which corresponds to the Cauchy–Riemann system in the theory of elliptic equations. The continuously differentiable solution $w = f(z)$ of the complex equation (1.6) in D is called a hyperbolic regular function in D.

If the function $w(z)$ is defined and continuous in the neighborhood of a point z_0, and the following limit exists and is finite:

$$w'(z_0) = \lim_{z \to z_0} \frac{w(z) - w(z_0)}{z - z_0}$$

$$= \lim_{\mu \to \mu_0, \nu \to \nu_0} \left[\frac{\xi(z) - \xi(z_0)}{\mu - \mu_0} e_1 + \frac{\eta(z) - \eta(z_0)}{\nu - \nu_0} e_2 \right]$$

$$= [\xi_\mu e_1 + \eta_\nu e_2]|_{\mu = \mu_0, \nu = \nu_0} = w_z(z_0),$$

then we say that $w(z)$ possesses the derivative $w'(z_0)$ at z_0. From the above formula, we see that $w(z)$ possesses a derivative at z_0 if and only if $\xi(z) = \operatorname{Re} w(z) + \operatorname{Im} w(z)$, $\eta(z) = \operatorname{Re} w(z) - \operatorname{Im} w(z)$ possess derivatives at $\mu_0 = x_0 + y_0, \nu_0 = x_0 - y_0$ respectively.

Now, we can define some elementary hyperbolic regular functions according to series representations in their convergent domains as follows:

$$z^n = [\mu e_1 + \nu e_2]^n = \mu^n e_1 + \nu^n e_2 = \frac{(x+y)^n + (x-y)^n}{2} + j \frac{(x+y)^n - (x-y)^n}{2},$$

$$e^z = 1 + z + \frac{z^2}{2!} + \cdots + \frac{z^n}{n!} + \cdots = e^\mu e_1 + e^\nu e_2 = \frac{e^{x+y} + e^{x-y}}{2} + j \frac{e^{x+y} - e^{x-y}}{2},$$

$$\ln z = \ln \mu \, e_1 + \ln \nu \, e_2 = \frac{\ln(x+y) + \ln(x-y)}{2} + j \frac{\ln(x+y) - \ln(x-y)}{2},$$

$$\sin z = z - \frac{z^3}{3!} + \cdots + (-1)^n \frac{z^{2n+1}}{(2n+1)!} + \cdots = \sin \mu \, e_1 + \sin \nu \, e_2,$$

$$\cos z = 1 - \frac{z^2}{2!} + \cdots + (-1)^n \frac{z^{2n}}{(2n)!} + \cdots = \cos \mu \, e_1 + \cos \nu \, e_2,$$

$$\operatorname{tg} z = \frac{\sin z}{\cos z} = (\sin \mu e_1 + \sin \nu e_2) \left(\frac{1}{\cos \mu} e_1 + \frac{1}{\cos \nu} e_2 \right) = \operatorname{tg} \mu \, e_1 + \operatorname{tg} \nu \, e_2,$$

$$\operatorname{ctg} z = \frac{\cos z}{\sin z} = (\cos \mu e_1 + \cos \nu e_2) \left(\frac{1}{\sin \mu} e_1 + \frac{1}{\sin \nu} e_2 \right) = \operatorname{ctg} \mu \, e_1 + \operatorname{ctg} \nu \, e_2,$$

$$(1+z)^\alpha = 1 + \alpha z + \cdots + \frac{\alpha(\alpha-1) \dots (\alpha-n+1)}{n!} z^n + \cdots = (1+\mu)^\alpha e_1 + (1+\nu)^\alpha e_2,$$

where n is a positive integer and α is a positive number. Moreover we can define the series expansion of hyperbolic regular functions and discuss its convergency.

1.2 Hyperbolic continuous functions and their integrals

Suppose that $w = f(z) = u(x, y) + jv(x, y)$ is any hyperbolic complex function in a domain D and possesses continuous partial derivatives of first order in D. Then for any point $z_0 \in D$, we have

$$\Delta w = f_x'(z_0)\Delta x + f_y'(z_0)\Delta y + \varepsilon(\Delta z),$$

where

$$\Delta w = f(z) - f(z_0),\ f_x'(z_0) = u_x(z_0) + jv_x(z_0),\ f_y'(z_0) = u_y(z_0) + jv_y(z_0),$$

and $z = z_0 + \Delta z$, ε is a function of Δz and

$$\lim_{\Delta z \to 0} \varepsilon(\Delta z) \to 0.$$

Suppose that C is a piecewise smooth curve in the domain D, and $w = f(z) = u + jv = \xi e_1 + \eta e_2$ is a continuous function in D. Then the integral of $f(z)$ along C and D are defined by

$$\int_C f(z)dz = \int_C udx + vdy + j[\int_C vdx + udy] = \int_C [\xi d\mu e_1 + \eta dv e_2],$$

$$\iint_D f(z)dxdy = \iint_D udxdy + j\iint_D vdxdy.$$

We easily obtain some properties of integrals of $f(z)$ as follows.

Theorem 1.2   (1) *If $f(z), g(z)$ are continuous functions in D and C is a piecewise smooth curve in D, then*

$$\int_C [f(z) + g(z)]dz = \int_C f(z)dz + \int_C g(z)dz,$$

$$\iint_D [f(z) + g(z)]dxdy = \iint_D f(z)dxdy + \iint_D g(z)dxdy.$$

(2) *Under the conditions in (1) and denoting $M_1 = \max_{z \in C} \| f(z) \|$, $M_2 = \sup_{z \in D} \| f(z) \|$, the length of C by l and the area of D by S, then*

$$\| \int_C f(z)dz \| \le \sqrt{2}M_1 l,$$

$$\| \iint_D f(z)dxdy \| \le \sqrt{2}M_2 S.$$

(3) *If C is a piecewise smooth closed curve and G is the finite domain bounded by C, $f(z)$ is continuously differentiable in $\bar{G}$, then we have Green's formulas*

$$\iint_G [f(z)]_{\bar{z}}dxdy = \frac{j}{2}\int_C f(z)dz,$$

$$\iint_G [f(z)]_z dxdy = -\frac{j}{2}\int_C f(z)d\bar{z}.$$

(4) *Under the conditions as in* (3) *of Theorem* 1.1 *and* $w = f(z)$ *is a hyperbolic regular function in* $\bar{G}$, *then*

$$\int_C f(z)dz = 0.$$

In the following, we introduce the definition of hyperbolic pseudoregular functions, and prove some properties of hyperbolic pseudoregular functions.

1.3 Hyperbolic pseudoregular functions and their properties

Let $w(z), F(z), G(z)$ be continuous functions in a domain D and $G(z), F(z)$ satisfy the conditions

$$\text{Im } F(z)\overline{G(z)} \neq 0 \text{ in } D. \tag{1.9}$$

Then for every point $z_0 \in D$, we can obtain a unique pair of real numbers δ_0 and γ_0, such that

$$w(z_0) = \delta_0 F(z_0) + \gamma_0 G(z_0). \tag{1.10}$$

Setting

$$W(z) = w(z) - \delta_0 F(z) - \gamma_0 G(z), \tag{1.11}$$

it is easy to see that

$$W(z_0) = 0. \tag{1.12}$$

If the following limit exists and is finite:

$$\dot{w}(z_0) = \lim_{z \to z_0} \frac{w(z) - \delta_0 F(z) - \gamma_0 G(z)}{z - z_0} = \lim_{z \to z_0} \frac{W(z) - W(z_0)}{z - z_0}, \text{ i.e.}$$

$$\dot{w}(z_0) = [\text{Re } W(z_0) + \text{Im } W(z_0)]_\mu e_1 + [\text{Re } W(z_0) - \text{Im } W(z_0)]_\nu e_2 = W'(z_0), \tag{1.13}$$

where $\mu = x + y, \nu = x - y$, then we say that $\dot{w}(z)$ is a (F, G)-derivative of $w(z)$ at z_0. In order to express the existence of (1.13) by partial differential equations, we suppose again that

$$F_z(z), F_{\bar{z}}(z), G_z(z) \text{ and } G_{\bar{z}}(z) \text{ exist and are continuous} \tag{1.14}$$

in a neighborhood of z_0. According to the definition of $W(z)$, if $W_z, W_{\bar{z}}$ exist, then

$$W_z(z) = w_z(z) - \delta_0 F_z(z) - \gamma_0 G_z(z),$$
$$W_{\bar{z}}(z) = w_{\bar{z}}(z) - \delta_0 F_{\bar{z}}(z) - \gamma_0 G_{\bar{z}}(z). \tag{1.15}$$

From (1.13), (1.15) and Theorem 1.1, we see that if $\dot{w}(z_0)$ exists, then $W_z(z_0)$ exists,

$$W_z(z_0) = \dot{w}(z_0), \tag{1.16}$$

and

$$W_{\bar{z}}(z_0) = 0; \tag{1.17}$$

and if $w_z(z)$, $w_{\bar{z}}(z)$ are continuous in a neighborhood of z_0, and (1.17) holds, then we have (1.13) and (1.16). Since

$$W(z) = \frac{\begin{vmatrix} w(z) & w(z_0) & \overline{w(z_0)} \\ F(z) & F(z_0) & \overline{F(z_0)} \\ G(z) & G(z_0) & \overline{G(z_0)} \end{vmatrix}}{\begin{vmatrix} F(z_0) & \overline{F(z_0)} \\ G(z_0) & \overline{G(z_0)} \end{vmatrix}}, \tag{1.18}$$

(1.17) can be rewritten as

$$\begin{vmatrix} w_{\bar{z}}(z_0) & w(z_0) & \overline{w(z_0)} \\ F_{\bar{z}}(z_0) & F(z_0) & \overline{F(z_0)} \\ G_{\bar{z}}(z_0) & G(z_0) & \overline{G(z_0)} \end{vmatrix} = 0. \tag{1.19}'$$

If (1.13) exists, then (1.16) can be written as

$$\dot{w}(z_0) = \frac{\begin{vmatrix} w_z(z_0) & w(z_0) & \overline{w(z_0)} \\ F_z(z_0) & F(z_0) & \overline{F(z_0)} \\ G_z(z_0) & G(z_0) & \overline{G(z_0)} \end{vmatrix}}{\begin{vmatrix} F(z_0) & \overline{F(z_0)} \\ G(z_0) & \overline{G(z_0)} \end{vmatrix}}. \tag{1.20}$$

Unfolding (1.19) and (1.20) respectively, and arranging them, we obtain

$$w_{\bar{z}} = aw + b\bar{w}, \tag{1.21}$$

$$\dot{w} = w_z - Aw - B\bar{w}, \tag{1.22}$$

where

$$a = -\frac{\bar{F}G_{\bar{z}} - F_{\bar{z}}\bar{G}}{F\bar{G} - \bar{F}G}, \quad b = \frac{FG_{\bar{z}} - F_{\bar{z}}G}{F\bar{G} - \bar{F}G},$$

$$A = -\frac{\bar{F}G_z - F_z\bar{G}}{F\bar{G} - \bar{F}G}, \quad B = \frac{FG_z - F_zG}{F\bar{G} - \bar{F}G}, \tag{1.23}$$

here $a(z), b(z), A(z)$ and $B(z)$ are called the characteristic coefficients of the generating pair (F, G). Obviously $\dot{F} = \dot{G} = 0$, and

$$F_{\bar{z}} = aF + b\bar{F}, \quad G_{\bar{z}} = aG + b\bar{G},$$

$$F_z = AF + B\bar{F}, \quad G_z = AG + B\bar{G}$$

uniquely determine a, b, A and B. Denote them by $a_{(F,G)}, b_{(F,G)}, A_{(F,G)}$ and $B_{(F,G)}$ respectively.

From the above discussion, we see that if $\dot{w}(z_0)$ exists, then w_z at z_0 exists and (1.21), (1.22) are true. If w_z and $w_{\bar{z}}(z)$ exist and are continuous in a neighborhood $z_0 \in D$, and (1.21) holds at z_0, then $\dot{w}(z_0)$ exists, and (1.22) is true.

For any function $w(z)$, if $\dot{w}(z)$ exists and is continuous in the domain D, then $w(z)$ is called the first-class (F, G) hyperbolic pseudoregular function or hyperbolic pseudoregular function for short. It is clear that the following theorem holds.

Theorem 1.3 $w(z)$ *is a hyperbolic pseudoregular function if and only if* $w_z(z)$ *and* $w_{\bar{z}}(z)$ *exist and are continuous, and* (1.21) *holds.*

By (1.9), it is easy to see that every function $w(z)$ has a unique expression

$$w(z) = \phi(z)F(z) + \psi(z)G(z), \tag{1.24}$$

where $\phi(z)$ and $\psi(z)$ are two real-valued functions. Let

$$K(z) = \phi(z) + j\psi(z). \tag{1.25}$$

Then we can give the following definition.

If $w(z)$ is the first-class (F, G) hyperbolic pseudoregular complex function, then $K(z) = \phi(z) + j\psi(z)$ is called the second-class (F, G) hyperbolic pseudoregular function.

Theorem 1.4 $K(z) = \phi(z) + j\psi(z)$ *is a second-class* (F, G) *hyperbolic pseudoregular function if and only if* ϕ *and* ψ *have continuous partial derivatives and*

$$F\phi_{\bar{z}} + G\psi_{\bar{z}} = 0. \tag{1.26}$$

Under this condition,

$$\dot{w}(z) = F\phi_z + G\psi_z \tag{1.27}$$

holds, where

$$
\begin{aligned}
\phi_z &= (e_1\phi_\mu + e_2\phi_\nu) = [(\phi_x + \phi_y)e_1 + (\phi_x - \phi_y)e_2]/2, \\
\phi_{\bar{z}} &= (e_1\phi_\nu + e_2\phi_\mu) = [(\phi_x - \phi_y)e_1 + (\phi_x + \phi_y)e_2]/2, \\
\psi_z &= (e_1\psi_\mu + e_2\psi_\nu) = [(\psi_x + \psi_y)e_1 + (\psi_x - \psi_y)e_2]/2, \\
\psi_{\bar{z}} &= (e_1\psi_\nu + e_2\psi_\mu) = [(\psi_x - \psi_y)e_1 + (\psi_x + \psi_y)e_2]/2.
\end{aligned}
\tag{1.28}
$$

Proof From

$$W(z) = [\phi(z) - \phi(z_0)]F(z) + [\psi(z) - \psi(z_0)]G(z),$$

it follows that

$$W_z(z_0) = F(z_0)\phi_z(z_0) + G(z_0)\psi_z(z_0),$$
$$W_{\bar{z}}(z_0) = F(z_0)\phi_{\bar{z}}(z_0) + G(z_0)\psi_{\bar{z}}(z_0).$$

Thus the proof can be immediately obtained.

Setting

$$-\frac{G}{F} = \sigma + j\tau = (\sigma + \tau)e_1 + (\sigma - \tau)e_2,$$

$$-\frac{F}{G} = \tilde{\sigma} + j\tilde{\tau} = (\tilde{\sigma} + \tilde{\tau})e_1 + (\tilde{\sigma} - \tilde{\tau})e_2,$$

(1.29)

where $\sigma, \tau, \tilde{\sigma}$ and $\tilde{\tau}$ are real-valued functions, and $\tau \neq 0$, $\tilde{\tau} \neq 0$. Hence (1.26) is equivalent to the system of equations

$$\phi_x = \sigma\psi_x - \tau\psi_y, \quad \phi_y = -\tau\psi_x + \sigma\psi_y.$$

(1.30)

If ϕ and ψ have continuous partial derivatives up to second order, we find the derivatives with respect to x and y in (1.30), and then obtain

$$\phi_{xx} - \phi_{yy} + \tilde{\delta}\phi_x + \tilde{\gamma}\phi_y = 0,$$

$$\psi_{xx} - \psi_{yy} + \delta\psi_x + \gamma\psi_y = 0,$$

(1.31)

where

$$\tilde{\delta} = \frac{\tilde{\sigma}_y + \tilde{\tau}_x}{\tilde{\tau}}, \qquad \tilde{\gamma} = -\frac{\tilde{\sigma}_x + \tilde{\tau}_y}{\tilde{\tau}},$$

$$\delta = \frac{\sigma_y + \tau_x}{\tau}, \qquad \gamma = -\frac{\sigma_x + \tau_y}{\tau}.$$

(1.32)

In accordance with the following theorem, the elimination is reasonable.

Theorem 1.5 *Let $\delta, \gamma, \tilde{\delta}$ and $\tilde{\gamma}$ be determined by (1.32). Then the real-valued function ϕ (ψ) is the real (imaginary) part of a second-class hyperbolic pseudoregular function if and only if it has continuous partial derivatives up to second order, and satisfies the first (second) equation in (1.31).*

Proof From the second formula in (1.31), we see that the function

$$\phi(z) = \int_{z_0}^z [(\sigma\psi_x - \tau\psi_y)dx + (-\tau\psi_x + \sigma\psi_y)dy], \quad z_0, z \in D$$

is single-valued, and $\phi(z), \psi(z)$ satisfy system (1.30). The part of necessity can be derived from Theorem 2.3, Chapter II below.

1.4 Existence of a generating pair (F,G)

Theorem 1.6 *Let $a(z)$ and $b(z)$ be two continuous functions in a bounded and closed domain $D = \{\mu_0 \leq \mu \leq \mu_0 + R_1, \nu_0 \leq \nu \leq \nu_0 + R_2\}$, where R_1, R_2 are positive constants, and denote $z_0 = \mu_0 e_1 + \nu_0 e_2$. Then there exists a unique continuously differentiable hyperbolic pseudoregular function $w(z)$ satisfying the complex equation*

$$w_{\tilde{z}} = a(z)w(z) + b(z)\overline{w(z)},$$

(1.33)

and the boundary conditions

$$w(z) = c_1(\mu)e_1 + c_2(\nu_0)e_2, \quad \text{when} \quad z \in L_1,$$
$$w(z) = c_1(\mu_0)e_1 + c_2(\nu)e_2, \quad \text{when} \quad z \in L_2, \quad (1.34)$$

where $c_1(\mu)$ and $c_2(\nu)$ are two real continuous functions on L_1, L_2 respectively, $L_1 = \{\mu_0 \le \mu \le \mu_0 + R_1, \nu = \nu_0\}$ and $L_2 = \{\mu = \mu_0, \nu_0 \le \nu \le \nu_0 + R_2\}$.

The theorem is a special case of Theorems 3.3 and 3.4 below.

Theorem 1.7 *Let $a(z)$ and $b(z)$ be two continuous complex functions in the domain D as stated in Theorem 1.6. Then there exists a generating pair (F, G) in D, such that*

$$a = a_{(F,G)} \quad \text{and} \quad b = b_{(F,G)}. \quad (1.35)$$

Proof Denote by $F(z)$ and $G(z)$ two solutions of the complex equation (1.33) satisfying the boundary conditions

$$w(z) = e_1 + e_2 = 1, \quad \text{when} \quad z \in L_1 \cup L_2$$

and

$$w(z) = e_1 - e_2 = j, \quad \text{when} \quad z \in L_1 \cup L_2,$$

respectively. Then by Theorem 1.6, $F(z)$ and $G(z)$ have continuous partial derivatives, and

$$F_{\bar{z}} = aF + b\bar{F}, \quad \text{and} \quad G_{\bar{z}} = aG + b\bar{G},$$

$$F = 1, \ G = j, \quad \text{when} \quad z \in L_1 \cup L_2.$$

Hence $a = a_{(F,G)}$ and $b = b_{(F,G)}$. Whether $\text{Im } F(z)\overline{G(z)} \ne 0$ in D remains to be discussed.

Theorem 1.8 *Under the same conditions as in Theorem 1.7, and letting*

$$b = -a \quad \text{or} \quad b = a, \qquad z \in D, \quad (1.36)$$

there exists a generating pair (F, G) in D satisfying the complex equation (1.33), and

$$F(z) = 1 \quad \text{or} \quad G(z) = j \quad \text{in} \quad D. \quad (1.37)$$

Proof By the hypotheses in Theorem 1.7 and equation (1.36), there exists a unique generating pair (F, G) in D satisfying the conditions

$$F_{\bar{z}} = a(F - \bar{F}), \quad \text{or} \quad G_{\bar{z}} = a(G + \bar{G}), \qquad z \in D,$$

$$F(z) = 1 \quad \text{or} \quad G(z) = j, \quad \text{when} \quad z \in L_1 \cup L_2.$$

Hence we have (1.37). The above results are similar to those in [9]1) (see [89]).

2 Complex Forms of Linear and Nonlinear Hyperbolic Systems of First Order Equations

In this section, we transform linear and nonlinear hyperbolic systems of first order equations into complex forms.

2.1 Complex forms of linear hyperbolic systems of first order equations

We consider the linear hyperbolic system of first order partial differential equations

$$\begin{cases} a_{11}u_x + a_{12}u_y + b_{11}v_x + b_{12}v_y = a_1 u + b_1 v + c_1, \\ a_{21}u_x + a_{22}u_y + b_{21}v_x + b_{22}v_y = a_2 u + b_2 v + c_2, \end{cases} \tag{2.1}$$

where the coefficients $a_{kl}, b_{kl}, a_k, b_k, c_k$ $(k, l = 1, 2)$ are known functions in D, in which D is a bounded domain. System (2.1) is called hyperbolic at a point in D, if at the point, the inequality

$$I = (K_2 + K_3)^2 - 4K_1 K_4 > 0 \tag{2.2}$$

holds, in which

$$K_1 = \begin{vmatrix} a_{11} & b_{11} \\ a_{21} & b_{21} \end{vmatrix}, \quad K_2 = \begin{vmatrix} a_{11} & b_{12} \\ a_{21} & b_{22} \end{vmatrix}, \quad K_3 = \begin{vmatrix} a_{12} & b_{11} \\ a_{22} & b_{21} \end{vmatrix}, \quad K_4 = \begin{vmatrix} a_{12} & b_{12} \\ a_{22} & b_{22} \end{vmatrix}.$$

If the inequality (2.2) at every point (x, y) in D holds, then (2.1) is called a hyperbolic system in D. We can verify that (2.2) can be rewritten as

$$I = (K_2 - K_3)^2 - 4K_5 K_6 > 0, \tag{2.3}$$

where

$$K_5 = \begin{vmatrix} a_{11} & a_{12} \\ a_{21} & a_{22} \end{vmatrix}, \quad K_6 = \begin{vmatrix} b_{11} & b_{12} \\ b_{21} & b_{22} \end{vmatrix}.$$

If the coefficients a_{kl}, b_{kl} $(k, l = 1, 2)$ in D are bounded, and the condition

$$I = (K_2 + K_3)^2 - 4K_1 K_4 = (K_2 - K_3)^2 - 4K_5 K_6 \geq I_0 > 0 \tag{2.4}$$

holds, in which I_0 is a positive constant, then (2.1) is called uniformly hyperbolic in D. In the following, we reduce system (2.1) to complex form.

1) If K_2, K_3 are of same signs and $K_6 \neq 0$ at the point $(x, y) \in D$, then we can solve (2.1) for v_y, $-v_x$ and obtain the system of first order equations

$$\begin{cases} v_y = a u_x + b u_y + a_0 u + b_0 v + f_0, \\ -v_x = d u_x + c u_y + c_0 u + d_0 v + g_0, \end{cases} \tag{2.5}$$

where a, b, c, d are known functions of $a_{kl}, b_{kl}(k, l = 1, 2)$ and $a_0, b_0, c_0, d_0, f_0, g_0$ are known functions of $b_{kl}, a_k, b_k, c_k(k, l = 1, 2)$, and

$$a = K_1/K_6, \quad b = K_3/K_6, \quad c = K_4/K_6, \quad d = K_2/K_6.$$

The condition (2.2) of hyperbolic type is reduced to the condition

$$\Delta = I/4K_6^2 = \frac{(b+d)^2}{4} - ac > 0. \tag{2.6}$$

There is no harm in assuming that $a - c \geq 0$, because otherwise let y be replaced by $-y$, this requirement can be realized. If a, c are not of the same sign or one of them is equal to zero, then $-ac \geq 0$, $bd \geq 0$, and may be such that $a \geq 0$, $-c \geq 0$; or a, c are of same signs, then we may assume that $a > 0$, $c > 0$, because otherwise if v is replaced by $-v$, this requirement can be realized. Moreover, we can assume that $0 < c < 1$, otherwise, setting $v = h\tilde{v}$, herein h is a positive constant such that $h \geq c + 1$, we have $\tilde{K}_4 = hK_4$, $\tilde{K}_6 = h^2 K_6$, and $\tilde{c} = \tilde{K}_4/\tilde{K}_6 = c/h < 1$, $\tilde{b}\tilde{d} \geq 0$. Multiply the first formula of (2.5) by $-j$ and then subtract the second formula of (2.5). This gives

$$v_x - jv_y = -j(au_x + bu_y + a_0u + b_0v + f_0)$$
$$- du_x - cu_y - c_0u - d_0v - g_0.$$

Noting $z = x + jy$, $w = u + jv$, and using the relations

$$\begin{cases} u_x = (w_{\bar{z}} + \bar{w}_{\bar{z}} + w_z + \bar{w}_z)/2, & u_y = j(-w_{\bar{z}} + \bar{w}_{\bar{z}} + w_z - \bar{w}_z)/2, \\ v_x = j(w_{\bar{z}} - \bar{w}_{\bar{z}} + w_z - \bar{w}_z)/2, & v_y = (-w_{\bar{z}} - \bar{w}_{\bar{z}} + w_z + \bar{w}_z)/2, \end{cases}$$

we get

$$j(w_{\bar{z}} - \bar{w}_{\bar{z}}) = -(aj + d)(w_{\bar{z}} + \bar{w}_{\bar{z}} + w_z + \bar{w}_z)/2$$
$$-(c + bj)j(-w_{\bar{z}} + \bar{w}_{\bar{z}} + w_z - \bar{w}_z)/2$$
$$+\text{lower order terms,}$$

namely

$$(1 + q_1)w_{\bar{z}} + q_2\bar{w}_z = -q_2 w_z + (1 - q_1)\bar{w}_{\bar{z}} + \text{lower order terms}, \tag{2.7}$$

in which

$$q_1 = [a - c + (d - b)j]/2, \qquad q_2 = [a + c + (d + b)j]/2.$$

Noting

$$q_0 = (1 + q_1)(1 + \bar{q}_1) - q_2\bar{q}_2$$
$$= [(2 + a - c)^2 - (d - b)^2 - (a + c)^2 + (d + b)^2]/4$$
$$= 1 + a - c - (d - b)^2/4 + (d + b)^2/4 - ac$$
$$= 1 + a - c - (d - b)^2/4 + \Delta$$
$$= 1 + a - c + \sigma = 1 + a - c + bd - ac$$
$$= (1 + a)(1 - c) + bd > 0,$$

where $\sigma = \Delta - (b-d)^2/4 = bd - ac \geq 0$, thus we can solve (2.7) for $w_{\bar{z}}$, giving

$$w_{\bar{z}} - Q_1(z)w_z - Q_2(z)\bar{w}_{\bar{z}} = A_1(z)w + A_2(z)\bar{w} + A_3(z), \qquad (2.8)$$

in which

$$Q_1(z) = \frac{-2q_2(z)}{q_0(z)}, \qquad Q_2(z) = \frac{[q_2\bar{q}_2 - (q_1-1)(\bar{q}_1+1)]}{q_0}.$$

For the complex equation (2.8), if $(a-c)^2 - 4\Delta \geq 0$, $(1+\sigma)^2 - 4\Delta \geq 0$, i.e.

$$\begin{aligned}
(K_1 - K_4)^2 - (K_2 - K_3)^2 + 4K_5 K_6 \geq 0, \\
(K_6^2 + K_2 K_3 - K_1 K_4)^2 - (K_2 - K_3)^2 + 4K_5 K_6 \geq 0,
\end{aligned} \qquad (2.9)$$

then we can prove

$$|Q_1| + |Q_2| = |Q_1\overline{Q_1}|^{1/2} + |Q_2\overline{Q_2}|^{1/2} < 1, \qquad (2.10)$$

where $|Q_1| = |Q_1\overline{Q_1}|^{1/2}$ is the absolute value of Q_1. In fact,

$$|2q_2| = |(a+c)^2 - (d+b)^2|^{1/2} = |(a-c)^2 - 4\Delta|^{1/2},$$

$$|q_2\bar{q}_2 - (q_1-1)(\bar{q}_1+1)|$$

$$= |(a-c)^2/4 - \Delta - [a-c+(b-d)j-2]$$

$$\times[a-c-(d-b)j+2]/4| = |(1+\sigma)^2 - 4\Delta|^{1/2},$$

$$(1+\sigma)^2 + (a-c)^2 + 2(1+\sigma)(a-c) = [1+\sigma+(a-c)]^2 > 0,$$

$$(1+\sigma)^2(a-c)^2 + (4\Delta)^2 - 4\Delta(1+\sigma)^2 - 4\Delta(a-c)^2$$

$$< (4\Delta)^2 + (1+\sigma)^2(a-c)^2 + 8\Delta(1+\sigma)(a-c),$$

$$[(a-c)^2 - 4\Delta][(1+\sigma)^2 - 4\Delta]$$

$$< (4\Delta)^2 + (1+\sigma)^2(a-c)^2 + 8\Delta(1+\sigma)(a-c),$$

$$2\{[(a-c)^2 - 4\Delta][(1+\sigma)^2 - 4\Delta]\}^{1/2} < 8\Delta + 2(1+\sigma)(a-c),$$

$$(a-c)^2 - 4\Delta + (1+\sigma)^2 - 4\Delta$$

$$+2\{[(a-c)^2 - 4\Delta][(1+\sigma)^2 - 4\Delta]\}^{1/2}$$

$$< (1+\sigma+a-c)^2,$$

and then

$$|(a-c)^2 - 4\Delta|^{1/2} + |(1+\sigma)^2 - 4\Delta|^{1/2} < 1+\sigma+a-c,$$

thus, we can derive (2.10).

2) K_2, K_3 at $(x,y) \in D$ have same signs, $K_6 = 0$, $K_5 \neq 0$; by using similar methods, we can transform (2.1) into a complex equation in the form (2.8).

Now, we discuss the case:

3) K_2, K_3 are not of same signs, $K_4 \neq 0$ at the point $(x, y) \in D$, then we can solve (2.1) for u_y, v_y and obtain the system of first order equations

$$\begin{cases} v_y = au_x + bv_x + a_0 u + b_0 v + f_0, \\ -u_y = du_x + cv_x + c_0 u + d_0 v + g_0, \end{cases} \tag{2.11}$$

where a, b, c, d are known functions of a_{kl}, b_{kl} $(k, l = 1, 2)$ and $a_0, b_0, c_0, d_0, f_0, g_0$ are known functions of $a_{k2}, b_{k2}, a_k, b_k, c_k$ $(k, l = 1, 2)$, and

$$a = K_5/K_4, \quad b = -K_3/K_4, \quad c = K_6/K_4, \quad d = K_2/K_4.$$

The condition (2.2) of hyperbolic type is reduced to the condition

$$\Delta = I/4K_4^2 = (b+d)^2/4 - ac > 0. \tag{2.12}$$

Similarly to (2.5), multiply the second formula of (2.11) by j, and then subtract the first formula of (2.11), we get

$$-v_y - ju_y = w_{\bar{z}} - w_z = -(a - dj)u_x - (b - cj)v_x + \text{lower order terms}$$

$$= -(a - dj)(w_{\bar{z}} + \bar{w}_z + w_z + \bar{w}_{\bar{z}})/2$$

$$+ (c - bj)(w_{\bar{z}} - \bar{w}_z + w_z - \bar{w}_{\bar{z}})/2 + \text{lower order terms},$$

namely

$$(1 + q_1)w_{\bar{z}} + q_2\bar{w}_z = (1 - q_1)w_z - q_2\bar{w}_{\bar{z}} + \text{lower order terms}, \tag{2.13}$$

where

$$q_1 = \frac{[a - c - (d - b)j]}{2}, \quad q_2 = \frac{[a + c - (d + b)j]}{2}.$$

It is clear that

$$q_0 = (1 + q_1)(1 + \bar{q}_1) - q_2\bar{q}_2$$

$$= \frac{[(2 + a - c)^2 - (d - b)^2 - (a + c)^2 + (d + b)^2]}{4}$$

$$= (1 + a)(1 - c) + bd > 0,$$

thus we can solve (2.13) for $w_{\bar{z}}$, i.e.

$$w_{\bar{z}} - Q_1(z)w_z - Q_2(z)\bar{w}_{\bar{z}} = A_1(z)w + A_2\bar{w} + A_3(z), \tag{2.14}$$

in which

$$Q_1(z) = \frac{[(1 - q_1)(1 + \bar{q}_1) + |q_2|^2]}{q_0}, \quad Q_2 = \frac{-2q_2(z)}{q_0}.$$

4) K_2, K_3 are not of same signs, $K_4 = 0, K_1 \neq 0$, by using similar methods as in 3), we can transform (2.1) into the complex equation of the form (2.14).

2.2 Complex forms of nonlinear hyperbolic systems of first order equations

Next, we consider the general nonlinear hyperbolic system of first order partial differential equations

$$F_k(x, y, u, v, u_x, u_y, v_x, v_y) = 0, \quad k = 1, 2, \tag{2.15}$$

where the real functions $F_k(k = 1, 2)$ are defined and continuous at every point (x, y) in D and possess continuous partial derivatives in u_x, u_y, v_x, v_y. For system (2.15), its condition of hyperbolic type can be defined by the inequality (2.2) or (2.3), but in which

$$K_1 = \frac{D(F_1, F_2)}{D(u_x, v_x)}, \quad K_2 = \frac{D(F_1, F_2)}{D(u_x, v_y)}, \quad K_3 = \frac{D(F_1, F_2)}{D(u_y, v_x)},$$

$$K_4 = \frac{D(F_1, F_2)}{D(u_y, v_y)}, \quad K_5 = \frac{D(F_1, F_2)}{D(u_x, u_y)}, \quad K_6 = \frac{D(F_1, F_2)}{D(v_x, v_y)}, \tag{2.16}$$

where $F_{ku_x}, F_{ku_y}, F_{kv_x}, F_{kv_y}(k = 1, 2)$ can be found as follows

$$F_{ku_x} = \int_0^1 F_{ktu_x}(x, y, u, v, tu_x, tu_y, tv_x, tv_y)dt,$$

$$F_{ku_y} = \int_0^1 F_{ktu_y}(x, y, u, v, tu_x, tu_y, tv_x, tv_y)dt,$$

$$F_{kv_x} = \int_0^1 F_{ktv_x}(x, y, u, v, tu_x, tu_y, tv_x, tv_y)dt, \tag{2.17}$$

$$F_{kv_y} = \int_0^1 F_{ktv_y}(x, y, u, v, tu_x, tu_y, tv_x, tv_y)dt.$$

By using the method in Subsection 2.1, for cases 1) K_2, K_3 are of same signs, K_5 or $K_6 \neq 0$; 2) K_2, K_3 are not of same signs, K_1 or $K_4 \neq 0$; then system (2.15) can be reduced to the complex form

$$w_{\bar{z}} - Q_1 w_z - Q_2 \bar{w}_{\bar{z}} = A_1 w + A_2 \bar{w} + A_3, \tag{2.18}$$

where $z = x + jy$, $w = u + jv$ and

$$Q_k = Q_k(z, w, w_z, w_{\bar{z}}), \quad k = 1, 2, \qquad A_k = A_k(z, w, w_z, w_{\bar{z}}), \quad k = 1, 2, 3.$$

In particular, if (2.9) holds, from the condition of hyperbolic type in (2.2), it follows that (2.10) holds.

Theorem 2.1 *Let system (2.15) satisfy the condition of hyperbolic type as in (2.2) and the conditions from the existence theorem for implicit functions. Then (2.15) is solvable with respect to $w_{\bar{z}}$, and the corresponding hyperbolic complex equation of first order: (2.18) can be obtained.*

As for the cases 3) $K_1 = K_4 = 0$, K_2, K_3 are not of same signs, $K_5 \neq 0$ or $K_6 \neq 0$, and 4) $K_5 = K_6 = 0$, K_2, K_3 are of same signs, $K_1 \neq 0$ or $K_4 \neq 0$, we can transform the quasilinear case of hyperbolic system: (2.15) into the complex forms by using a similar method in the next subsection.

2.3 Complex forms of quasilinear hyperbolic systems of first order equations

Finally we discuss the quasilinear hyperbolic system of first order partial differential equations

$$\begin{cases} a_{11}u_x + a_{12}u_y + b_{11}v_x + b_{12}v_y = a_1u + b_1v + c_1, \\ a_{21}u_x + a_{22}u_y + b_{21}v_x + b_{22}v_y = a_2u + b_2v + c_2, \end{cases} \quad (2.19)$$

where the coefficients $a_{kl}, b_{kl}(k,l = 1,2)$ are known functions in $(x,y) \in D$ and $a_k, b_k, c_k(k = 1,2)$ are known functions of $(x,y) \in D$ and $u, v \in \mathbb{R}$. The hyperbolicity condition of (2.19) is the same as for system (2.1), i.e. for any point $(x,y) \in D$, the inequality

$$I = (K_2 + K_3)^2 - 4K_1K_4 = (K_2 - K_3)^2 - 4K_5K_6 > 0 \quad (2.20)$$

holds, in which

$$K_1 = \begin{vmatrix} a_{11} & b_{11} \\ a_{21} & b_{21} \end{vmatrix}, \quad K_2 = \begin{vmatrix} a_{11} & b_{12} \\ a_{21} & b_{22} \end{vmatrix}, \quad K_3 = \begin{vmatrix} a_{12} & b_{11} \\ a_{22} & b_{21} \end{vmatrix},$$

$$K_4 = \begin{vmatrix} a_{12} & b_{12} \\ a_{22} & b_{22} \end{vmatrix}, \quad K_5 = \begin{vmatrix} a_{11} & a_{12} \\ a_{21} & a_{22} \end{vmatrix}, \quad K_6 = \begin{vmatrix} b_{11} & b_{12} \\ b_{21} & b_{22} \end{vmatrix}.$$

We first consider the case: 1) $K_1 = K_4 = 0$, K_2, K_3 are not of same signs, $K_6 \neq 0$ at the point $(x,y) \in D$. From $K_1 = K_4 = 0$, there exist real constants λ, μ, such that

$$a_{11} = \lambda b_{11}, \quad a_{21} = \lambda b_{21}, \quad a_{12} = \mu b_{12}, \quad a_{22} = \mu b_{22},$$

thus

$$K_2 = \lambda K_6, \quad K_3 = -\mu K_6, \quad K_5 = \lambda\mu K_6,$$

and then

$$I = (K_2 - K_3)^2 - 4K_5K_6 = [(\lambda + \mu)^2 - 4\lambda\mu]K_6^2$$

$$= (K_2 + K_3)^2 - 4K_1K_4 = (\lambda - \mu)^2 K_6^2 > 0.$$

It is easy to see that $\lambda \neq \mu$, i.e. $K_2 \neq -K_3$, in this case, system (2.19) becomes the form

$$\begin{cases} b_{11}(\lambda u + v)_x + b_{12}(\mu u + v)_y = a_1u + b_1v + c_1, \\ b_{21}(\lambda u + v)_x + b_{22}(\mu u + v)_y = a_2u + b_2v + c_2. \end{cases} \quad (2.21)$$

Setting $U = \lambda u + v$, $V = \mu u + v$, and noting

$$\begin{vmatrix} U_u & U_v \\ V_u & V_v \end{vmatrix} = \begin{vmatrix} \lambda & 1 \\ \mu & 1 \end{vmatrix} = \lambda - \mu \neq 0,$$

and

$$u = \frac{U - V}{\lambda - \mu}, \quad v = \frac{\mu U - \lambda V}{\mu - \lambda},$$

system (2.21) can be written in the form

$$\begin{cases} b_{11}U_x + b_{12}V_y = a_1'U + b_1'V + c_1, \\ b_{21}U_x + b_{22}V_y = a_2'U + b_2'V + c_2, \end{cases} \tag{2.22}$$

where

$$a_1' = \frac{a_1 - \mu b_1}{\lambda - \mu}, \quad b_1' = \frac{-a_1 + \lambda b_1}{\lambda - \mu}, \quad a_2' = \frac{a_2 - \mu b_2}{\lambda - \mu}, \quad b_2' = \frac{-a_2 + \lambda b_2}{\lambda - \mu},$$

thus

$$\begin{cases} U_x = [(a_1'b_{22} - a_2'b_{12})U + (b_1'b_{22} - b_2'b_{12})V + (c_1b_{22} - c_2b_{12})]/K_6, \\ V_y = [(a_2'b_{11} - a_1'b_{21})U + (b_2'b_{11} - b_1'b_{21})V + (c_2b_{11} - c_1b_{21})]/K_6. \end{cases} \tag{2.23}$$

Subtracting the first equation from the second equation, the complex equation of $W = U + jV$

$$W_{\bar{z}} + \overline{W}_z = A_1(z, W)W + A_2(z, W)\overline{W} + A_3(z, W) \tag{2.24}$$

can be derived, where A_1, A_2, A_3 are known functions of $b_{kl}, a_k, b_k, c_k (k, l = 1, 2)$.

Moreover, we consider system (2.19) with the condition 2) $K_5 = K_6 = 0$, K_2, K_3 are of same signs, and $K_4 \neq 0$ at the point $(x, y) \in D$. In this case, due to $K_5 = K_6 = 0$ at the point $(x, y) \in D$, there exist real constants λ, μ, such that

$$a_{11} = \lambda a_{12}, \quad a_{21} = \lambda a_{22}, \quad b_{11} = \mu b_{12}, \quad b_{21} = \mu b_{22},$$

thus

$$K_1 = \lambda \mu K_4, \quad K_2 = \lambda K_4, \quad K_3 = \mu K_4,$$

and then

$$I = (K_2 + K_3)^2 - 4K_1 K_4 = [(\lambda + \mu)^2 - 4\lambda\mu]K_4^2 = (\lambda - \mu)^2 K_4^2 > 0.$$

It is clear that if $\lambda \neq \mu$, i.e. $K_2 \neq K_3$, then system (2.19) can become the form

$$\begin{cases} a_{12}(\lambda u_x + u_y) + b_{12}(\mu v_x + v_y) = a_1 u + b_1 v + c_1, \\ a_{22}(\lambda u_x + u_y) + b_{22}(\mu v_x + v_y) = a_2 u + b_2 v + c_2. \end{cases} \tag{2.25}$$

Let

$$\xi = \frac{x - \mu y}{\lambda - \mu}, \quad \eta = \frac{-x + \lambda y}{\lambda - \mu},$$

it is easy to see that

$$\begin{vmatrix} \xi_x & \xi_y \\ \eta_x & \eta_y \end{vmatrix} = \frac{1}{(\lambda - \mu)^2} \begin{vmatrix} 1 & -\mu \\ -1 & \lambda \end{vmatrix} = \frac{1}{\lambda - \mu} \neq 0,$$

and $x = \lambda\xi + \mu\eta$, $y = \xi + \eta$. Thus system (2.25) can be rewritten in the form

$$\begin{cases} a_{12}u_\xi + b_{12}v_\eta = a_1u + b_1v + c_1, \\ a_{22}u_\xi + b_{22}v_\eta = a_2u + b_2v + c_2. \end{cases} \tag{2.26}$$

This system can be solved for u_ξ, v_η, namely

$$\begin{cases} u_\xi = a_1'u + b_1'v + c_1', \\ v_\eta = a_2'u + b_2'v + c_2', \end{cases} \tag{2.27}$$

where

$$a_1' = (a_1b_{22} - a_2b_{12})/K_4, \quad a_2' = (a_2a_{12} - a_1a_{22})/K_4,$$

$$b_1' = (b_1b_{22} - b_2b_{12})/K_4, \quad b_2' = (b_2a_{12} - b_1a_{22})/K_4,$$

$$c_1' = (c_1b_{22} - c_2b_{12})/K_4, \quad c_2' = (c_2a_{12} - c_1a_{22})/K_4.$$

Denoting $\zeta = \xi + j\eta$, then system (2.27) can be written in the complex form

$$w_{\bar\zeta} + \bar{w}_\zeta = A_1'(\zeta, w)w + A_2'(\zeta, w)\bar{w} + A_3'(\zeta, w), \tag{2.28}$$

in which A_1', A_2', A_3' are known functions of $a_{k2}, b_{k2}, a_k, b_k, c_k (k = 1, 2)$.

For 3) $K_1 = K_4 = 0$, K_2, K_3 are not of same signs, and $K_5 \neq 0$, and 4) $K_5 = K_6 = 0$, K_2, K_3 are of same signs, and $K_1 \neq 0$, by using a similar method, we can transform (2.19) into the complex equations (2.24) and (2.28) respectively. We mention that it is possible that the case:

$$b_{11} = \lambda a_{11}, \quad b_{21} = \lambda a_{21}, \quad b_{12} = \mu a_{12}, \quad b_{22} = \mu a_{22}$$

occurs for 1) and 2), and

$$a_{12} = \lambda a_{11}, \quad a_{22} = \lambda a_{21}, \quad b_{12} = \mu b_{11}, \quad b_{22} = \mu b_{21}$$

occurs for 3) and 4), then we can similarly discuss. In addition, if $\lambda(x,y), \mu(x,y)$ are known functions of (x,y) in D, then in the left-hand sides of the two equations in (2.21) should be added $b_{11}\lambda_x u + b_{12}\mu_y u$ and $b_{21}\lambda_x u + b_{22}\mu_y u$. It is sufficient to modify the coefficient of u and the system is still hyperbolic. For 2)–4), we can similarly handle.

The complex equations as stated in (2.24) and (2.28) can be written in the form

$$w_{\bar{z}} + \bar{w}_z = A(z,w)w + B(z,w)\bar{w} + C(z,w), \tag{2.29}$$

which is equivalent to the system of first order equations

$$u_x = au + bv + f, \quad v_y = cu + dv + g, \tag{2.30}$$

where $z = x + jy$, $w = u + jv$, $A = (a + jb - c - jd)/2$, $B = (a - jb - c + jd)/2$, $C = f - g$. Let

$$W(z) = ve_1 + ue_2, \quad Z = xe_1 + ye_2, \tag{2.31}$$

where $e_1 = (1+j)/2$, $e_2 = (1-j)/2$. From (2.30), we can obtain the complex equation

$$W_{\bar{Z}} = v_y e_1 + u_x e_2 = A_1 W + A_2 \overline{W} + A_3 = F(Z, W), \tag{2.32}$$

in which $A_1(Z, W) = de_1 + ae_2$, $A_2(Z, W) = ce_1 + be_2$, $A_3(Z, W) = ge_1 + fe_2$ [92].

3   Boundary Value Problems of Linear Hyperbolic Complex Equations of First Order

In this section, we mainly discuss the Riemann–Hilbert boundary value problem for linear hyperbolic complex equations of first order in a simply connected domain. We first give a representation of solutions for the above boundary value problem, and then prove the uniqueness and existence of solutions for the above problem by using the successive iteration.

3.1   Formulation of the Riemann–Hilbert problem and uniqueness of its solutions for simplest hyperbolic complex equations

Let D be a simply connected bounded domain in the $x + jy$-plane with the boundary $\Gamma = L_1 \cup L_2 \cup L_3 \cup L_4$, where $L_1 = \{x = -y, 0 \leq x \leq R_1\}$, $L_2 = \{x = y + 2R_1, R_1 \leq x \leq R_2\}$, $L_3 = \{x = -y - 2R_1 + 2R_2, R = R_2 - R_1 \leq x \leq R_2\}$, $L_4 = \{x = y, 0 \leq x \leq R_2 - R_1\}$, and denote $z_0 = 0$, $z_1 = (1 - j)R_1$, $z_2 = R_2 + j(R_2 - 2R_1)$, $z_3 = (1 + j)(R_2 - R_1) = (1 + j)R$ and $L = L_1 \cup L_4$, where j is the hyperbolic unit. For convenience we only discuss the case $R_2 \geq 2R_1$, the other case can be discussed by a similar method. We consider the simplest hyperbolic complex equation of first order:

$$w_{\bar{z}} = 0 \text{ in } D. \tag{3.1}$$

The Riemann–Hilbert boundary value problem for the complex equation (3.1) may be formulated as follows:

Problem A   Find a continuous solution $w(z)$ of (3.1) in $\bar{D}$ satisfying the boundary conditions

$$\text{Re}\,[\lambda(z)w(z)] = r(z), \quad z \in L, \qquad \text{Im}\,[\lambda(z_0)w(z_0)] = b_1, \tag{3.2}$$

where $\lambda(z) = a(z) + jb(z) \neq 0$, $z \in L$, and $\lambda(z)$, $r(z)$, b_1 satisfy the conditions

$$C_\alpha[\lambda(z), L] = C_\alpha[\text{Re}\,\lambda, L] + C_\alpha[\text{Im}\,\lambda, L] \leq k_0, \ C_\alpha[r(z), L] \leq k_2, \tag{3.3}$$

$$|b_1| \leq k_2, \quad \max_{z \in L_1} \frac{1}{|a(z) - b(z)|}, \quad \max_{z \in L_4} \frac{1}{|a(z) + b(z)|} \leq k_0, \tag{3.4}$$

in which b_1 is a real constant and $\alpha \, (0 < \alpha < 1)$, k_0, k_2 are non-negative constants. In particular, when $a(z) = 1$, $b(z) = 0$, i.e. $\lambda(z) = 1$, $z \in L$, Problem A is the Dirichlet problem (Problem D), whose boundary condition is

$$\text{Re}\,[w(z)] = r(z), \quad z \in L, \qquad \text{Im}\,[w(z_0)] = b_1. \tag{3.5}$$

Problem A with the conditions $r(z) = 0, z \in L, b_1 = 0$ is called Problem A_0.

On the basis of Theorem 1.1, it is clear that the complex equation (3.1) can be reduced to the form

$$\xi_\nu = 0, \quad \eta_\mu = 0, \quad (\mu, \nu) \in Q = \{0 \leq \mu \leq 2R, 0 \leq \nu \leq 2R_1\}, \tag{3.6}$$

where $\mu = x + y$, $\nu = x - y$, $\xi = u + v$, $\eta = u - v$. Hence the general solution of system (3.6) can be expressed as

$$\xi = u + v = f(\mu) = f(x+y), \quad \eta = u - v = g(\nu) = g(x-y), \quad \text{i.e.}$$
$$u = [f(x+y) + g(x-y)]/2, \quad v = [f(x+y) - g(x-y)]/2, \tag{3.7}$$

in which $f(t), g(t)$ are two arbitrary real continuous functions on $[0, 2R]$, $[0, 2R_1]$ respectively. From the boundary condition (3.2), we have

$$a(z)u(z) + b(z)v(z) = r(z) \quad \text{on} \quad L, \quad \lambda(z_0)w(z_0) = r(z_0) + jb_1, \quad \text{i.e.}$$

$$[a((1-j)x) + b((1-j)x)]f(0) + [a((1-j)x) - b((1-j)x)]$$
$$\times g(2x) = 2r((1-j)x) \quad \text{on} \quad [0, R_1],$$
$$[a((1+j)x) + b((1+j)x)]f(2x) + [a((1+j)x) - b((1+j)x)] \tag{3.8}$$
$$\times g(0) = 2r((1+j)x) \quad \text{on} \quad [0, R],$$

$$f(0) = u(0) + v(0) = \frac{r(0) + b_1}{a(0) + b(0)}, \quad g(0) = u(0) - v(0) = \frac{r(0) - b_1}{a(0) - b(0)}.$$

The above formulas can be rewritten as

$$[a((1-j)t/2) + b((1-j)t/2)]f(0) + [a((1-j)t/2) - b((1-j)t/2)]$$
$$\times g(t) = 2r((1-j)t/2), \quad t \in [0, 2R_1],$$
$$[a((1+j)t/2) + b((1+j)t/2)]f(t) + [a((1+j)t/2) - b((1+j)t/2)]$$
$$\times g(0) = 2r((1+j)t/2), \quad t \in [0, 2R], \quad \text{i.e.}$$

$$f(x+y) = \frac{2r((1+j)(x+y)/2)}{a((1+j)(x+y)/2) + b((1+j)(x+y)/2)} \tag{3.9}$$
$$- \frac{[a((1+j)(x+y)/2) - b((1+j)(x+y)/2)]g(0)}{a((1+j)(x+y)/2) + b((1+j)(x+y)/2)},$$
$$0 \le x+y \le 2R,$$

$$g(x-y) = \frac{2r((1-j)(x-y)/2)}{a((1-j)(x-y)/2) - b((1-j)(x-y)/2)}$$
$$- \frac{[a((1-j)(x-y)/2) + b((1-j)(x-y)/2)]f(0)}{a((1-j)(x-y)/2) - b((1-j)(x-y)/2)},$$
$$0 \le x-y \le 2R_1.$$

Thus the solution $w(z)$ of (3.1) can be expressed as

$$w(z) = f(x+y)e_1 + g(x-y)e_2$$
$$= \frac{1}{2}\{f(x+y) + g(x-y) + j[f(x+y) - g(x-y)]\}, \tag{3.10}$$

where $f(x+y)$, $g(x-y)$ are as stated in (3.9) and $f(0), g(0)$ are as stated in (3.8). It is not difficult to see that $w(z)$ satisfies the estimate

$$C_\alpha[w(z), \bar{D}] \le M_1 = M_1(\alpha, k_0, k_2, D), \quad C_\alpha[w(z), \bar{D}] \le M_2 k_2 = M_2(\alpha, k_0, D)k_2, \quad (3.11)$$

where M_1, M_2 are two non-negative constants only dependent on α, k_0, k_2, D and α, k_0, D respectively. The above results can be written as a theorem.

Theorem 3.1 *Any solution $w(z)$ of Problem A for the complex equation (3.1) possesses the representation (3.10), which satisfies the estimate (3.11).*

3.2 Uniqueness of solutions of the Riemann–Hilbert problem for linear hyperbolic complex equations

Now we discuss the linear case of the complex equation (2.32), namely

$$w_{\bar{z}} = A_1(z)w + A_2(z)\bar{w} + A_3(z), \qquad (3.12)$$

and suppose that the complex equation (3.12) satisfies the following conditions:

Condition C $A_l(z)$ $(l = 1, 2, 3)$ are continuous in $\bar{D}$ and satisfy

$$C[A_l, \bar{D}] = C[\operatorname{Re} A_l, \bar{D}] + C[\operatorname{Im} A_l, \bar{D}] \le k_0, \quad l = 1, 2, \qquad C[A_3, \bar{D}] \le k_1. \qquad (3.13)$$

Due to $w = u + jv = \xi e_1 + \eta e_2$, $w_z = \xi_\mu e_1 + \eta_\nu e_2$, $w_{\bar{z}} = \xi_\nu e_1 + \eta_\mu e_2$, from the formulas in Section 1, equation (3.12) can be rewritten in the form

$$\xi_\nu e_1 + \eta_\mu e_2 = [A(z)\xi + B(z)\eta + E(z)]e_1$$
$$+[C(z)\xi + D(z)\eta + F(z)]e_2, \quad z \in D, \quad \text{i.e.}$$
$$\begin{cases} \xi_\nu = A(z)\xi + B(z)\eta + E(z), \\ \eta_\mu = C(z)\xi + D(z)\eta + F(z), \end{cases} z \in D, \qquad (3.14)$$

in which

$$A = \operatorname{Re} A_1 + \operatorname{Im} A_1, \quad B = \operatorname{Re} A_2 + \operatorname{Im} A_2, \quad C = \operatorname{Re} A_2 - \operatorname{Im} A_2,$$
$$D = \operatorname{Re} A_1 - \operatorname{Im} A_1, \quad E = \operatorname{Re} A_3 + \operatorname{Im} A_3, \quad F = \operatorname{Re} A_3 - \operatorname{Im} A_3.$$

The boundary condition (3.2) can be reduced to

$$\operatorname{Re}[\lambda(\xi e_1 + \eta e_2)] = r(z), \quad \operatorname{Im}[\lambda(\xi e_1 + \eta e_2)]|_{z=z_0} = b_1, \qquad (3.15)$$

where $\lambda = (a+b)e_1 + (a-b)e_2$. Moreover, the domain D is transformed into $Q = \{0 \le \mu \le 2R, 0 \le \nu \le 2R_1, R = R_2 - R_1\}$, which is a rectangle and A, B, C, D, E, F

are known functions of $(\mu, \nu) \in Q$. There is no harm in assuming that $w(z_0) = 0$, otherwise through the transformation

$$W(z) = w(z) - [a(z_0) - jb(z_0)]\frac{[r(z_0) + jb_1]}{[a^2(z_0) - b^2(z_0)]}, \tag{3.16}$$

the requirement can be realized. For convenience, sometimes we write $z \in D$ or $z \in Q$ and denote $L_1 = \{\mu = 0, 0 \le \nu \le 2R_1\}$, $L_4 = \{0 \le \mu \le 2R, \nu = 0\}$.

Now we give a representation of solutions of Problem A for equation (3.12).

Theorem 3.2 *If equation (3.12) satisfies Condition C, then any solution $w(z)$ of Problem A for (3.12) can be expressed as*

$$w(z) = w_0(z) + \Phi(z) + \Psi(z) \quad in \quad D,$$

$$w_0(z) = f(x+y)e_1 + g(x-y)e_2, \qquad \Phi(z) = \tilde{f}(x+y)e_1 + \tilde{g}(x-y)e_2, \tag{3.17}$$

$$\Psi(z) = \int_0^{x-y}[A\xi + B\eta + E]d(x-y)e_1 + \int_0^{x+y}[C\xi + D\eta + F]d(x+y)e_2,$$

where $f(x+y)$, $g(x-y)$ are as stated in (3.9) and $\tilde{f}(x+y)$, $\tilde{g}(x-y)$ are similar to $f(x+y)$, $g(x-y)$ in (3.9), but $\Phi(z)$ satisfies the boundary condition

$$\operatorname{Re}[\lambda(z)\Phi(z)] = -\operatorname{Re}[\lambda(z)\Psi(z)], \quad z \in L, \qquad \operatorname{Im}[\lambda(z_0)\Phi(z_0)] = -\operatorname{Im}[\lambda(z_0)\Psi(z_0)]. \tag{3.18}$$

Proof It is not difficult to see that the functions $w_0(z)$, $\Phi(z)$ are solutions of the complex equation (3.1) in $\bar{D}$, which satisfy the boundary conditions (3.2) and (3.18) respectively, and $\Psi(z)$ satisfies the complex equation

$$[\Psi]_{\bar{z}} = [A\xi + B\eta + E]e_1 + [C\xi + D\eta + F]e_2, \tag{3.19}$$

and $\Phi(z) + \Psi(z)$ satisfies the boundary condition of Problem A_0. Hence $w(z) = w_0(z) + \Phi(z) + \Psi(z)$ satisfies the boundary condition (3.2) and is a solution of Problem A for (3.12).

Theorem 3.3 *Suppose that Condition C holds. Then Problem A for the complex equation (3.12) has at most one solution.*

Proof Let $w_1(z), w_2(z)$ be any two solutions of Problem A for (3.12) and substitute them into equation (3.12) and the boundary condition (3.2). It is clear that $w(z) = w_1(z) - w_2(z)$ satisfies the homogeneous complex equation and boundary conditions

$$w_{\bar{z}} = A_1 w + A_2 \overline{w} \quad in \quad D, \tag{3.20}$$

$$\operatorname{Re}[\lambda(z)w(z)] = 0, \qquad if \ (x, y) \in L, \quad w(z_0) = 0. \tag{3.21}$$

On the basis of Theorem 3.2, the function $w(z)$ can be expressed in the form

$$w(z) = \Phi(z) + \Psi(z),$$

$$\Psi(z) = \int_0^{x-y}[A\xi + B\eta]e_1 d(x-y) + \int_0^{x+y}[C\xi + D\eta]e_2 d(x+y). \tag{3.22}$$

Suppose $w(z) \not\equiv 0$ in the neighborhood $(\subset \bar{D})$ of the point $z_0 = 0$. We may choose a sufficiently small positive number $R_0 < 1$, such that $8M_3MR_0 < 1$, where $M_3 = \max\{C[A, Q_0], C[B, Q_0], C[C, Q_0], C[D, Q_0]\}$, $M = 1 + 4k_0^2(1 + k_0^2)$ is a positive constant, and $m = C[w(z), \overline{Q_0}] > 0$, herein $Q_0 = \{0 \leq \mu \leq R_0\} \cap \{0 \leq \nu \leq R_0\}$. From (3.9),(3.10),(3.17),(3.18),(3.22) and Condition C, we have

$$\| \Psi(z) \| \leq 8M_3 m R_0, \quad \| \Phi(z) \| \leq 32M_3 k_0^2(1 + k_0^2)m R_0,$$

thus an absurd inequality $m \leq 8M_3 M m R_0 < m$ is derived. It shows $w(z) = 0, (x, y) \in Q_0$. Moreover, we extend along the positive directions of $\mu = x + y$ and $\nu = x - y$ successively, and finally obtain $w(z) = 0$ for $(x, y) \in D$, i.e. $w_1(z) - w_2(z) = 0$ in D. This proves the uniqueness of solutions of Problem A for (3.12).

3.3   Solvability of Problem A for linear hyperbolic complex equations of first order

Theorem 3.4   *If the complex equation* (3.12) *satisfies Condition C, then Problem A for* (3.12) *has a solution.*

Proof   In order to find a solution $w(z)$ of Problem A in D, we can express $w(z)$ in the form (3.17). In the following, by using successive iteration we can find a solution of Problem A for the complex equation (3.12). First of all, substituting $w_0(z) = \xi_0 e_1 + \eta_0 e_2$ of Problem A for (3.1) into the position of $w = \xi e_1 + \eta e_2$ in the right-hand side of (3.12), the function

$$w_1(z) = w_0(z) + \Phi_1(z) + \Psi_1(z),$$

$$\Psi_1(z) = \int_0^\nu [A\xi_0 + B\eta_0 + E]e_1 d\nu + \int_0^\mu [C\xi_0 + D\eta_0 + F]e_2 d\mu, \tag{3.23}$$

can be determined, where $\mu = x + y$, $\nu = x - y$, $\Phi_1(z)$ is a solution of (3.1) in $\bar{D}$ satisfying the boundary conditions

$$\mathrm{Re}\,[\lambda(z)\Phi_1(z)] = -\mathrm{Re}\,[\lambda(z)\Psi_1(z)], \quad z \in L,$$

$$\mathrm{Im}\,[\lambda(z_0)\Phi_1(z_0)] = -\mathrm{Im}\,[\lambda(z_0)\Psi_1(z_0)]. \tag{3.24}$$

Thus from (3.23), (3.24), we have

$$\| w_1(z) - w_0(z) \| = C[w_1(z) - w_0(z), \bar{D}] \leq 2M_4 M(4m + 1)R', \tag{3.25}$$

where $M_4 = \max_{z \in \bar{D}}(|A|, |B|, |C|, |D|, |E|, |F|)$, $m = \| w_0 \|_{C(\bar{D})}$, $R' = \max(2R_1, 2R)$, $M = 1 + 4k_0^2(1 + k_0^2)$ is a positive constant as in the proof of Theorem 3.3. Moreover, we substitute $w_1(z) = w_0(z) + \Phi_1(z) + \Psi_1(z)$ and the corresponding functions $\xi_1(z) = \mathrm{Re}\,w_1(z) + \mathrm{Im}\,w_1(z)$, $\eta_1(z) = \mathrm{Re}\,w_1(z) - \mathrm{Im}\,w_1(z)$ into the positions of $w(z), \xi(z), \eta(z)$ in (3.17), and similarly to (3.23)–(3.25), we can find the corresponding functions $\Psi_2(z), \Phi_2(z)$ in $\bar{D}$ and the function

$$w_2(z) = w_0(z) + \Phi_2(z) + \Psi_2(z) \quad \text{in} \quad \bar{D}.$$

It is clear that the function $w_2(z) - w_1(z)$ satisfies the equality

$$w_2(z) - w_1(z) = \Phi_2(z) - \Phi_1(z) + \Psi_2(z) - \Psi_1(z)$$

$$= \Phi_2(z) - \Phi_1(z) + \int_0^\nu [A(\xi_1 - \xi_0) + B(\eta_1 - \eta_0)]e_1 d\nu$$

$$+ \int_0^\mu [C(\xi_1 - \xi_0) + D(\eta_1 - \eta_0)]e_2 d\mu,$$

and

$$\| w_2 - w_1 \| \le [2M_3 M(4m+1)]^2 \int_0^{R'} R' dR' \le \frac{[2M_3 M(4m+1)R']^2}{2!},$$

where M_3 is a constant as stated in the proof of Theorem 3.3. Thus we can find a sequence of functions $\{w_n(z)\}$ satisfying

$$w_n(z) = w_0(z) + \Phi_n(z) + \Psi_n(z),$$

$$\Psi_n(z) = + \int_0^\nu [A\xi_n + B\eta_n + E]e_1 d\nu + \int_0^\mu [C\xi_n + D\eta_n + F]e_2 d\mu, \tag{3.26}$$

and $w_n(z) - w_{n-1}(z)$ satisfies

$$w_n(z) - w_{n-1}(z) = \Phi_n(z) - \Phi_{n-1}(z) + \Psi_n(z) - \Psi_{n-1}(z)$$

$$= \Phi_n(z) - \Phi_{n-1}(z) + \int_0^\nu [A(\xi_{n-1} - \xi_{n-2})$$

$$+ B(\eta_{n-1} - \eta_{n-2})]e_1 d\nu$$

$$+ \int_0^\mu [C(\xi_{n-1} - \xi_{n-2}) + D(\eta_{n-1} - \eta_{n-2})]e_2 d\mu, \tag{3.27}$$

and then

$$\| w_n - w_{n-1} \| \le [2M_3 M(4m+1)]^n \int_0^{R'} \frac{R'^{n-1}}{(n-1)!} dR' \le \frac{[2M_3 M(4m+1)R']^n}{n!}. \tag{3.28}$$

From the above inequality, it is seen that the sequence of functions $\{w_n(z)\}$, i.e.

$$w_n(z) = w_0(z) + [w_1(z) - w_0(z)] + \cdots + [w_n(z) - w_{n-1}(z)] (n = 1, 2, \ldots) \tag{3.29}$$

uniformly converges to a function $w_*(z)$, and $w_*(z)$ satisfies the equality

$$w_*(z) = w_0(z) + \Phi_*(z) + \Psi_*(z),$$

$$\Psi_*(z) = \int_0^\nu [A\xi_* + B\eta_* + E]e_1 d\nu + \int_0^\mu [C\xi_* + D\eta_* + F]e_2 d\mu. \tag{3.30}$$

It is easy to see that $w_*(z)$ satisfies equation (3.12) and the boundary condition (3.2), hence it is just a solution of Problem A for the complex equation (3.12) in the closed domain $\bar{D}$ ([87]2).

3.4 Another boundary value problem for linear hyperbolic complex equations of first order

Now we introduce another boundary value problem for equation (3.12) in D with the boundary conditions

$$\text{Re}\,[\lambda(z)w(z)] = r(z) \ \text{on} \ L_1 \cup L_5, \ \text{Im}\,[\lambda(z_1)w(z_1)] = b_1, \qquad (3.31)$$

where $L_1 = \{y = -x, 0 \le x \le R\}, L_5 = \{y = (R+R_1)[x/(R-R_1) - 2R_1R/(R^2 - R_1^2)],\ R_1 \le x \le R = R_2 - R_1\}, R_2 \ge 2R_1,\ \lambda(z) = a(z) + jb(z), z \in L_1, \lambda(z) = a(z) + jb(z) = 1 + j, z \in L_5$ and $\lambda(z), r(z), b_1$ satisfy the conditions

$$C_\alpha[\lambda(z), L_1] \le k_0, C_\alpha[r(z), L_1 \cup L_5] \le k_2, |b_1| \le k_2, \max_{z \in L_1} \frac{1}{|a(z) - b(z)|} \le k_0, \quad (3.32)$$

in which b_1 is a real constant and $\alpha\,(0 < \alpha < 1), k_0, k_2$ are non-negative constants. The boundary value problem is called Problem A_1.

On the basis of Theorem 1.1, it is clear that the complex equation (3.1) can be reduced to the form (3.6) in $\bar{D}$. The general solution of system (3.6) can be expressed as

$$
\begin{aligned}
w(z) &= u(z) + jv(z) \\
&= [u(z) + v(z)]e_1 + [u(z) - v(z)]e_2 \\
&= f(x+y)e_1 + g(x-y)e_2 \\
&= \frac{1}{2}\{f(x+y) + g(x-y) + j[f(x+y) - g(x-y)]\},
\end{aligned}
\qquad (3.33)
$$

where $f(t)\,(0 \le t \le 2R), g(t)\,(0 \le t \le 2R_1)$ are two arbitrary real continuous functions. Noting that the boundary condition (3.31), namely

$$a(z)u(z) + b(z)v(z) = r(z) \ \text{on} \ L_1 \cup L_5, \ \lambda(z_1)w(z_1) = r(z_1) + jb_1, \ \text{i.e.}$$

$$[a((1-j)x) + b((1-j)x)]f(0) + [a((1-j)x) - b((1-j)x)]g(2x)$$

$$= 2r((1-j)x) \ \text{on} \ [0, R_1], \quad f(z_1) = u(z_1) + v(z_1) = \frac{r(z_1) + b_1}{a(z_1) + b(z_1)} \qquad (3.34)$$

$$\text{Re}\,[\lambda(z)w(z)] = u(z) + v(z) = r\left[\left(1 + \frac{R+R_1}{R-R_1}j\right)x - j\frac{2RR_1}{R-R_1}\right] \ \text{on} \ [R_1, R].$$

It is easy to see that the above formulas can be rewritten as

$$[a((1-j)t/2) + b((1-j)t/2)]f(0) + [a((1-j)t/2) - b((1-j)t/2)]$$

$$\times g(t) = 2r((1-j)t/2), \quad t \in [0, 2R_1], \quad f(x+y) = f\left[\left(\frac{2R}{R-R_1}\right)x - \frac{2RR_1}{R-R_1}\right]$$

$$f(t) = r\left[((1+j)R - (1-j)R_1)\frac{t}{2R} + (1-j)R_1\right], \quad t \in [0, 2R],$$

and then

$$g(x-y) = \frac{2r((1-j)(x-y)/2)}{a((1-j)(x-y)/2) - b((1-j)(x-y)/2)}$$
$$-\frac{[a((1-j)(x-y)/2)+b((1-j)(x-y)/2)]f(0)}{a((1-j)(x-y)/2)-b((1-j)(x-y)/2)}, \tag{3.35}$$
$$0 \le x-y \le 2R_1,$$

$$f(x+y) = r[((1+j)R - (1-j)R_1)\frac{x+y}{2R} + (1-j)R_1], \quad 0 \le x+y \le 2R.$$

Substitute the above function $f(x+y), g(x-y)$ into (3.33), the solution $w(z)$ of (3.6) is obtained. We are not difficult to see that $w(z)$ satisfies the estimate

$$C_\alpha[w(z), \bar{D}] \le M_1, \quad C_\alpha[w(z), \bar{D}] \le M_2 k_2, \tag{3.36}$$

where $M_1 = M_1(\alpha, k_0, k_2, D)$, $M_2 = M_2(\alpha, k_0, D)$ are two non-negative constants.

Next we consider Problem A_1 for equation (3.12). Similarly to before, we can derive the representation of solutions $w(z)$ of Problem A_1 for (3.12) as stated in (3.17), where $f(x+y), g(x-y)$ possess the form (3.35), and $L = L_1 \cup L_2$, z_0 in the formula (3.18) should be replaced by $L_1 \cup L_5$, z_1. Moreover applying the successive iteration, the uniqueness and existence of solutions of Problem A_1 for equation (3.12) can be proved, but L, z_0 in the formulas (3.21) and (3.24) are replaced by $L_1 \cup L_5$, z_1. We write the results as a theorem.

Theorem 3.5 *Suppose that equation (3.12) satisfies Condition C. Then Problem A_1 for (3.12) has a unique solution $w(z)$, which can be expressed in the form (3.17), where $f(x+y), g(x-y)$ possess the form (3.35).*

4 Boundary Value Problems of Quasilinear Hyperbolic Complex Equations of First Order

In this section, we mainly discuss the Riemann–Hilbert boundary value problem for quasilinear hyperbolic complex equations of first order in a simply connected domain. We first prove the uniqueness of solutions for the above boundary value problem, and then give a priori estimates of solutions of the problem, moreover, by using the successive iteration, the existence of solutions for the above problem is proved. Finally, we also discuss the solvability of the above boundary value problem in general domains.

4.1   Uniqueness of solutions of the Riemann–Hilbert problem for quasilinear hyperbolic complex equations

In the subsection, we first discuss the quasilinear hyperbolic complex equation

$$w_{\bar{z}} = F(z, w), \quad F = A_1(z, w)w + A_2(z, w)\overline{w} + A_3(z, w) \text{ in } \bar{D}, \qquad (4.1)$$

where D is a simply connected bounded domain in the $x+jy$-plane with the boundary $\Gamma = L_1 \cup L_2 \cup L_3 \cup L_4$ as stated in Section 3.

Suppose that the complex equation (4.1) satisfies the following conditions:

Condition C

1) $A_l(z, w)\,(l = 1, 2, 3)$ are continuous in $z \in \bar{D}$ for any continuous complex function $w(z)$ and satisfy

$$C[A_l, \bar{D}] = C[\operatorname{Re} A_l, \bar{D}] + C[\operatorname{Im} A_l, \bar{D}] \leq k_0, \quad l = 1, 2, \ C[A_3, \bar{D}] \leq k_1. \qquad (4.2)$$

2) For any continuous complex functions $w_1(z), w_2(z)$ in $\bar{D}$, the equality

$$F(z, w_1) - F(z, w_2) = \tilde{A}_1(z, w_1, w_2)(w_1 - w_2) + \tilde{A}_2(z, w_1, w_2)(\overline{w_1} - \overline{w_2}) \text{ in } D \qquad (4.3)$$

holds, where $C[\tilde{A}_l, \bar{D}] \leq k_0$, $l = 1, 2$ and k_0, k_1 are non-negative constants. In particular, when (4.1) is a linear equation, the condition (4.3) obviously holds.

In order to give an a priori $C_\alpha(\bar{D})$-estimate of solutions for Problem A, we need the following conditions: For any hyperbolic numbers $z_1, z_2(\in \bar{D})$, w_1, w_2, the above functions satisfy

$$\| \tilde{A}_l(z_1, w_1) - A_l(z_2, w_2) \| \leq k_0[\| z_1 - z_2 \|^\alpha + \| w_1 - w_2 \|], l = 1, 2,$$
$$\| A_3(z_1, w_1) - A_3(z_2, w_2) \| \leq k_2[\| z_1 - z_2 \|^\alpha + \| w_1 - w_2 \|], \qquad (4.4)$$

in which $\alpha(0 < \alpha < 1), k_0, k_2$ are non-negative constants.

Similarly to (3.12) and (3.14), due to $w = u + jv = \xi e_1 + \eta e_2 = \zeta$, $w_z = \xi_\mu e_1 + \eta_\nu e_2$, $w_{\bar{z}} = \xi_\nu e_1 + \eta_\mu e_2$, the quasilinear hyperbolic complex equation (4.1) can be rewritten in the form

$$\xi_\nu e_1 + \eta_\mu e_2 = [A(z, \zeta)\xi + B(z, \zeta)\eta + E(z, \zeta)]e_1$$
$$+[C(z, \zeta)\xi + D(z, \zeta)\eta + F(z, \zeta)]e_2, \ z \in D, \quad \text{i.e.}$$
$$\begin{cases} \xi_\nu = A(z, \zeta)\xi + B(z, \zeta)\eta + E(z, \zeta), \\ \eta_\mu = C(z, \zeta)\xi + D(z, \zeta)\eta + F(z, \zeta), \end{cases} z \in D, \qquad (4.5)$$

in which

$$A = \operatorname{Re} A_1 + \operatorname{Im} A_1, \ B = \operatorname{Re} A_2 + \operatorname{Im} A_2, \ C = \operatorname{Re} A_2 - \operatorname{Im} A_2,$$
$$D = \operatorname{Re} A_1 - \operatorname{Im} A_1, \ E = \operatorname{Re} A_3 + \operatorname{Im} A_3, \ F = \operatorname{Re} A_3 - \operatorname{Im} A_3.$$

Obviously, any solution of Problem A for equation (4.1) possesses the same representation (3.17) as stated in Theorem 3.2. In the following, we prove the existence and uniqueness of solutions for Problem A for (4.1) with Condition C.

Theorem 4.1 *If Condition C holds, then Problem A for the quasilinear complex equation* (4.1) *has at most one solution.*

Proof Let $w_1(z), w_2(z)$ be any two solutions of Problem A for (4.1) and substitute them into equation (4.1) and boundary condition (3.2). By Condition C, we see that $w(z) = w_1(z) - w_2(z)$ satisfies the homogeneous complex equation and boundary conditions

$$w_{\bar{z}} = \tilde{A}_1 w + \tilde{A}_2 \bar{w} \ \text{in} \ D, \tag{4.6}$$

$$\text{Re}\,[\lambda(z)w(z)] = 0, \ z \in L, \ \text{Im}\,[\lambda(z_0)w(z_0)] = 0. \tag{4.7}$$

On the basis of Theorem 3.2, the function $w(z)$ can be expressed in the form

$$w(z) = \tilde{\Phi}(z) + \tilde{\Psi}(z),$$

$$\tilde{\Psi}(z) = \int_0^{x-y} [\tilde{A}\xi + \tilde{B}\eta] e_1 d(x-y) + \int_0^{x+y} [\tilde{C}\xi + \tilde{D}\eta] e_2 d(x+y), \tag{4.8}$$

where the relation between the coefficients $\tilde{A}, \tilde{B}, \tilde{C}, \tilde{D}$ and $\tilde{A}_1, \tilde{A}_2$ is the same with that between A, B, C, D and A_1, A_2 in (4.5). Suppose $w(z) \not\equiv 0$ in the neighborhood $Q_0 (\subset \bar{D})$ of the point $z_0 = 0$, we may choose a sufficiently small positive number $R_0 < 1$, such that $8M_5 M R_0 < 1$, where $M_5 = \max\{C[\tilde{A}, Q_0], C[\tilde{B}, Q_0], C[\tilde{C}, Q_0], C[\tilde{D}, Q_0]\}$. Similarly to the proof of Theorem 3.3, we can derive a contradiction. Hence $w_1(z) = w_2(z)$ in D.

4.2 Solvability of Problem A for quasilinear hyperbolic complex equations

Theorem 4.2 *If the quasilinear complex equation* (4.1) *satisfies Condition C, then Problem A for* (4.1) *has a solution.*

Proof Similarly to the proof of Theorem 3.4, we use the successive iteration to find a solution of Problem A for the complex equation (4.1). Firstly, substituting $w_0(z) = \xi_0 e_1 + \eta_0 e_2$ of Problem A for (3.1) into the position of $w = \xi e_1 + \eta e_2$ in the right-hand side of (4.1), the function

$$w_1(z) = w_0(z) + \Phi_1(z) + \Psi_1(z),$$

$$\Psi_1(z) = \int_0^{\nu} [A\xi_0 + B\eta_0 + E] e_1 d\nu + \int_0^{\mu} [C\xi_0 + D\eta_0 + F] e_2 d\mu, \tag{4.9}$$

can be determined, where $\mu = x+y$, $\nu = x-y$, $\Phi_1(z)$ is a solution of (3.1) in $\bar{D}$ satisfying the boundary conditions

$$\text{Re}\,[\lambda(z)\Phi_1(z)] = -\text{Re}\,[\lambda(z)\Psi_1(z)], \quad z \in L,$$

$$\text{Im}\,[\lambda(z_0)\Phi_1(z_0)] = -\text{Im}\,[\lambda(z_0)\Psi_1(z_0)]. \tag{4.10}$$

Moreover, we can find a sequence of functions $\{w_n(z)\}$ satisfying

$$w_n(z) = w_0(z) + \Phi_n(z) + \Psi_n(z),$$

$$\Psi_n(z) = \int_0^\nu [A\xi_{n-1} + B\eta_{n-1} + E]e_1 d\nu + \int_0^\mu [C\xi_{n-1} + D\eta_{n-1} + F]e_2 d\mu, \qquad (4.11)$$

and $w_n(z) - w_{n-1}(z)$ satisfies

$$\begin{aligned}
w_n(z) - w_{n-1}(z) &= \Phi_n(z) - \Phi_{n-1}(z) + \Psi_n(z) - \Psi_{n-1}(z) \\
&= \Phi_n(z) - \Phi_{n-1}(z) \\
&\quad + \int_0^\nu [\tilde{A}(\xi_{n-1} - \xi_{n-2}) + \tilde{B}(\eta_{n-1} - \eta_{n-2})]e_1 d\nu \qquad (4.12) \\
&\quad + \int_0^\mu [\tilde{C}(\xi_{n-1} - \xi_{n-2}) + \tilde{D}(\eta_{n-1} - \eta_{n-2})]e_2 d\mu.
\end{aligned}$$

Denoting $M_5 = \max_{\bar{D}}(|\tilde{A}|, |\tilde{B}|, |\tilde{C}|, |\tilde{D}|)$, we can obtain

$$\| w_n - w_{n-1} \| \le [2M_5 M (4m + 1)]^n \int_0^{R'} \frac{R'^{n-1}}{(n-1)!} dR' \le \frac{[2M_5 M (4m + 1) R']^n}{n!}, \qquad (4.13)$$

in which $m = \| w_0 \|_{C(Q)}$, $R' = \max(2R_1, 2R)$, $M = 1 + 4k_0^2(1 + k_0^2)$ is a positive constant as in the proof of Theorem 3.3. The remained proof is identical with the proof of Theorem 3.4.

4.3 A priori estimates of solutions of the Riemann–Hilbert problem for hyperbolic complex equations

We first give the boundedness estimate of solutions for Problem A.

Theorem 4.3 *If Condition C holds, then any solution $u(z)$ of Problem A for the hyperbolic equation* (4.1) *satisfies the estimates*

$$C[w(z), \bar{D}] \le M_6, \ C[w(z), \bar{D}] \le M_7 k, \qquad (4.14)$$

in which $M_6 = M_6(\alpha, k_0, k_1, k_2, D)$, $k = k_1 + k_2$, $M_7 = M_7(\alpha, k_0, D)$ are non-negative constants.

Proof On the basis of Theorems 4.1 and 4.2, we see that under Condition C, Problem A for equation (4.1) has a unique solution $w(z)$, which can be found by using successive iteration. Due to the functions $w_{n+1}(z) - w_n(z) \, (n = 1, 2, \ldots)$ in $\bar{D}$ are continuous, the limit function $w(z)$ of the sequence $\{w_n(z)\}$ in $\bar{D}$ is also continuous, and satisfies the estimate

$$C[w(z), \bar{D}] \le \sum_{n=0}^\infty \frac{[2M_5 M (4m + 1) R']^n}{n!} = e^{2M_5 M (4m+1) R'} = M_6, \qquad (4.15)$$

where $R' = \max(2R_1, 2R)$. This is the first estimate in (4.14). As for the second estimate in (4.14), if $k = k_1 + k_2 = 0$, then it is true from Theorem 4.1. If $k = k_1 + k_2 > 0$, let the solution $w(z)$ of Problem A be substituted into (4.1) and (3.2), and dividing them by k, we obtain the equation and boundary conditions for $\tilde{w}(z) = w(z)/k$:

$$\tilde{w}_{\bar{z}} = A_1\tilde{w} + \tilde{A}_2\tilde{w} + A_3/k, \quad z \in \bar{D},$$

$$\text{Re}\,[\lambda(z)\tilde{w}(z)] = r(z)/k, \quad z \in \bar{D}, \quad \text{Im}\,[\lambda(z_0)\tilde{w}(z_0)] = b_1/k. \tag{4.16}$$

Noting that A_3/k, r/k, b_1/k satisfy the conditions

$$C[A_3/k, \bar{D}] \leq 1, \; C[r/k, L] \leq 1, \; |b_1/k| \leq 1,$$

by using the method of deriving $C[w(z), \bar{D}]$ in (4.15), we can obtain the estimate

$$C[\tilde{w}(z), \bar{D}] \leq M_7 = M_7(\alpha, k_0, D).$$

From the above estimate, the second estimate in (4.14) is immediately derived.

Here we mention that in the proof of the estimate (4.14), we have not required that the coefficients $\lambda(z)$, $r(z)$ of (3.2) satisfy a Hölder (continuous) condition and only require that they are continuous on L.

Next, we shall give the $C_\alpha(\bar{D})$-estimates of solutions of Problem A for (4.1), and first discuss the linear hyperbolic complex equation (3.12) or (3.14).

Theorem 4.4 *Suppose that the linear complex equation (3.12) satisfies the conditions (3.13) and (4.4), i.e. the coefficents of (3.12) satisfies the conditions*

$$C_\alpha[A_l, \bar{D}] \leq k_0, \quad l = 1, 2, \quad C_\alpha[A_3, \bar{D}] \leq k_2, \tag{4.17}$$

in which $\alpha(0 < \alpha < 1)$, k_0, k_2 are non-negative constants. Then the solution $w(z) = w_0(z) + \Phi(z) + \Psi(z)$ satisfies the following estimates

$$C_\alpha[w_0(z), \bar{D}] \leq M_8, \; C_\alpha[\Psi(z), \bar{D}] \leq M_8,$$

$$C_\alpha[\Phi(z), \bar{D}] \leq M_8, \; C_\alpha[w(z), \bar{D}] \leq M_8, \tag{4.18}$$

where $w_0(z)$ is a solution of (3.1) as stated in (3.10), $M_8 = M_8(\alpha, k_0, k_1, k_2, D)$ is a non-negative constant.

Proof As stated before $w_0(z)$ is the function as in (3.10), which satisfies the estimate (3.11), namely the first estimate in (4.18). In order to prove that $\Psi(z) = \Psi_1(z) = \Psi_1^1(z)e_1 + \Psi_1^2(z)e_2$ satisfies the second estimate in (4.18), from

$$\Psi_1^1(z) = \int_0^{x-y} G_1(z)d(x-y), \quad \Psi_1^2(z) = \int_0^{x+y} G_2(z)d(x+y),$$

$$G_1(z) = A(z)\xi + B(z)\eta + E(z), \quad G_2(z) = C(z)\xi + D(z)\eta + F(z), \tag{4.19}$$

and (4.17), we see that $\Psi_1^1(z) = \Psi_1^1(\mu, \nu)$, $\Psi_1^2(z) = \Psi_1^2(\mu, \nu)$ in $\bar{D}$ with respect to $\nu = x - y$, $\mu = x + y$ satisfy the estimates

$$C_\alpha[\Psi_1^1(\cdot, \nu), \bar{D}] \leq M_9 R', \; C_\alpha[\Psi_1^2(\mu, \cdot), \bar{D}] \leq M_9 R', \tag{4.20}$$

respectively, where $R' = \max(2R, 2R_1)$, $M_9 = M_9(\alpha, k_0, k_1, k_2, D)$ is a non-negative constant. If we substitute the solution $w_0 = w_0(z) = \xi_0 e_1 + \eta_0 e_2$ of Problem A of (3.1) into the position of $w = \xi e_1 + \eta e_2$ in (4.19), and $\xi_0 = \text{Re}\, w_0 + \text{Im}\, w_0$, $\eta_0 = \text{Re}\, w_0 - \text{Im}\, w_0$, from (4.17) and (3.11), we obtain

$$C_\alpha[G_1(\mu, \cdot), \bar{D}] \le M_{10}, \ C_\alpha[G_2(\cdot, \nu), \bar{D}] \le M_{10},$$
$$C_\alpha[\Psi_1^1(\mu, \cdot), \bar{D}] \le M_{10} R', \ C_\alpha[\Psi_1^2(\cdot, \nu), \bar{D}] \le M_{10} R', \tag{4.21}$$

in which $M_{10} = M_{10}(\alpha, k_0, k_1, k_2, D)$ is a non-negative constant. Due to $\Phi(z) = \Phi_1(z)$ satisfies the complex equation (3.1) and boundary condition (3.18), and $\Phi_1(z)$ possesses a representation similar to that in (3.17), the estimate

$$C_\alpha[\Phi_1(z), \bar{D}] \le M_{11} R' = R' M_{11}(\alpha, k_0, k_1, k_2, D) \tag{4.22}$$

can be derived. Thus setting $w_1(z) = w_0(z) + \Phi_1(z) + \Psi_1(z)$, $\tilde{w}_1(z) = w_1(z) - w_0(z)$, it is clear that the functions $\tilde{w}_1^1(z) = \text{Re}\, \tilde{w}_1(z) + \text{Im}\, \tilde{w}_1(z)$, $\tilde{w}_1^2(z) = \text{Re}\, \tilde{w}_1(z) - \text{Im}\, \tilde{w}_1(z)$ satisfy as functions of $\mu = x + y$, $\nu = x - y$ respectively the estimates:

$$C_\alpha[\tilde{w}_1^1(\cdot, \nu), \bar{D}] \le M_{12} R', \ C_\alpha[\tilde{w}_1^1(\mu, \cdot), \bar{D}] \le M_{12} R',$$
$$C_\alpha[\tilde{w}_1^2(\mu, \cdot), \bar{D}] \le M_{12} R', \ C_\alpha[\tilde{w}_1^2(\cdot, \nu), \bar{D}] \le M_{12} R', \tag{4.23}$$

where $M_{12} = 2M_{13} M(4m + 1)$, $M = 1 + 4k_0^2(1 + k_0)$, $m = C_\alpha[w_0, \bar{D}]$, $M_{13} = \max_{\bar{D}}[|A|, |B|, |C|, |D|, |E|, |F|]$. By using successive iteration, we obtain the sequence of functions: $w_n(z)$ $(n = 1, 2, \ldots)$, and the corresponding functions $\tilde{w}_n^1 = \text{Re}\, \tilde{w}_n + \text{Im}\, \tilde{w}_n$, $\tilde{w}_n^2 = \text{Re}\, \tilde{w}_n - \text{Im}\, \tilde{w}_n$ satisfying the estimates

$$C_\alpha[\tilde{w}_n^1(\cdot, \nu), \bar{D}] \le \frac{(M_{12} R')^n}{n!}, \ C_\alpha[\tilde{w}_n^1(\mu, \cdot), \bar{D}] \le \frac{(M_{12} R')^n}{n!},$$
$$C_\alpha[\tilde{w}_n^2(\mu, \cdot), \bar{D}] \le \frac{(M_{12} R')^n}{n!}, \ C_\alpha[\tilde{w}_n^2(\cdot, \nu), \bar{D}] \le \frac{(M_{12} R')^n}{n!}, \tag{4.24}$$

and denote by $w(z)$ the limit function of $w_n(z) = \sum_{m=0}^n \tilde{w}_n(z)$ in $\bar{D}$, the corresponding functions $\tilde{w}^1 = \text{Re}\, w(z) + \text{Im}\, w(z)$, $\tilde{w}^2 = \text{Re}\, w(z) - \text{Im}\, w(z)$ satisfy the estimates

$$C_\alpha[\tilde{w}^1(\cdot, \nu), \bar{D}] \le e^{M_{12} R'}, \ C_\alpha[\tilde{w}^1(\mu, \cdot), \bar{D}] \le e^{M_{12} R'},$$
$$C_\alpha[\tilde{w}^2(\mu, \cdot), \bar{D}] \le e^{M_{12} R'}, \ C_\alpha[\tilde{w}^2(\cdot, \nu), \bar{D}] \le e^{M_{12} R'}.$$

Combining the first formula in (4.18), (4.20)–(4.24) and the above formulas, the last three estimates in (4.18) are derived.

Theorem 4.5   *Let the quasilinear complex equation (4.1) satisfy Condition C and (4.4). Then the solution $w(z)$ of Problem A for (4.1) satisfies the following estimates*

$$C_\alpha[w(z), \bar{D}] \le M_{14}, \ C_\alpha[w(z), \bar{D}] \le M_{15} k, \tag{4.25}$$

where $k = k_1 + k_2$, $M_{14} = M_{14}(\alpha, k_0, k_1, k_2, D)$, $M_{15} = M_{15}(\alpha, k_0, D)$ are non-negative constants.

Proof According to the proof of Theorem 4.3, from the first formula in (4.25), the second formula in (4.25) is easily derived. Hence we only prove the first formula in (4.25). Similarly to the proof of Theorem 4.4, we see that the function $\Psi_1(z) = \Psi_1^1(\mu, \nu)e_1 + \Psi_1^2(\mu, \nu)e_2$ still possesses the estimate (4.20). Noting that the coefficients are the functions of $z \in \bar{D}$ and w, and applying the condition (4.4), we can derive similar estimates as in (4.21). Hence we also obtain estimates similar to (4.22) and (4.23), and the constant M_{12} in (4.23) can be chosen as $M_{12} = 2M_{13}M(4m+1)$, $m = C_\alpha[w_0(z), \bar{D}]$. Thus the first estimate in (4.25) can be derived.

Moreover, according to the above methods, we can obtain estimates for $[\text{Re}\,w + \text{Im}\,w]_\nu$, $[\text{Re}\,w - \text{Im}\,w]_\mu$ analogous to those in (4.14) and (4.25).

4.4 The boundary value problem for quasilinear hyperbolic equations of first order in general domains

In this subsection, we shall generalize the domain D to general cases.

1. The boundary L_1 of the domain D is replaced by a curve L_1', and the boundary of the domain D' is $L_1' \cup L_2' \cup L_3 \cup L_4$, where the parameter equations of the curves L_1', L_2' are as follows:

$$L_1' = \{\gamma_1(x) + y = 0, 0 \le x \le l\}, \ L_2' = \{x - y = 2R_1, l \le x \le R_2\}, \qquad (4.26)$$

in which $\gamma_1(x)$ on $0 \le x \le l = 2R_1 - \gamma_1(l)$ is continuous and $\gamma_1(0) = 0$, $\gamma_1(x) > 0$ on $0 < x \le l$, and $\gamma_1(x)$ is differentiable except at isolated points on $0 \le x \le l$ and $1 + \gamma_1'(x) > 0$. By this condition, the inverse function $x = \sigma(\nu)$ of $x + \gamma_1(x) = \nu$ can be found, and $\sigma'(\nu) = 1/[1 + \gamma_1'(x)]$, hence the curve L_1' can be expressed by $x = \sigma(\nu) = (\mu + \nu)/2$, i.e. $\mu = 2\sigma(\nu) - \nu$, $0 \le \nu \le 2R_1$. We make a transformation

$$\tilde{\mu} = \frac{2R[\mu - 2\sigma(\nu) + \nu]}{2R - 2\sigma(\nu) + \nu}, \quad \tilde{\nu} = \nu, \quad 2\sigma(\nu) - \nu \le \mu \le 2R, 0 \le \nu \le 2R_1, \qquad (4.27)$$

its inverse transformation is

$$\mu = \frac{1}{2R}[2R - 2\sigma(\nu) + \nu]\tilde{\mu} + 2\sigma(\nu) - \nu, \quad \nu = \tilde{\nu}, \quad 0 \le \tilde{\mu} \le 2R, 0 \le \tilde{\nu} \le 2R_1. \qquad (4.28)$$

The transformation (4.27) can be expressed by

$$
\begin{cases}
\tilde{x} = \dfrac{1}{2}(\tilde{\mu} + \tilde{\nu}) \\
\quad = \dfrac{2R(x+y) + 2R(x-y) - (2R + x - y)[2\sigma(x + \gamma_1(x)) - x - \gamma_1(x)]}{4R - 4\sigma(x + \gamma_1(x)) + 2x + 2\gamma_1(x)}, \\
\tilde{y} = \dfrac{1}{2}(\tilde{\mu} - \tilde{\nu}) \\
\quad = \dfrac{2R(x+y) - 2R(x-y) - (2R - x + y)[2\sigma(x + \gamma_1(x)) - x - \gamma_1(x)]}{4R - 4\sigma(x + \gamma_1(x)) + 2x + 2\gamma_1(x)},
\end{cases}
\qquad (4.29)
$$

where $\gamma_1(x) = -y$, and its inverse transformation is

$$
\begin{cases}
x = \dfrac{1}{2}(\mu + \nu) = \dfrac{[2R - 2\sigma(x + \gamma_1(x)) + x + \gamma_1(x)](\tilde{x} + \tilde{y})}{4R} \\
\qquad\qquad\qquad + \sigma(x + \gamma_1(x)) - \dfrac{x + \gamma_1(x) - \tilde{x} + \tilde{y}}{2}, \\
y = \dfrac{1}{2}(\mu - \nu) = \dfrac{[2R - 2\sigma(x + \gamma_1(x)) + x + \gamma_1(x)](\tilde{x} + \tilde{y})}{4R} \\
\qquad\qquad\qquad + \sigma(x + \gamma_1(x)) - \dfrac{x + \gamma_1(x) + \tilde{x} - \tilde{y}}{2}.
\end{cases}
\tag{4.30}
$$

Denote by $\tilde{z} = \tilde{x} + j\tilde{y} = f(z)$, $z = x + jy = f^{-1}(\tilde{z})$ the transformation (4.29) and the inverse transformation (4.30) respectively. In this case, the system of equations and boundary conditions are

$$
\xi_\nu = A\xi + B\eta + E, \quad \eta_\mu = C\xi + D\eta + F, \quad z \in \overline{D'}, \tag{4.31}
$$

$$
\mathrm{Re}\,[\lambda(z)w(z)] = r(z), \quad z \in L_1' \cup L_2', \quad \mathrm{Im}\,[\lambda(z_0)w(z_0)] = b_1, \tag{4.32}
$$

in which $z_0' = l - j\gamma_1(l)$, $\lambda(z)$, $r(z)$, b_1 on $L_1' \cup L_2'$ satisfy the conditions (3.3),(3.4). Suppose system (4.31) in $\overline{D'}$ satisfies Condition C, through the transformation (4.28) and $\xi_{\tilde{\nu}} = \xi_\nu$, $\eta_{\tilde{\mu}} = [2R - 2\sigma(\nu) + \nu]\eta_\mu/2R$, system (4.31) is reduced to

$$
\xi_{\tilde{\nu}} = A\xi + B\eta + E, \quad \eta_{\tilde{\mu}} = [2R - 2\sigma(\nu) + \nu]\frac{[C\xi + D\eta + F]}{2R}. \tag{4.33}
$$

Moreover, through the transformation (4.30), i.e. $z = f^{-1}(\tilde{z})$, the boundary condition (4.32) is reduced to

$$
\mathrm{Re}\,[\lambda(f^{-1}(\tilde{z}))w(f^{-1}(\tilde{z}))] = r(f^{-1}(\tilde{z})), \quad \tilde{z} \in L_1 \cup L_2,
$$
$$
\mathrm{Im}\,[\lambda(f^{-1}(\tilde{z}_0))w(f^{-1}(\tilde{z}_0))] = b_1,
\tag{4.34}
$$

in which $\tilde{z}_0 = f(z_0) = 0$. Therefore the boundary value problem (4.31),(4.32) is transformed into the boundary value problem (4.33),(4.34). On the basis of Theorems 4.1 and 4.2, we see that the boundary value problem (4.33), (4.34) has a unique solution $w(\tilde{z})$, and then $w[f(z)]$ is just a solution of the boundary value problem (4.31),(4.32) in D'.

Theorem 4.6 *If the complex equation* (4.1) *satisfies Condition C in the domain D' with the boundary $L_1' \cup L_2' \cup L_3 \cup L_4$, where L_1', L_2' are as stated in* (4.26), *then Problem A with the boundary conditions*

$$
\mathrm{Re}\,[\lambda(z)w(z)] = r(z),\ z \in L_1' \cup L_2',\ \mathrm{Im}\,[\lambda(z_0)w(z_0)] = b_1
$$

has a unique solution $w(z)$.

2. The boundary L_1, L_4 of the domain D are replaced by the two curves L_1'', L_4'' respectively, and the boundary of the domain D'' is $L_1'' \cup L_2'' \cup L_3'' \cup L_4''$, where the parameter equations of the curves L_1'', L_2'', L_3'', L_4'' are as follows:

$$L_1'' = \{\gamma_1(x) + y = 0, 0 \le x \le l_1\}, \quad L_2'' = \{x - y = 2R_1, l_1 \le x \le R_2\},$$
$$L_3'' = \{x + y = 2R, l_2 \le x \le R_2\}, \quad L_4'' = \{-\gamma_4(x) + y = 0, 0 \le x \le l_2\}, \tag{4.35}$$

in which and $\gamma_1(0) = 0$, $\gamma_4(R_2) = 2R - R_2$; $\gamma_1(x) > 0, 0 \le x \le l_1$; $\gamma_4(x) > 0, 0 \le x \le l_2$; $\gamma_1(x)$ on $0 \le x \le l_1$, $\gamma_4(x)$ on $0 \le x \le l_2$ are continuous, and $\gamma_1(x)$, $\gamma_4(x)$ possess derivatives except at finite points on $0 \le x \le l_1$, $0 \le x \le l_2$ respectively, and $1 + \gamma_1'(x) > 0$, $1 + \gamma_4'(x) > 0$, $z_1'' = x - j\gamma_1(l_1) \in L_2$, $z_3'' = x + j\gamma_2(l_2) \in L_3$. By the conditions, the inverse functions $x = \sigma(\nu)$, $x = \tau(\mu)$ of $x + \gamma_1(x) = \nu$, $x + \gamma_4(x) = \mu$ can be found respectively, namely

$$\mu = 2\sigma(\nu) - \nu, 0 \le \nu \le l_1 + \gamma_1(l_1), \quad \nu = 2\tau(\mu) - \mu, l_2 + \gamma_4(l_2) \le \mu \le 2R_1. \tag{4.36}$$

We make a transformation

$$\tilde{\mu} = \mu, \quad \tilde{\nu} = \frac{2R_1(\nu - 2\tau(\mu) + \mu)}{2R_1 - 2\tau(\mu) + \mu}, \quad 0 \le \mu \le 2R, 2\tau(\mu) - \mu \le \mu \le 2R_1, \tag{4.37}$$

its inverse transformation is

$$\mu = \tilde{\mu}, \quad \nu = \frac{2R_1 - 2\tau(\mu) + \mu}{2R_1}\tilde{\nu} + 2\tau(\mu) - \nu, \quad 0 \le \tilde{\mu} \le 2R, \quad 0 \le \tilde{\nu} \le 2R_1. \tag{4.38}$$

Hence we have

$$\tilde{x} = \frac{1}{2}(\tilde{\mu} + \tilde{\nu}) = \frac{4R_1 x - [2\tau(x + \gamma_1(x)) - x - \gamma_1(x)](2R_1 + x + y)}{4R_1 - 4\tau(x + \gamma_1(x)) + 2x + 2\gamma_1(x)},$$
$$\tilde{y} = \frac{1}{2}(\tilde{\mu} - \tilde{\nu}) = \frac{2R_1 y + [2\tau(x + \gamma_1(x)) - x - \gamma_1(x)](2R_1 - x - y)}{4R_1 - 4\tau(x + \gamma_1(x)) + 2x + 2\gamma_1(x)}, \tag{4.39}$$

and its inverse transformation is

$$x = \frac{1}{2}(\mu + \nu) = \frac{4R_1 \tilde{x} + [2\tau(x + \gamma_1(x)) - x - \gamma_1(x)](2R_1 - \tilde{x} + \tilde{y})}{4R_1},$$
$$y = \frac{1}{2}(\mu - \nu) = \frac{4R_1 \tilde{y} - [2\tau(x + \gamma_1(x)) - x - \gamma_1(x)](2R_1 + \tilde{x} - \tilde{y})}{4R_1}. \tag{4.40}$$

Denote by $\tilde{z} = \tilde{x} + j\tilde{y} = g(z)$, $z = x + jy = g^{-1}(\tilde{z})$ the transformation (4.39) and its inverse transformation (4.40) respectively. Through the transformation (4.38), we have

$$(u + v)_{\tilde{\nu}} = \frac{2R_1 - 2\tau(\mu) + \mu}{2R_1}(u + v)_\nu, \quad (u - v)_{\tilde{\mu}} = (u - v)_\mu,$$

system (4.31) in $\overline{D''}$ is reduced to

$$\xi_{\tilde{\nu}} = \frac{2R_1 - 2\tau(\mu) + \mu}{2R_1}[A\xi + B\eta + E], \quad \eta_{\tilde{\mu}} = C\xi + D\eta + F, \quad z \in D', \tag{4.41}$$

where D' is a bounded domain with boundary $L_1' \cup L_2' \cup L_3 \cup L_4$ and $L_1' = L_1''$. Moreover, through the transformation (4.40), the boundary condition (4.32) on $L_1'' \cup L_4''$ is reduced to

$$\text{Re}\,[\lambda(g^{-1}(\tilde{z}))w(g^{-1}(\tilde{z}))] = r(g^{-1}(\tilde{z})),\ \tilde{z} \in L_1' \cup L_4,$$

$$\text{Im}\,[\lambda(g^{-1}(\tilde{z}_0))w(g^{-1}(\tilde{z}_0))] = b_1,$$

(4.42)

in which $\tilde{z}_0 = g(z_0) = 0$. Therefore the boundary value problem (4.31),(4.32) in D'' is transformed into the boundary value problem (4.41),(4.42). On the basis of Theorem 4.6, we see that the boundary value problem (4.41),(4.42) has a unique solution $w(\tilde{z})$, and then $w[g(z)]$ is just a solution of the boundary value problem (4.31),(4.32).

Theorem 4.7   *If the complex equation* (4.1) *satisfies Condition C in the domain D'' with the boundary $L_1'' \cup L_2'' \cup L_3'' \cup L_4''$, then Problem A with the boundary conditions*

$$\text{Re}\,[\lambda(z)w(z)] = r(z),\ z \in L_1'' \cup L_4'',\ \text{Im}\,[\lambda(z_0)w(z_0)] = b_1$$

has a unique solution $w(z)$, where $z_1'' = x - j\gamma_1(l_1) \in L_2$, $z_3'' = x + j\gamma_4(l_2) \in L_3$.

Now, we give an example to illustrate the above results. When $R_2 = 2R_1$, the boundary of the domain D is $L_1 = \{x = -y, 0 \le x \le R_1\}$, $L_2 = \{x = y + 2R_1, R_1 \le x \le R_1\}$, $L_3 = \{x = -y + 2R_1, R_1 \le x \le 2R_1\}$, $L_4 = \{x = y, 0 \le x \le R_1\}$. We replace $L_1 \cup L_4$ by a left semi-circumference $L_1'' \cup L_4''$ with the center R_1 and the radius R_1, namely

$$L_1'' = \{x - y = \nu,\ y = -\gamma_1(x) = -\sqrt{R_1^2 - (x - R_1)^2},\ 0 \le x \le R_1\},$$

$$L_4'' = \{x + y = \mu,\ y = \gamma_4(x) = \sqrt{R_1^2 - (x - R_1)^2},\ 0 \le x \le R_1\},$$

where $1 + \gamma_1'(x) > 0$, $0 < x \le R_1$, $1 + \gamma_4'(x) > 0$, $0 \le x < R_1$, and

$$x = \sigma(\nu) = \frac{R_1 + \nu - \sqrt{R_1^2 + 2R_1\nu - \nu^2}}{2},$$

$$x = \tau(\mu) = \frac{R_1 + \mu - \sqrt{R_1^2 + 2R_1\mu - \mu^2}}{2}.$$

It is clear that according to the above method, the domain D can be generalized to a general domain D'', namely its boundary consists of the general curves $L_1'', L_2'', L_3'', L_4''$ with some conditions, which includes the circumference $L = \{|z - R_1| = R_1\}$.

Finally, we mention that some boundary value problems for equations (3.12) and (4.1) with one of the boundary conditions

$$\text{Re}\,[\lambda(z)w(z)] = r(z),\ z \in L_1 \cup L_2,\ \text{Im}\,[\lambda(z_1)w(z_1)] = b_1,$$

$$\text{Re}\,[\lambda(z)w(z)] = r(z),\ z \in L_3 \cup L_4,\ \text{Im}\,[\lambda(z_3)w(z_3)] = b_1,$$

$$\text{Re}\,[\lambda(z)w(z)] = r(z),\ z \in L_1 \cup L_5,\ \text{Im}\,[\lambda(z_0)w(z_0)] = b_1,$$

$$\text{Re}\,[\lambda(z)w(z)] = r(z),\ z \in L_1 \cup L_6,\ \text{Im}\,[\lambda(z_0)w(z_0)] = b_1,$$

can be discussed, where $\lambda(z), r(z), b_1$ satisfy the conditions similar to those in (3.3),(3.4),(3.31) and $L_j(j = 1,\ldots,5)$, $z_j(j = 0,1,3)$ are as stated in Section 3, $L_6 = \{y = \dfrac{R_2 - 2R_1}{R_2 x}, 0 \le x \le R_2\}$, and $\lambda(z) = 1 + j, z \in L_5$, $\lambda(z) = 1 - j$, $z \in L_6$.

For corresponding boundary value problems of hyperbolic systems of first order complex equations, whether there are similar results as before? The problem needs to be investigated.

5 Hyperbolic Mappings and Quasi-hyperbolic Mappings

Now, we introduce the definitions of hyperbolic mappings and quasi-hyperbolic mappings, and prove some properties of quasi-hyperbolic mappings.

5.1 Hyperbolic mappings

A so-called hyperbolic mapping in a domain D is a univalent mapping given by a hyperbolic continuously differentiable function $w = f(z) = u + jv$ satisfying the simplest hyperbolic system of first order equations

$$u_x = v_y, \quad v_x = u_y, \tag{5.1}$$

which maps D onto a domain G in the w-plane. By Theorem 1.1, system (5.1) is equivalent to the system

$$\xi_\nu = 0, \quad \eta_\mu = 0, \tag{5.2}$$

where $\xi = u + v$, $\eta = u - v$, $\mu = x + y$, $\nu = x - y$. Noting

$$\begin{vmatrix} \mu_x & \nu_x \\ \mu_y & \nu_y \end{vmatrix} = \begin{vmatrix} 1 & 1 \\ 1 & -1 \end{vmatrix} = -2, \quad \begin{vmatrix} \xi_u & \eta_u \\ \xi_v & \eta_v \end{vmatrix} = \begin{vmatrix} 1 & 1 \\ 1 & -1 \end{vmatrix} = -2, \tag{5.3}$$

if we find a homeomorphic solution of (5.2), then the solution

$$w = u + jv = \xi[(\mu + \nu)/2, (\mu - \nu)/2]e_1 + \eta[(\mu + \nu)/2, (\mu - \nu)/2]e_2 \tag{5.4}$$

of the corresponding system (5.1) is also homeomorphic. In fact,

$$\xi = \xi(\mu), \quad \eta = \eta(\nu) \tag{5.5}$$

is a homeomorphic solution of (5.2), if $\xi(\mu)$ and $\eta(\nu)$ are strictly monotonous continuous functions of $\mu(\mu_0 \le \mu \le \mu_1)$ and $\nu(\nu_0 \le \nu \le \nu_1)$ respectively. When $[\xi(\mu), \eta(\nu)]$ is univalent continuous in $\Delta = \{\mu_0 \le \mu \le \mu_1, \nu_0 \le \nu \le \nu_1\}$ and $D(\Delta)$ is the closed domain in the $z = x + jy$-plane corresponding to Δ, it is easy to see that $[u(x,y), v(x,y)]$ is a homeomorphic solution of (5.1) in $D(\Delta)$.

5.2   Quasi-hyperbolic mappings

In this subsection, we first discuss the uniformly hyperbolic system in the complex form

$$w_{\bar{z}} = Q(z)w_z, \quad Q(z) = a(z) + jb(z), \tag{5.6}$$

where $Q(z)$ is a continuous function satisfying the condition $|Q(z)| \le q_0 < 1$, here q_0 is a non-negative constant. On the basis of the representation

$$w_{\bar{z}} = \xi_\nu e_1 + \eta_\mu e_2, \; w_z = \xi_\mu e_1 + \eta_\nu e_2, \; Q = q_1 e_1 + q_2 e_2,$$

from (5.6), it follows that

$$\xi_\nu = q_1 \xi_\mu, \quad \eta_\mu = q_2 \eta_\nu. \tag{5.7}$$

Due to $Q = a + jb = q_1 e_1 + q_2 e_2$, here $q_1 = a + b$, $q_2 = a - b$, thus $|Q|^2 = |Q\bar{Q}| = |a^2 - b^2| = |q_1 q_2| \le q_0^2 < 1$, and a representation theorem of solutions for (5.6) can be obtained.

Theorem 5.1 *Let $\chi(z)$ be a homeomorphic solution of* (5.6) *in a domain D, and $w(z)$ be a solution of* (5.6) *in D. Then $w(z)$ can be expressed as*

$$w(z) = \Phi[\chi(z)], \tag{5.8}$$

where $\Phi(\chi)$ is a hyperbolic regular function in the domain $G = \chi(D)$.

Proof Suppose that $z(\chi)$ is the inverse function of $\chi(z)$, we can find

$$\{w[z(\chi)]\}_{\bar{\chi}} = w_z z_{\bar{\chi}} + w_{\bar{z}} \bar{z}_{\bar{\chi}} = w_z[z_{\bar{\chi}} + Q\bar{z}_{\bar{\chi}}],$$

and

$$\chi_\chi = 1 = \chi_z z_\chi + \chi_{\bar{z}} \bar{z}_\chi = \chi_z[z_\chi + Q\bar{z}_\chi],$$

$$\chi_{\bar{\chi}} = 0 = \chi_z z_{\bar{\chi}} + \chi_{\bar{z}} \bar{z}_{\bar{\chi}} = \chi_z[z_{\bar{\chi}} + Q\bar{z}_{\bar{\chi}}].$$

From the above equalities, we see $\chi_z \ne 0$, $z_{\bar{\chi}} + Q\bar{z}_{\bar{\chi}} = 0$, consequently $\Phi(\chi) = w[z(\chi)]$ satisfies

$$[\Phi(\chi)]_{\bar{\chi}} = 0, \quad \chi \in G = \chi(D).$$

This shows that $\Phi(\chi)$ is a hyperbolic regular function in $G = \chi(D)$, therefore the representation (5.8) holds.

Next we prove the existence of a homeomorphic solution of equation (5.6) with some conditions for the coefficient of (5.6).

From (5.7), we see that the complex equation (5.6) can be written in the form

$$\xi_\nu = (a + b)\xi_\mu, \quad \eta_\mu = (a - b)\eta_\nu. \tag{5.9}$$

Let $\Delta = \{\mu_0 \le \mu \le \mu_0 + R_1, \nu_0 \le \nu \le \nu_0 + R_2\}$, in which μ_0, ν_0 are two real numbers and R_1, R_2 are two positive numbers, if a, b possess continuously differentiable derivatives with respect to μ, ν in Δ, then the solution $w = \xi e_1 + \eta e_2$ of (5.9) in Δ is a homeomorphism, provided that one of the following sets of conditions

$$b > 0, \quad -b < a < b, \tag{5.10}$$

$$b < 0, \quad b < a < -b, \tag{5.11}$$

holds and ξ and η are strictly monotonous continuous functions of $\mu \, (\mu_0 \leq \mu \leq \mu_0 + R_1)$ and $\nu \, (\nu_0 \leq \nu \leq \nu_0 + R_2)$ respectively.

Thus we have the following theorem.

Theorem 5.2 *Denote by D the corresponding domain of Δ in the $(x + jy)$-plane and let $w(z)$ be a continuous solution of (5.6) in D. If (5.10) or (5.11) in D holds, $\xi_\mu(\xi_\nu) > 0$ and $\eta_\mu(\eta_\nu) > 0$ except some possible isolated points on $\mu \, (\mu_0 \leq \mu \leq \mu_0 + R_1) \, (\nu \, (\nu_0 \leq \nu \leq \nu_0 + R_2))$, then the solution $w(z)$ in D is a homeomorphism.*

In particular, if the coefficient $Q(z) = a + jb$ of the complex equation (5.6) is a hyperbolic constant, which satisfies the condition

$$|Q(z)|^2 = |Q\overline{Q}| = |a^2 - b^2| = |q_1 q_2| \leq q_0^2 < 1, \quad z \in D,$$

we make a nonsingular transformation

$$\mu = -(a + b)\sigma + \tau, \quad \nu = \sigma - (a - b)\tau. \tag{5.12}$$

Thus system (5.9) can be transformed into the system

$$\xi_\sigma = 0, \quad \eta_\tau = 0, \quad (\sigma, \tau) \in G, \tag{5.13}$$

where the domain G is the corresponding domain of Δ under the transformation (5.12). According to the discussion of hyperbolic mappings, we see that system (5.13) in G possesses a homeomorphic solution, hence system (5.9) in Δ has a homeomorphic solution and then the complex equation (5.6) in D has a homeomorphic solution. The above result can be written as a theorem.

Theorem 5.3 *Suppose that $Q(z) = a + jb$ is a hyperbolic constant and $|Q(z)| < 1$. Then the complex equation (5.6) in D has a homeomorphic solution.*

5.3 Other Quasi-hyperbolic mappings

Now, we consider the hyperbolic complex equation

$$\bar{w}_z = Q(z)w_z, \tag{5.14}$$

where $Q(z)$ is a continuous function in $D = \{0 \leq x \leq R_1, 0 \leq y \leq R_2\}$ satisfying the condition $|Q(z)| \leq q_0 < 1$. If $Q(z)$ in D is a hyperbolic regular function of $\bar{z}$, we introduce a transformation of functions

$$W = \bar{w} - Qw, \quad \text{i.e. } w = \frac{\overline{W} + \overline{Q(z)}W}{1 - |Q(z)|^2}. \tag{5.15}$$

Then (5.14) is reduced to the complex equation

$$W_z = 0. \tag{5.16}$$

The solution $W(z)$ of (5.16) in D is a hyperbolic regular function of $\bar{z}$. As stated before, the complex equation (5.16) possesses a homeomorphic solution, which realizes a hyperbolic mapping in D. Moreover, if $Q(z)$ is a hyperbolic regular function in D, we find the partial derivative with respect to $\bar{z}$ in (5.14), and obtain

$$\bar{w}_{z\bar{z}} = Q(z)w_{z\bar{z}}, \quad \text{i.e.} \quad w_{z\bar{z}} = \overline{Q(z)}\bar{w}_{z\bar{z}}, \tag{5.17}$$

from $|Q(z)| < 1$, it follows that

$$(1 - |Q(z)|^2)w_{z\bar{z}} = 0, \quad \text{i.e.} \quad w_{z\bar{z}} = 0, \tag{5.18}$$

the solution of the above complex equation (5.18) is called a hyperbolic harmonic complex function.

A hyperbolic harmonic complex function $w(z)$ can be expressed as

$$w(z) = u(z) + jv(z) = \phi(z) + \overline{\phi(z)} + \psi(z) - \overline{\psi(z)}$$
$$= \phi(z) + \psi(z) + \overline{\phi(z)} - \overline{\psi(z)} = f(z) + \overline{g(z)},$$

in which $\phi(z), \psi(z)$ are hyperbolic regular functions, hence $f(z) = \phi(z) + \psi(z)$, $g(z) = \phi(z) - \psi(z)$ are hyperbolic regular functions. This is a representation of hyperbolic harmonic functions through hyperbolic regular functions. Hence in order to find a hyperbolic harmonic function, it is sufficient to find solutions of the following two boundary value problems with the boundary conditions

$$\text{Re}\, f(x) = \text{Re}\, \phi_0(x), \ \text{Re}\, f(jy) = \text{Re}\, \phi_1(y),$$

and

$$\text{Im}\, g(x) = \text{Im}\, \phi_0(x), \ \text{Im}\, g(jy) = \text{Im}\, \phi_1(y),$$

respectively, where $\phi_0(x), \phi_1(y)$ are given hyperbolic complex functions on $\{0 \leq x \leq R_1\}$, $\{0 \leq y \leq R_2\}$ respectively, and R_1, R_2 are two positive constants.

At last we mention that the notations of hyperbolic numbers and hyperbolic complex functions are mainly used in this and next chapters. From Chapter III to Chapter VI except in Section 5, Chapter V, we do not use them.

The references for this chapter are [5],[9],[12],[19],[26],[29],[32],[34],[38],[44],[51], [59],[68],[74],[80],[83],[85],[87],[89],[92],[97].

CHAPTER II

HYPERBOLIC COMPLEX EQUATIONS OF SECOND ORDER

In this chapter, we mainly discuss oblique derivative boundary value problems for linear and quasilinear hyperbolic equations of second order in a simply connected domain. Firstly, we transform some linear and nonlinear uniformly hyperbolic equations of second order with certain conditions into complex forms, give the uniqueness theorem of solutions for the above boundary value problems. Moreover by using the successive iteration, the existence of solutions for several oblique derivative problems is proved. Finally we introduce some boundary value problems for degenerate hyperbolic equations of second order with certain conditions.

1 Complex Form of Hyperbolic Equations of Second Order

This section deals with hyperbolic equations of second order in the plane domains, we first transform some linear and nonlinear uniformly hyperbolic equations of second order with certain conditions into complex forms, and then we state the conditions of some hyperbolic complex equations of second order.

1.1 Reduction of linear and nonlinear hyperbolic equations of second order

Let D be a bounded domain, we consider the linear hyperbolic partial differential equation of second order

$$au_{xx} + 2bu_{xy} + cu_{yy} + du_x + eu_y + fu = g, \qquad (1.1)$$

where the coefficients a, b, c, d, e, f, g are known continuous functions of $(x, y) \in D$, in which D is a bounded domain. The condition of hyperbolic type for (1.1) is that for any point (x, y) in D, the inequality

$$I = ac - b^2 < 0, \quad a > 0, \qquad (1.2)$$

holds. If a, b, c are bounded in D and

$$I = ac - b^2 \leq I_0 < 0, \quad a > 0 \qquad (1.3)$$

in D, where I_0 is a negative constant, then equation (1.1) is called uniformly hyperbolic in D. Introduce the notations as follows

$$(\ )_z = \frac{(\ )_x + j(\ )_y}{2}, \ (\ )_{\bar z} = \frac{(\ )_x - j(\ )_y}{2}, \ (\ )_{z\bar z} = \frac{(\ )_{xx} - (\ )_{yy}}{4},$$

$$(\ )_{zz} = \frac{(\ )_{xx} + (\ )_{yy} + 2j(\ )_{xy}}{4}, \ (\ )_{\bar z \bar z} = \frac{(\ )_{xx} + (\ )_{yy} - 2j(\ )_{xy}}{4}, \tag{1.4}$$

$$(\ )_x = (\ )_z + (\ )_{\bar z}, \ (\ )_y = j[(\ )_z - (\ )_{\bar z}], \ (\ )_{xy} = j[(\ )_{zz} - (\ )_{\bar z \bar z}],$$

$$(\ )_{xx} = (\ )_{zz} + (\ )_{\bar z \bar z} + 2(\ )_{z\bar z}, \ (\ )_{yy} = (\ )_{zz} + (\ )_{\bar z \bar z} - 2(\ )_{z\bar z},$$

equation (1.1) can be written in the form

$$2(a - c)u_{z\bar z} + (a + c + 2bj)u_{zz} + (a + c - 2bj)u_{\bar z \bar z}$$
$$+ (d + ej)u_z + (d - ej)u_{\bar z} + fu = g \text{ in } D. \tag{1.5}$$

If $a \neq c$ in D, then equation (1.5) can be reduced to the complex form

$$u_{z\bar z} - \text{Re}\,[Q(z)u_{zz} + A_1(z)u_z] - A_2(z)u = A_3(z) \text{ in } D, \tag{1.6}$$

in which

$$Q = \frac{a + c + 2bj}{a - c}, \ A_1 = \frac{d + ej}{a - c}, \ A_2 = \frac{f}{2(a - c)}, \ A_3 = \frac{g}{2(a - c)}.$$

If $(a + c)^2 \geq 4b^2$, then the conditions of hyperbolic type and uniformly hyperbolic are transformed into

$$|Q(z)| < 1 \text{ in } D, \tag{1.7}$$

and

$$|Q(z)| \leq q_0 < 1 \text{ in } D, \tag{1.8}$$

respectively.

For the nonlinear hyperbolic equation of second order

$$\Phi(x, y, u, u_x, u_y, u_{xx}, u_{xy}, u_{yy}) = 0 \text{ in } D, \tag{1.9}$$

from (1.4) we have $\Phi = F(z, u, u_z, u_{zz}, u_{z\bar z})$. Under certain conditions, equation (1.9) can be reduced to the real form

$$au_{xx} + 2bu_{xy} + cu_{yy} + du_x + eu_y + fu = g \text{ in } D, \tag{1.10}$$

and its complex form is as follows

$$a_0 u_{z\bar z} - \text{Re}\,[qu_{zz} + a_1 u_z] - a_2 u = a_3 \text{ in } D, \tag{1.11}$$

in which

$$a = \int_0^1 \Phi_{\tau u_{xx}}(x,y,u,u_x,u_y,\tau u_{xx},\tau u_{xy},\tau u_{yy})d\tau = a(x,y,u,u_x,u_y,u_{xx},u_{xy},u_{yy}),$$

$$2b = \int_0^1 \Phi_{\tau u_{xy}}(x,y,u,u_x,u_y,\tau u_{xx},\tau u_{xy},\tau u_{yy})d\tau = 2b(x,y,u,u_x,u_y,u_{xx},u_{xy},u_{yy}),$$

$$c = \int_0^1 \Phi_{\tau u_{yy}}(x,y,u,u_x,u_y,\tau u_{xx},\tau u_{xy},\tau u_{yy})d\tau = c(x,y,u,u_x,u_y,u_{xx},u_{xy},u_{yy}),$$

$$d = \int_0^1 \Phi_{\tau u_x}(x,y,u,\tau u_x,\tau u_y,0,0,0)d\tau = d(x,y,u,u_x,u_y),$$

$$e = \int_0^1 \Phi_{\tau u_y}(x,y,u,\tau u_x,\tau u_y,0,0,0)d\tau = e(x,y,u,u_x,u_y),$$

$$f = \int_0^1 \Phi_{\tau u}(x,y,\tau u,0,0,0,0,0)d\tau = f(x,y,u),$$

$$g = -\Phi(x,y,0,0,0,0,0,0) = g(x,y),$$

and

$$a_0 = 2(a-c) = \int_0^1 F_{\tau u_{z\bar{z}}}(z,u,u_z,\tau u_{zz},\tau u_{z\bar{z}})d\tau = a_0(z,u,u_z,u_{zz},u_{z\bar{z}}),$$

$$q = 2(a+c+2bj) = -2\int_0^1 F_{\tau u_{zz}}(z,u,u_z,\tau u_{zz},\tau u_{z\bar{z}})d\tau = q(z,u,u_z,u_{zz},u_{z\bar{z}}),$$

$$a_1 = 2(d+ej) = -2\int_0^1 F_{\tau u_z}(z,u,\tau u_z,0,0,0)d\tau = a_1(z,u,u_z), \qquad (1.12)$$

$$a_2 = f = -\int_0^1 F_{\tau u}(z,\tau u,0,0,0,0)d\tau = a_2(z,u),$$

$$a_3 = -F(z,0,0,0,0) = a_3(z).$$

The condition of uniformly hyperbolic type for equation (1.10) is the same with (1.3). If $a \neq c$ in D, the complex equation (1.11) can be rewritten in the form

$$u_{z\bar{z}} - \operatorname{Re}\left[Qu_{zz} + A_1 u_z\right] - A_2 u = A_3 \ \text{ in } D, \qquad (1.13)$$

where

$$Q = q/a_0, \quad A_1 = a_1/a_0, \quad A_2 = a_2/a_0, \quad A_3 = a_3/a_0$$

are functions of $z \in D, u, u_z, u_{zz}, u_{z\bar{z}}$, the condition of uniformly hyperbolic type for (1.13) is as stated in the form (1.8).

As stated in [12] 3), for the linear hyperbolic equation (1.1) or its complex form (1.6), if the coefficients a, b, c are sufficiently smooth, through a nonsingular transformation of z, equation (1.1) can be reduced to the standard form

$$u_{xx} - u_{yy} + du_x + eu_y + fu = g \qquad (1.14)$$

or its complex form

$$u_{z\bar{z}} - \operatorname{Re}\left[A_1(z)u_z\right] - A_2(z)u = A_3(z). \qquad (1.15)$$

1.2   Conditions of some hyperbolic equations of second order

Let D be a simply connected bounded domain with the boundary $\Gamma = L_1 \cup L_2 \cup L_3 \cup L_4$ as stated in Chapter I, where $L_1 = \{x = -y, 0 \leq x \leq R_1\}$, $L_2 = \{x = y + 2R_1, R_1 \leq x \leq R_2\}$, $L_3 = \{x = -y - 2R_1 + 2R_2, R_2 - R_1 \leq x \leq R_2\}$, $L_4 = \{x = y, 0 \leq x \leq R_2 - R_1\}$, and denote $z_0 = 0$, $z_1 = (1-j)R_1$, $z_2 = R_2 + j(R_2 - 2R_1)$, $z_3 = (1+j)(R_2 - R_1)$, $L = L_3 \cup L_4$, here there is no harm in assuming that $R_2 \geq 2R_1$. In the following, we mainly consider second order quasilinear hyperbolic equation in the form

$$u_{z\bar{z}} - \mathrm{Re}\,[A_1(z, u, u_z)u_z] - A_2(z, u, u_z)u = A_3(z, u, u_z), \qquad (1.16)$$

whose coefficients satisfy the following conditions: **Condition C**

1) $A_l(z, u, u_z)(l = 1, 2, 3)$ are continuous in $z \in \bar{D}$ for all continuously differentiable functions $u(z)$ in $\bar{D}$ and satisfy

$$C[A_l, \bar{D}] = C[\mathrm{Re}\,A_l, \bar{D}] + C[\mathrm{Im}\,A_l, \bar{D}] \leq k_0, \ l = 1, 2, \ C[A_3, \bar{D}] \leq k_1. \qquad (1.17)$$

2) For any continuously differentiable functions $u_1(z), u_2(z)$ in $\bar{D}$, the equality

$$F(z, u_1, u_{1z}) - F(z, u_2, u_{2z}) = \mathrm{Re}\,[\tilde{A}_1(u_1 - u_2)_z] + \tilde{A}_2(u_1 - u_2) \text{ in } \bar{D} \qquad (1.18)$$

holds, where

$$C[\tilde{A}_l(z, u_1, u_2), \bar{D}] \leq k_0, \ l = 1, 2, \qquad (1.19)$$

in (1.17),(1.19), k_0, k_1 are non-negative constants. In particular, when (1.16) is a linear equation, from (1.17) it follows that the conditions (1.18), (1.19) hold.

In order to give a priori estimates in $C_\alpha(\bar{D})$ of solutions for some boundary value problems, we need to add the following conditions: For any two real numbers u_1, u_2 and hyperbolic numbers $z_1, z_2 \in \bar{D}$, w_1, w_2, the above functions satisfy

$$\begin{aligned}
&\| A_l(z_1, u_1, w_1) - A_l(z_2, u_2, w_2) \| \\
&\leq k_0[\| z_1 - z_2 \|^\alpha + \| u_1 - u_2 \|^\alpha + \| w_1 - w_2 \|], l = 1, 2, \\
&\| A_3(z_1, u_1, w_1) - A_3(z_2, u_2, w_2) \| \\
&\leq k_1[\| z_1 - z_2 \|^\alpha + \| u_1 - u_2 \|^\alpha + \| w_1 - w_2 \|],
\end{aligned} \qquad (1.20)$$

where $\alpha\,(0 < \alpha < 1)$, k_0, k_1 are non-negative constants.

It is clear that (1.16) is the complex form of the following real equation of second order

$$u_{xx} - u_{yy} = au_x + bu_y + cu + d \text{ in } D, \qquad (1.21)$$

in which a, b, c, d are functions of $(x, y)(\in D)$, $u, u_x, u_y(\in \mathbb{R})$, and

$$A_1 = \frac{a + jb}{2}, \quad A_2 = \frac{c}{4}, \quad A_3 = \frac{d}{4} \text{ in } D.$$

Due to $z = x + jy = \mu e_1 + \nu e_2$, $w = u_z = \xi e_1 + \eta e_2$, and

$$w_z = \frac{w_x + j w_y}{2} = \xi_\mu e_1 + \eta_\nu e_2, \; w_{\bar{z}} = \frac{w_x - j w_y}{2} = \xi_\nu e_1 + \eta_\mu e_2,$$

the quasilinear hyperbolic equation (1.16) can be rewritten in the form

$$\xi_\nu e_1 + \eta_\mu e_2 = [A(z, u, w)\xi + B(z, u, w)\eta + C(z, u, w)u + D(z, u, w)]e_1$$
$$+ [A(z, u, w)\xi + B(z, u, w)\eta + C(z, u, w)u + D(z, u, w)]e_2, \text{ i.e.}$$

$$\begin{cases} \xi_\nu = A(z, u, w)\xi + B(z, u, w)\eta + C(z, u, w)u + D(z, u, w) \\ \eta_\mu = A(z, u, w)\xi + A(z, u, w)\eta + C(z, u, w)u + D(z, u, w) \end{cases} \text{ in } D, \quad (1.22)$$

in which

$$A = \frac{a+b}{4}, \; B = \frac{a-b}{4}, \; C = \frac{c}{4}, \; D = \frac{d}{4}.$$

In the following, we mainly discuss the oblique derivative problem for linear hyperbolic equation (1.6) and quasilinear hyperbolic equation (1.16) in the simply connected domain. We first prove that there exists a unique solution of the boundary value problem and give a priori estimates of their solutions, and then prove the solvability of the boundary value problem for general hyperbolic equations.

2 Oblique Derivative Problems for Quasilinear Hyperbolic Equations of Second Order

Here we first introduce the oblique derivative problem for quasilinear hyperbolic equations of second order in a simply connected domain, and give the representation theorem of solutions for hyperbolic equations of second order.

2.1 Formulation of the oblique derivative problem and the representation of solutions for hyperbolic equations

The oblique derivative problem for equation (1.16) may be formulated as follows:

Problem P Find a continuously differentiable solution $u(z)$ of (1.16) in $\bar{D}$ satisfying the boundary conditions

$$\frac{1}{2}\frac{\partial u}{\partial l} = \text{Re}\,[\lambda(z)u_z] = r(z), \quad z \in L = L_3 \cup L_4,$$

$$u(0) = b_0, \quad \text{Im}\,[\lambda(z)u_z]|_{z=z_3} = b_1, \quad (2.1)$$

where l is a given vector at every point on L, $\lambda(z) = a(z) + jb(z) = \cos(l, x) + j\cos(l, y)$, $z \in L$, b_0, b_1 are real constants, and $\lambda(z)$, $r(z)$, b_0, b_1 satisfy the conditions

$$C_\alpha[\lambda(z), L] \leq k_0, \ C_\alpha[r(z), L] \leq k_2, \quad |b_0|, |b_1| \leq k_2,$$

$$\max_{z \in L_3} \frac{1}{|a(z) - b(z)|}, \max_{z \in L_4} \frac{1}{|a(z) + b(z)|} \leq k_0, \tag{2.2}$$

in which $\alpha(0 < \alpha < 1)$, k_0, k_2 are non-negative constants. The above boundary value problem for (1.16) with $A_3(z, u, w) = 0$, $z \in D$, $u \in \mathbb{R}$, $w \in \mathbb{C}$ and $r(z) = b_0 = b_1 = 0$, $z \in L$ will be called Problem P_0.

By $z = x + jy = \mu e_1 + \nu e_2$, $w = u_z = \xi e_1 + \eta e_2$, the boundary condition (2.1) can be reduced to

$$\text{Re} \left[\lambda(z)(\xi e_1 + \eta e_2)\right] = r(z), \ u(0) = b_0, \ \text{Im} \left[\lambda(z)(\xi e_1 + \eta e_2)\right]|_{z=z_3} = b_1, \tag{2.3}$$

where $\lambda(z) = (a + b)e_1 + (a - b)e_2$. Moreover, the domain D is transformed into $Q = \{0 \leq \mu \leq 2R, \ 0 \leq \nu \leq 2R_1\}$, $R = R_2 - R_1$, which is a rectangle and A, B, C, D are known functions of (μ, ν) and unknown continuous functions u, w, and they satisfy the condition

$$\begin{cases} \lambda(z_3)w(z_3) = \lambda(z)[\xi e_1 + \eta e_2]|_{z=z_3} = r(z_3) + jb_1, \ u(0) = b_0, \\[2mm] \text{Re} \left[\lambda(z)w(2Re_1 + \nu e_2)\right] = r(z), \\[2mm] \quad \text{if } (x, y) \in L_3 = \{\mu = 2R, \ 0 \leq \nu \leq 2R_1\}, \\[2mm] \text{Re} \left[\lambda(z)w(\mu e_1 + 0e_2)\right] = r(z), \\[2mm] \quad \text{if } (x, y) \in L_4 = \{0 \leq \mu \leq 2R, \ \nu = 0\}, \end{cases} \tag{2.4}$$

where $\lambda(z)$, $r(z)$, b_0, b_1 are as stated in (2.1). We can assume that $w(z_3) = 0$, otherwise through the transformation $W(z) = \{w(z) - [a(z_3) - jb(z_3)][r(z_3) + jb_1]\}/[a^2(z_3) - b^2(z_3)]$, the requirement can be realized.

It is not difficult to see that the oblique derivative boundary value problem (Problem P) includes the Dirichlet boundary value problem (Problem D) as a special case. In fact, the boundary condition of Dirichlet problem (Problem D) for equation (1.21) is as follows

$$u(z) = \phi(z) \text{ on } L = L_3 \cup L_4. \tag{2.5}$$

We find the derivative with respect to the tangent direction $s = (x \mp jy)/\sqrt{2}$ for (2.5), in which $\mp$ are determined by L_3 and L_4 respectively, it is clear that the following equalities hold:

$$\text{Re} \left[\lambda(z)u_z\right] = r(z), \quad z \in L, \quad \text{Im} \left[\lambda(z)u_z\right]|_{z=z_3} = b_1, \tag{2.6}$$

in which

$$\lambda(z) = a + jb = \begin{cases} \dfrac{1-j}{\sqrt{2}} \text{ on } L_3, \\ \dfrac{1+j}{\sqrt{2}} \text{ on } L_4, \end{cases} \quad r(z) = \dfrac{\phi_x}{\sqrt{2}} \text{ on } L = L_3 \cup L_4,$$

$$b_1^+ = \text{Im}\left[\frac{1-j}{\sqrt{2}}u_z(z_3)\right] = -\frac{\phi_x + \phi_x}{2\sqrt{2}}\Big|_{z=z_3-0} = -\sqrt{2}\phi_x\big|_{z=z_3-0},$$

$$b_1^- = \text{Im}\left[\frac{1+j}{\sqrt{2}}u_z(z_3)\right] = \sqrt{2}\phi_x\big|_{z=z_3+0}, \quad b_0 = \phi(0),$$

(2.7)

in which $a = 1/\sqrt{2} \neq b = -1/\sqrt{2}$ on L_3 and $a = 1/\sqrt{2} \neq -b = -1/\sqrt{2}$ on L_4.

Noting that Problem P for (1.16) is equivalent to the Riemann–Hilbert problem (Problem A) for the complex equation of first order and boundary conditions:

$$w_{\bar{z}} = \text{Re}\,[A_1 w] + A_2 u + A_3 \text{ in } D,$$

(2.8)

$$\text{Re}\,[\lambda(z)w(z)] = r(z), \quad z \in L, \quad \text{Im}\,[\lambda(z)w(z)]|_{z=z_3} = b_1,$$

(2.9)

and the relation

$$u(z) = 2\text{Re}\int_0^z w(z)dz + b_0 \text{ in } D,$$

(2.10)

from Theorem 1.2 (3), Chapter I, and equation (2.8), we see that

$$2\text{Re}\int_\Gamma w(z)dz = \int_\Gamma w(z)dz + \int_\Gamma \overline{w(z)}d\bar{z}$$

$$= 2j\iint_D [w_{\bar{z}} - \bar{w}_z]dxdy = 4\iint_D \text{Im}\,[w_{\bar{z}}]dxdy = 0,$$

the above equality for any subdomain in D is also true, hence the function determined by the integral in (2.10) is independent of integral paths in $\bar{D}$. In this case we may choose that the integral path is along two family of characteristic lines, namely first along one of characteristic lines: $x + y = \mu\,(0 \le \mu \le 2R)$ and then along one of characteristic lines: $x - y = \nu\,(0 \le \nu \le 2R_1)$, for instance, the value of $u(z^*)(z^* = x^* + jy^* \in D, y^* \le 0)$ can be obtained by the integral

$$u(z^*) = 2\text{Re}\left[\int_{s_1} w(z)dz + \int_{s_2} w(z)dz\right] + b_0,$$

where $s_1 = \{x+y = 0, 0 \le x \le (x^*-y^*)/2\}$, $s_2 = \{x-y = x^*-y^*, (x^*-y^*)/2 \le x \le x^*-y^*\}$, in which x^*-y^* is the intersection of the characteristic line: $\{x-y = x^*-y^*\}$ through the point z^* and real axis. In particular, when $A_j = 0, j = 1, 2, 3$, equation (1.16) becomes the simplest hyperbolic complex equation

$$u_{z\bar{z}} = 0.$$

(2.11)

Problem P for (2.11) is equivalent to Problem A for the simplest hyperbolic complex equation of first order

$$w_{\bar{z}} = 0 \text{ in } D \tag{2.12}$$

with the boundary condition (2.9) and the relation (2.10). Hence similarly to Theorem 3.1, Chapter I, we can derive the representation and existence theorem of solutions of Problem A for the simplest equation (2.12), namely

Theorem 2.1 *Any solution $u(z)$ of Problem P for the hyperbolic equation (2.11) can be expressed as (2.10), where $w(z)$ is as follows*

$$w(z) = f(x+y)e_1 + g(x-y)e_2$$

$$= \frac{1}{2}\{f(x+y) + g(x-y) + j[f(x+y) - g(x-y)]\},$$

$$f(2R) = u(z_3) + v(z_3) = \frac{r((1+j)R) + b_1}{a((1+j)R) + b((1+j)R)}, \tag{2.13}$$

$$g(0) = u(z_3) - v(z_3) = \frac{r((1+j)R) - b_1}{a((1+j)R) - b((1+j)R)},$$

here $f(x+y)$, $g(x-y)$ possess the forms

$$g(x-y) = \frac{2r((1-j)(x-y)/2 + (1+j)R)}{a((1-j)(x-y)/2 + (1+j)R) - b((1-j)(x-y)/2 + (1+j)R)}$$
$$- \frac{[a((1-j)(x-y)/2+(1+j)R)+b((1-j)(x-y)/2+(1+j)R)]f(2R)}{a((1-j)(x-y)/2+(1+j)R)-b((1-j)(x-y)/2+(1+j)R)},$$
$$0 \le x - y \le 2R_1, \tag{2.14}$$

$$f(x+y) = \frac{2r((1+j)(x+y)/2) - [a((1+j)(x+y)/2) - b((1+j)(x+y)/2)]g(0)}{a((1+j)(x+y)/2) + b((1+j)(x+y)/2)}.$$
$$0 \le x + y \le 2R.$$

Moreover $u(z)$ satisfies the estimate

$$C_{\alpha}^1[u(z), \bar{D}] \le M_1 = M_1(\alpha, k_0, k_2, D), C_{\alpha}^1[u(z), \bar{D}] \le M_2 k_2 = M_2(\alpha, k_0, D)k_2, \tag{2.15}$$

where M_1, M_2 are two non-negative constants only dependent on α, k_0, k_2, D and α, k_0, D respectively.

Proof   Let the general solution $w(z) = u_z = \frac{1}{2}\{f(x+y) + g(x-y) + j[f(x+y) - g(x-y)]\}$ of (2.12) be substituted in the boundary condition (2.1), we obtain

$$a(z)u(z) + b(z)v(z) = r(z) \text{ on } L, \ \lambda(z_3)w(z_3) = r(z_3) + jb_1, \quad \text{i.e.}$$

$$[a((1-j)x + 2jR) + b((1-j)x + 2jR)]f(2R) + [a((1-j)x + 2jR)$$

$$-b((1-j)x + 2jR)]g(2x - 2R) = 2r((1-j)x + 2jR) \text{ on } L_3,$$

$$[a((1+j)x) + b((1+j)x)]f(2x) + [a((1+j)x)$$

$$-b((1+j)x)]g(0) = 2r((1+j)x) \text{ on } L_4,$$

$$f(2R) = u(z_3) + v(z_3) = \frac{r((1+j)R) + b_1}{a((1+j)R) + b((1+j)R)},$$

$$g(0) = u(z_3) - v(z_3) = \frac{r((1+j)R) - b_1}{a((1+j)R) - b((1+j)R)},$$

and the above formulas can be rewritten as

$$[a((1-j)t/2 + (1+j)R) + b((1-j)t/2 + (1+j)R)]f(2R)$$

$$+ [a((1-j)t/2 + (1+j)R) - b((1-j)t/2 + (1+j)R)]g(t)$$

$$= 2r((1-j)t/2 + (1+j)R), \quad t \in [0, 2R_1],$$

$$[a((1+j)t/2) + b((1+j)t/2)]f(t) + [a((1+j)t/2)$$

$$-b((1+j)t/2)]g(0) = 2r((1+j)t/2), \quad t \in [0, 2R],$$

thus the solution $w(z)$ can be expressed as (2.13),(2.14). From the condition (2.2) and the relation (2.10), we see that the estimate (2.15) of $u(z)$ for (2.11) is obviously true.

Next we give the representation of Problem P for the quasilinear equation (1.16).

Theorem 2.2 *Under Condition C, any solution $u(z)$ of Problem P for the hyperbolic equation (1.16) can be expressed as*

$$u(z) = 2\mathrm{Re} \int_0^z w(z)dz + b_0 \text{ in } D,$$

$$w(z) = W(z) + \Phi(z) + \Psi(z) \text{ in } D,$$

$$W(z) = f(\mu)e_1 + g(\nu)e_2, \quad \Phi(z) = \tilde{f}(\mu)e_1 + \tilde{g}(\nu)e_2, \qquad (2.16)$$

$$\Psi(z) = \int_0^\nu [A\xi + B\eta + Cu + D]e_1 d\nu + \int_{2R}^\mu [A\xi + B\eta + Cu + D]e_2 d\mu,$$

where $f(\mu)$, $g(\nu)$ are as stated in (2.14) and $\tilde{f}(\mu)$, $\tilde{g}(\nu)$ are similar to $f(\mu)$, $g(\nu)$ in (2.14), but $r(z), b_1$ are replaced by $-\mathrm{Re}\,[\lambda(z)\Psi(z)]$, $-\mathrm{Im}\,[\lambda(z_3)\Psi(z_3)]$, namely

$$\mathrm{Re}\,[\lambda(z)\Phi(z)] = -\mathrm{Re}\,[\lambda(x)\Psi(x)], \quad z \in L, \quad \mathrm{Im}\,[\lambda(z_3)\Phi(z_3)] = -\mathrm{Im}\,[\lambda(z_3)\Psi(z_3)].$$
$$(2.17)$$

Proof Let the solution $u(z)$ of Problem P be substituted into the coefficients of equation (1.16). Then the equation in this case can be seen as a linear hyperbolic equation (1.15). Due to Problem P is equivalent to the Problem A for the complex equation (2.8) with the relation (2.10), from Theorem 3.2, Chapter I, it is not difficult to see that the function $\Psi(z)$ satisfies the complex equation

$$[\Psi]_{\tilde{z}} = [A\xi + B\eta + Cu + D]e_1 + [A\xi + B\eta + Cu + D]e_2 \text{ in } D, \qquad (2.18)$$

and $\Phi(z) = w(z) - W(z) - \Psi(z)$ satisfies the complex equation and the boundary conditions

$$\xi_\nu e_1 + \eta_\mu e_2 = 0, \tag{2.19}$$

$$\operatorname{Re}\left[\lambda(z)(\xi e_1 + \eta e_2)\right] = -\operatorname{Re}\left[\lambda(z)\Psi(z)\right] \text{ on } L,$$

$$\operatorname{Im}\left[\lambda(z)(\xi e_1 + \eta e_2)\right]|_{z=z_3} = -\operatorname{Im}\left[\lambda(z_3)\Psi(z_3)\right]. \tag{2.20}$$

By the representation of solutions of Problem A for (1.16) as stated in (3.17), Chapter I, we can obtain the representation (2.16) of Problem P for (1.16).

2.2   Existence and uniqueness of solutions of Problem P for hyperbolic equations of second order

Theorem 2.3   *If the complex equation* (1.16) *satisfies Condition C, then Problem P for* (1.16) *has a solution.*

Proof   We consider the expression of $u(z)$ in the form (2.16). In the following, by using successive iteration we shall find a solution of Problem P for equation (1.16). Firstly, substitute

$$u_0(z) = 2\operatorname{Re}\int_0^z w_0(z)dz + b_0, \quad w_0(z) = W(z) = \xi_0 e_1 + \eta_0 e_2, \tag{2.21}$$

into the position of $u, w = \xi e_1 + \eta e_2$ in the right-hand side of (1.16), where $w_0(z)$ is the same function with $W(z)$ in (2.16) and satisfies the estimate (2.15). Moreover, we have

$$u_1(z) = 2\operatorname{Re}\int_0^z w_1(z)dz + b_0, \quad w_1(z) = w_0(z) + \Phi_1(z) + \Psi_1(z),$$

$$\Psi_1(z) = \int_0^{x-y}[A\xi_0 + B\eta_0 + Cu_0 + D]e_1 d(x-y) \tag{2.22}$$

$$+ \int_{2R}^{x+y}[A\xi_0 + B\eta_0 + Cu_0 + D]e_2 d(x+y),$$

from the first equality in (2.22), the estimate

$$C^1[u_0(z), \bar{D}] \le M_3 C[w_0(z), \bar{D}] + k_2, \tag{2.23}$$

can be derived, where $M_3 = M_3(D)$. From the third and second equalities in (2.22), we can obtain

$$C[\Psi_1(z), \bar{D}] \le 2M_4[(4 + M_3)m + k_2 + 1]R',$$

$$C[\Phi_1(z), \bar{D}] \le 8M_4 k_0^2(1 + k_0^2)[(4 + M_3)m + k_2 + 1]R', \tag{2.24}$$

$$C[w_1(z) - w_0(z), \bar{D}] \le 2M_4 M[(4 + M_3)m + k_2 + 1]R',$$

where $M_4 = \max_{\bar{D}}(|A|, |B|, |C|, |D|)$, $R' = \max(2R_1, 2R)$, $m = \| w_0(z) \|_{C(\bar{D})}$, $M = 1 + 4k_0^2(1 + k_0^2)$ is a positive constant. Thus we can find a sequence of functions $\{w_n(z)\}$ satisfying

$$u_{n+1}(z) = 2\mathrm{Re} \int_0^z w_{n+1}(z)dz + b_0,$$

$$w_{n+1}(z) = w_0(z) + \Phi_n(z) + \int_0^\nu [A\xi_n + B\eta_n + Cu_n + D]e_1 d\nu \qquad (2.25)$$

$$+ \int_{2R}^\mu [B\eta_n + A\eta_n + Cu_n + D]e_2 d\mu,$$

and then

$$\| w_n - w_{n-1} \| \leq \{2M_4 M[(4 + M_3)m + 1]\}^n \times \int_0^{R'} \frac{R'^{n-1}}{(n-1)!} dR'$$

$$\leq \frac{\{2M_4 M[(4 + M_3)m + 1)]R'\}^n}{n!}. \qquad (2.26)$$

From the above inequality, it is seen that the sequence of functions $\{w_n(z)\}$, i.e.

$$w_n(z) = w_0(z) + [w_1(z) - w_0(z)] + \cdots + [w_n(z) - w_{n-1}(z)](n = 1, 2, \ldots) \quad (2.27)$$

in $\bar{D}$ uniformly converges a function $w_*(z)$, and $w_*(z)$ satisfies the equality

$$w_*(z) = \xi_* e_1 + \eta_* e_2$$

$$= w_0(z) + \Phi_*(z) + \int_0^\nu [A\xi_* + B\eta_* + Cu_* + D]e_1 d\nu \qquad (2.28)$$

$$+ \int_{2R}^\mu [A\xi_* + B\eta_* + Cu_* + D]e_2 d\mu,$$

and the function

$$u_*(z) = 2\mathrm{Re} \int_0^z w_*(z)dz + b_0, \qquad (2.29)$$

is just a solution of Problem P for equation (1.16) in the closed domain $\bar{D}$.

Theorem 2.4 *Suppose that Condition C holds. Then Problem P for the complex equation (1.16) has at most one solution in $\bar{D}$.*

Proof Let $u_1(z), u_2(z)$ be any two solutions of Problem P for (1.16), we see that $u(z) = u_1(z) - u_2(z)$ and $w(z) = u_{1z}(z) - u_{2z}(z)$ satisfies the homogeneous complex equation and boundary conditions

$$w_{\bar{z}} = \mathrm{Re}\,[\tilde{A}_1 w] + \tilde{A}_2 u \text{ in } D, \qquad (2.30)$$

$$\mathrm{Re}\,[\lambda(z)w(z)] = 0 \text{ on } L, \quad \mathrm{Im}\,[\lambda(z_3)w(z_3)] = 0, \qquad (2.31)$$

and the relation

$$u(z) = 2\mathrm{Re} \int_0^z w(z)dz, \; z \in \bar{D}. \qquad (2.32)$$

From Theorem 2.2, we see that the function $w(z)$ can be expressed in the form

$$w(z) = \Phi(z) + \Psi(z),$$

$$\Psi(z) = \int_0^\nu [\tilde{A}\xi + \tilde{B}\eta + \tilde{C}u]e_1 d\nu + \int_{2R}^\mu [\tilde{A}\xi + \tilde{B}\eta + \tilde{C}u]e_2 d\mu,$$

(2.33)

moreover from (2.32),

$$C_1[u(z), \bar{D}] \le M_3 C[w(z), \bar{D}]$$

(2.34)

can be obtained, in which $M_3 = M_3(D)$ is a non-negative constant. By using the process of iteration similar to the proof of Theorem 2.3, we can get

$$\| w(z) \| = \| w_1 - w_2 \| \le \frac{\{2M_5 M[(4 + M_3)m + 1]R'\}^n}{n!},$$

where $M_5 = \max_{\bar{D}}(|\tilde{A}|, |\tilde{B}|, |\tilde{C}|)$. Let $n \to \infty$, it can be seen $w(z) = w_1(z) - w_2(z) = 0$, $\Psi(z) = \Phi(z) = 0$ in $\bar{D}$. This proves the uniqueness of solutions of Problem P for (1.16).

3 Oblique Derivative Problems for General Quasilinear Hyperbolic Equations of Second Order

In this section, we first give a priori estimates in $C^1(\bar{D})$ of solutions for the oblique derivative problem, moreover by using the estimates of solutions, the existence of solutions for the above problem for general quasilinear equation is proved. Finally we discuss the oblique derivative problem for hyperbolic equations of second order in general domains.

3.1 A priori estimates of solutions of Problem P for hyperbolic equations of second order

From Theorems 2.3 and 2.4, we see that under Condition C, Problem P for equation (1.16) has a unique solution $u(z)$, which can be found by using successive iteration. Noting that $w_{n+1}(z) - w_n(z)$ satisfy the estimate (2.26), the limit $w(z)$ of the sequence of functions $\{w_n(z)\}$ satisfies the estimate

$$\max_{z \in \bar{D}} \| w(z) \| = C[w(z), \bar{D}] \le e^{2M_5 M[(4+M_3)m+1]R'} = M_6,$$

(3.1)

and the solution $u(z)$ of Problem P is as stated in (2.10), which satisfies the estimate

$$C^1[u(z), \bar{D}] \le R^* M_6 + k_2 = M_7,$$

(3.2)

where $R^* = 2R_1 + 2R$. Thus we have

Theorem 3.1 *If Condition C holds, then any solution $u(z)$ of Problem P for the hyperbolic equation* (1.16) *satisfies the estimates*

$$C^1[u, \bar{D}] \le M_7, \ C^1[u, \bar{D}] \le M_8 k, \tag{3.3}$$

in which $M_7 = M_7(\alpha, k_0, k_1, k_2, D)$, $k = k_1 + k_2$, $M_8 = M_8(\alpha, k_0, D)$ *are non-negative constants.*

In the following, we give the $C_\alpha^1(\bar{D})$-estimates of solution $u(z)$ for Problem P for (1.16).

Theorem 3.2 *If Condition C and* (1.20) *hold, then any solution $u(z)$ of Problem P for the hyperbolic equation* (1.16) *satisfies the estimates*

$$C_\alpha[u_z, \bar{D}] \le M_9, \ C_\alpha^1[u, \bar{D}] \le M_{10}, \ C_\alpha^1[u, \bar{D}] \le M_{11} k, \tag{3.4}$$

in which $k = k_1 + k_2$, $M_j = M_j(\alpha, k_0, k_1, k_2, D)$, $j = 9, 10$, $M_{11} = M_{11}(\alpha, k_0, D)$ *are non-negative constants.*

Proof Similarly to Theorem 4.3, Chapter I, it suffices to prove the first estimate in (3.4). Due to the solution $u(z)$ of Problem P for (1.16) is found by the successive iteration through the integral expression (2.16), we first choose the solution

$$u_0(z) = 2\text{Re} \int_0^z w_0(z) dz + b_0, \ w_0(z) = W(z) = \xi_0 e_1 + \eta_0 e_2, \tag{3.5}$$

of Problem P for the equation

$$u_{z\bar{z}} = 0 \text{ in } D, \tag{3.6}$$

and substitute u_0, w_0 into the position of $u, w = \xi e_1 + \eta e_2$ on the right-hand side of (1.16), where $w_0(z)$ is the same function with $W(z)$ in (2.16) and $w_0(z), u_0(z)$ satisfy the first estimates

$$C_\alpha[w_0, \bar{D}] = C_\alpha[\text{Re}\, w_0 \bar{D}] + C_\alpha[\text{Im}\, w_0, \bar{D}] \le M_{12} k_2, \ C_\alpha^1[u_0, \bar{D}] \le M_{13} k_2 = M_{14}, \tag{3.7}$$

where $M_j = M_j(\alpha, k_0, D)$, $j = 12, 13$, and then we have

$$u_1(z) = 2\text{Re} \int_0^z w_1(z) dz + b_0, \ w_1(z) = w_0(z) + \Phi_1(z) + \Psi_1(z),$$

$$\Psi_1(z) = \Psi_1^1(z) e_1 + \Psi_1^2(z) e_2, \ \Psi_1^1(z) = \int_0^\nu G_1(z) d\nu, \tag{3.8}$$

$$\Psi_1^2(z) = \int_{2R}^\mu G_2(z) d\mu, \ G_1(z) = G_2(z) = A\xi_0 + B\eta_0 + Cu_0 + D.$$

From (3.7) and the last three equalities in (3.8), it is not difficult to see that $\Psi_1^1(z) = \Psi_1^1(\mu, \nu)$, $\Psi_1^2(z) = \Psi_1^2(\mu, \nu)$ satisfy the Hölder continuous estimates about ν, μ respectively, namely

$$C_\alpha[\Psi_1^1(\cdot, \nu), \bar{D}] \le M_{15} R', \ C_\alpha[\Psi_1^2(\mu, \cdot), \bar{D}] \le M_{15} R', \tag{3.9}$$

in which $M_{15} = M_{15}(\alpha, k_0, k_1, k_2, D)$, there is no harm assuming that $R' = \max(2R_1, 2R) \geq 1$. By Condition C and (1.20), we can see that $G_1(z)$, $\Psi_1^1(z)$ $G_2(z)$, $\Psi_1^2(z)$ about μ, ν satisfy the the Hölder conditions respectively, i.e.

$$C_\alpha[G_1(\mu, \cdot), \bar{D}] \leq M_{16}R', \ C_\alpha[\Psi_1^1(\mu, \cdot), \bar{D}] \leq M_{16}R',$$

$$C_\alpha[G_2(\cdot, \nu), \bar{D}] \leq M_{16}R', \ C_\alpha[\Psi_1^2(\cdot, \nu), \bar{D}] \leq M_{16}R', \tag{3.10}$$

where $M_{16} = M_{16}(\alpha, k_0, k_1, k_2, D)$. Moreover we can obtain the estimates of $\Psi_1(z)$, $\Phi_1(z)$ as follows

$$C_\alpha[\Psi_1(z), \bar{D}] \leq M_{17}R', \ C_\alpha[\Phi_1(z), \bar{D}] \leq M_{17}R', \tag{3.11}$$

in which $M_{17} = M_{17}(\alpha, k_0, k_1, k_2, D)$. Setting $\tilde{w}_1^1(z) = \operatorname{Re} \tilde{w}_1(z) + \operatorname{Im} \tilde{w}_1(z)$, $\tilde{w}_2^1(z) = \operatorname{Re} \tilde{w}_1(z) - \operatorname{Im} \tilde{w}_1(z)$, $\tilde{w}_1(z) = w_1(z) - w_0(z)$ and $\tilde{u}_1(z) = u_1(z) - u_0(z)$, from (3.8)–(3.11), it follows that

$$C_\alpha[\tilde{w}_1^1(z), \bar{D}] \leq M_{18}R', \ C_\alpha[\tilde{w}_1^2(z), \bar{D}] \leq M_{18}R',$$

$$C_\alpha[\tilde{w}_1(z), \bar{D}] \leq M_{18}R', \ C_\alpha^1[\tilde{u}_1(z), \bar{D}] \leq M_{18}R', \tag{3.12}$$

where $M_{18} = M_{18}(\alpha, k_0, k_1, k_2, D)$. According to the successive iteration, the estimates of functions $\tilde{w}_n^1(z) = \operatorname{Re} \tilde{w}_n(z) + \operatorname{Im} \tilde{w}_n(z)$, $\tilde{w}_n^2(z) = \operatorname{Re} \tilde{w}_n(z) - \operatorname{Im} \tilde{w}_n(z)$, $\tilde{w}_n(z) = w_n(z) - w_{n-1}(z)$ and the corresponding function $\tilde{u}_n(z) = u_n(z) - u_{n-1}(z)$ can be obtained, namely

$$C_\alpha[\tilde{w}_n^1(z), \bar{D}] \leq \frac{(M_{18}R')^n}{n!}, \ C_\alpha[\tilde{w}_n^2(z), \bar{D}] \leq \frac{(M_{18}R')^n}{n!},$$

$$C_\alpha[\tilde{w}_n(z), \bar{D}] \leq \frac{(M_{18}R')^n}{n!}, \ C_\alpha^1[\tilde{u}_n(z), \bar{D}] \leq \frac{(M_{18}R')^n}{n!}. \tag{3.13}$$

Therefore the sequences of functions

$$w_n(z) = w_0(z) + \sum_{m=1}^n \tilde{w}_m(z), \ u_n(z) = u_0(z) + \sum_{m=1}^n \tilde{u}_m(z) \ (n = 1, 2, \ldots)$$

uniformly converge to $w(z)$, $u(z)$ in $\bar{D}$ respectively, and $w(z)$, $u(z)$ satisfy the estimates

$$C_\alpha[w(z), \bar{D}] \leq M_9 = e^{M_{18}R'}, \ C_\alpha^1[u(z), \bar{D}] \leq M_{10}, \tag{3.14}$$

this is just the first estimate in (3.4).

3.2 The existence of solutions of Problem P for general hyperbolic equations of second order

Now, we consider the general quasilinear equation of second order

$$u_{z\bar{z}} = F(z, u, u_z) + G(z, u, u_z),$$

$$F = \operatorname{Re}[A_1 u_z] + A_2 u + A_3, \tag{3.15}$$

$$G = A_4 \|u_z\|^\sigma + A_5 |u|^\tau, \quad z \in D,$$

where $F(z, u, u_z)$ satisfies Condition C, σ, τ are positive constants, and $A_j(z, u, u_z)$ $(j = 4, 5)$ satisfy the conditions in Condition C, i.e.

$$C[A_j(z, u, u_z), \bar{D}] \leq k_0, \quad j = 4, 5,$$

and denote by Condition C' the above conditions.

Theorem 3.3 *Let the complex equation (3.15) satisfy Condition C'.*

(1) *When $0 < \max(\sigma, \tau) < 1$, Problem P for (3.15) has a solution $u(z) \in C_1^1(\bar{D})$.*

(2) *When $\min(\sigma, \tau) > 1$, Problem P for (3.15) has a solution $u(z) \in C_1^1(\bar{D})$, provided that*

$$M_{19} = k_1 + k_2 + |b_0| + |b_1| \tag{3.16}$$

is sufficiently small.

(3) *In general, the above solution of Problem P is not unique, if $0 < \max(\sigma, \tau) < 1$.*

Proof (1) Consider the algebraic equation for t :

$$M_8\{k_1 + k_2 + 2k_0 t^\sigma + 2k_0 t^\tau + |b_1| + |b_0|\} = t, \tag{3.17}$$

where M_8 is a non-negative constant in (3.3), it is not difficult to see that equation (3.17) has a unique solution $t = M_{20} \geq 0$. Now, we introduce a closed and convex subset B^* in the Banach space $C^1(\bar{D})$, whose elements are the functions $u(z)$ satisfying the conditions

$$u(z) \in C^1(\bar{D}), \quad C^1[u(z), \bar{D}] \leq M_{20}. \tag{3.18}$$

We arbitrarily choose a function $u_0(z) \in B^*$ for instance $u_0(z) = 0$ and substitute it into the position of u in coefficients of (3.15) and $G(z, u, u_z)$, from Theorems 2.3 and 2.4, it is clear that problem P for

$$u_{z\bar{z}} - \text{Re}\,[A_1(z, u_0, u_{0z})u_z] - A_2(z, u_0, u_{0z})u - A_3(z, u_0, u_{0z}) = G(z, u_0, u_{0z}), \tag{3.19}$$

has a unique solution $u_1(z) \in B^*$. By Theorem 3.1, we see that the solution $u_1(z)$ satisfies the estimate in (3.18). By using the successive iteration, we obtain a sequence of solutions $u_m(z)(m = 1, 2, ...) \in B^*$ of Problem P, which satisfy the equations

$$u_{m+1z\bar{z}} - \text{Re}\,[A_1(z, u_m, u_{mz})u_{m+1z}] - A_2(z, u_m, u_{mz})u_{m+1}$$
$$-A_3(z, u_m, u_{mz}) = G(z, u_m, u_{mz}) \text{ in } \bar{D}, \quad m = 1, 2, \ldots \tag{3.20}$$

and $u_{m+1}(z) \in B^*$. From (3.20), we see that $\tilde{u}_{m+1}(z) = u_{m+1}(z) - u_m(z)$ satisfies the equations and boundary conditions

$$\tilde{u}_{m+1z\bar{z}} - \text{Re}[\tilde{A}_1\tilde{u}_{m+1z}] - \tilde{A}_2\tilde{u}_{m+1}$$
$$= G(z, u_m, u_{mz}) - G(z, u_{m-1}, u_{m-1z}) \text{ in } \bar{D}, \quad m = 1, 2, \ldots, \tag{3.21}$$

$$\text{Re}\,[\lambda(z)(\tilde{u}_{m+1}(z)] = 0 \text{ on } L, \quad \text{Im}\,[\lambda(z_3)(\tilde{u}_{m+1}(z_3)] = 0.$$

Noting that $C[G(z, u_m, u_{mz}) - G(z, u_{m-1}, u_{m-1z}), \bar{D}] \leq 2k_0 M_{20}$, M_{20} is a solution of the algebraic equation (3.17) and according to Theorem 3.1,

$$\| \tilde{u}_{m+1} \| = C^1[\tilde{u}_{m+1}, \bar{D}] \leq M_{20} \tag{3.22}$$

can be obtained. Due to $\tilde{u}_{m+1}$ can be expressed as

$$\tilde{u}_{m+1}(z) = 2\text{Re} \int_0^z w_{m+1}(z)dz, \ w_{m+1}(z) = w_0(z) + \Phi_{m+1}(z) + \Psi_{m+1}(z),$$

$$\Psi_{m+1}(z) = \int_0^{x-y} [\tilde{A}\xi_m + \tilde{B}\eta_m + \tilde{C}u_m + \tilde{G}]e_1 d(x-y) \tag{3.23}$$

$$+ \int_{2R}^{x+y} [\tilde{A}\xi_m + \tilde{B}\eta_m + \tilde{C}u_m + \tilde{G}]e_2 d(x+y),$$

in which the relation between $\tilde{A}_1, \tilde{A}_2, \tilde{G}$ and $\tilde{A}, \tilde{B}, \tilde{C}, \tilde{G}$ is the same as that of A_1, A_2, A_3 and A, B, C, D in Section 1 and $\tilde{G} = G(z, u_m, u_{mz}) - G(z, u_{m-1}, u_{m-1z})$. By using the method in the proof of Theorem 2.3, we can obtain

$$\| u_{m+1} - u_m \| = C^1[\tilde{u}_{m+1}, \bar{D}] \leq \frac{(M_{21}R')^n}{n!},$$

where $M_{21} = 4M_{22}M(M_3 + 2)(2m_0 + 1)$, $R' = \max(2R_1, 2R)$, $m_0 = \| w_0(z) \|_{C(\bar{D})}$, herein $M_{22} = \max\{C[\tilde{A}, Q], C[\tilde{B}, Q], C[\tilde{C}, Q], C[\tilde{G}, Q]\}$, $M = 1 + 4k_0^2(1 + k_0^2)$. From the above inequality, it is seen that the sequences of functions: $\{u_m(z)\}$, $\{w_m(z)\}$, i.e.

$$u_m(z) = u_0(z) + [u_1(z) - u_0(z)] + \cdots + [u_m(z) - u_{m-1}(z)](m = 1, 2, \ldots),$$
$$w_m(z) = w_0(z) + [w_1(z) - w_0(z)] + \cdots + [w_m(z) - w_{m-1}(z)](m = 1, 2, \ldots), \tag{3.24}$$

uniformly converge to the functions $u_*(z)$, $w_*(z)$ respectively, and $w_*(z)$ satisfies the equality

$$w_*(z) = w_0(z) + \Phi_*(z)$$

$$+ \int_0^{x-y} [A\sigma_* + B\eta_* + Cu_* + D]e_1 d(x-y) \tag{3.25}$$

$$+ \int_{2R}^{x+y} [B\sigma_* + A\eta_* + Cu_* + D]e_2 d(x+y),$$

and the function

$$u_*(z) = 2\text{Re} \int_0^z w_*(z)dz + b_0, \tag{3.26}$$

is just a solution of Problem P for the nonlinear equation (3.15) in the closed domain $\bar{D}$.

(2) Consider the algebraic equation for t :

$$M_8\{k_1 + k_2 + 2k_0 t^\sigma + 2k_0 t^\tau + |b_0| + |b_1|\} = t, \tag{3.27}$$

it is not difficult to see that equation (3.27) has a solution $t = M_{20} \geq 0$, provided that M_{19} in (3.16) is small enough. If there exist two solutions, then we choose the minimum of both as M_{20}. Now, we introduce a closed and convex subset B_* of the Banach space $C^1(\bar{D})$, whose elements are of the functions $u(z)$ satisfying the conditions

$$u(z) \in C^1(\bar{D}), \ C^1[u(z), \bar{D}] \leq M_{20}. \tag{3.28}$$

By using the same method in (1), we can find a solution $u(z) \in B_*$ of Problem P for equation (3.15) with $\min(\sigma, \tau) > 1$.

(3) We can give an example to explain that there exist two solutions for equation (3.15) with $\sigma = 0$, $\tau = 1/2$, namely the equation

$$u_{xx} - u_{yy} = Au^{1/2}, \quad A = 8\text{sgn}(x^2 - y^2) \text{ in } D \tag{3.29}$$

has two solutions $u_1(x, y) = 0$ and $u_2(x, y) = (x^2 - y^2)^2/4$, and they satisfy the boundary conditions

$$\text{Re}\,[\lambda(z)u_z] = r(z), \quad z \in L, \quad u(0) = b_0, \ \text{Im}\,[\lambda(z)u_z]|_{z=0} = b_1, \tag{3.30}$$

where

$$\lambda(z) = 1 - i, \quad z \in L_1, \quad \lambda(x) = 1 + i, \quad z \in L_4,$$

$$b_0 = 0, \ r(z) = 0, \quad z \in L = L_1 \cup L_4, \quad b_1 = 0.$$

3.3 The existence of solutions of Problem P for hyperbolic equations of second order in general domains

In this subsection, we shall generalize the domain D to general cases.

1. The boundaries L_3, L_4 of the domain D are replaced by the curves L_3', L_4', and the boundary of the domain D' is $L_1 \cup L_2 \cup L_3' \cup L_4'$, where the parameter equations of the curves L_3', L_4' are as follows:

$$L_3' = \{x + y = 2R, l \leq x \leq R_2\}, L_4' = \{x + y = \mu, y = \gamma_1(x), 0 \leq x \leq l\}, \tag{3.31}$$

in which $\gamma_1(x)$ on $0 \leq x \leq l = \gamma_1(l)$ is continuous and $\gamma_1(0) = 0$, $\gamma_1(x) > 0$ on $0 < x \leq l$, and $\gamma_1(x)$ possesses the derivatives on $0 \leq x \leq l$ except some isolated points and $1 + \gamma_1'(x) > 0$. By the condition, we can find the inverse function $x = \tau(\mu) = (\mu + \nu)/2$ of $x + \gamma_1(x) = \mu$, and then $\nu = 2\tau(\mu) - \mu$, $0 \leq \mu \leq 2R$. We make a transformation

$$\tilde{\mu} = \mu, \ \tilde{\nu} = \frac{2R_1(\nu - 2\tau(\mu) + \mu)}{2R_1 - 2\tau(\mu) + \mu}, \quad 0 \leq \mu \leq 2R, \quad 2\tau(\mu) - \mu \leq \nu \leq 2R_1, \tag{3.32}$$

its inverse transformation is

$$\mu = \tilde{\mu}, \ \nu = \frac{2R_1 - 2\tau(\mu) + \mu}{2R_1}\tilde{\nu} + 2\tau(\mu) - \mu, \quad 0 \leq \tilde{\mu} \leq 2R, \quad 0 \leq \tilde{\nu} \leq 2R_1. \tag{3.33}$$

Hence we have

$$\tilde{x} = \frac{1}{2}(\tilde{\mu} + \tilde{\nu}) = \frac{4R_1 x - [2\tau(x + \gamma_1(x)) - x - \gamma_1(x)](2R_1 + x + y)}{4R_1 - 4\tau(x + \gamma_1(x)) + 2x + 2\gamma_1(x)},$$

$$\tilde{y} = \frac{1}{2}(\tilde{\mu} - \tilde{\nu}) = \frac{4R_1 y + [2\tau(x + \gamma_1(x)) - x - \gamma_1(x)](2R_1 - x - y)}{4R_1 - 4\tau(x + \gamma_1(x)) + 2x + 2\gamma_1(x)},$$

$$(3.34)$$

and

$$x = \frac{1}{2}(\mu + \nu) = \frac{4R_1 \tilde{x} + [2\tau(x + \gamma_1(x)) - x - \gamma_1(x)](2R_1 - \tilde{x} + \tilde{y})}{4R_1},$$

$$y = \frac{1}{2}(\mu - \nu) = \frac{4R_1 \tilde{y} - [2\tau(x + \gamma_1(x)) - x - \gamma_1(x)](2R_1 + \tilde{x} - \tilde{y})}{4R_1}.$$

$$(3.35)$$

Denote by $\tilde{z} = \tilde{x} + j\tilde{y} = f(z)$, $z = x + jy = f^{-1}(\tilde{z})$ the transformations (3.34) and (3.35) respectively. In this case, setting $w(z) = u_z$, equation (1.16) in $\bar{D}'$ and boundary condition (2.1) on $L_3' \cup L_4'$ can be rewritten in the form

$$\xi_\nu = A\xi + B\eta + Cu + D, \quad \eta_\mu = A\xi + B\eta + Cu + D, \quad z \in \bar{D}', \tag{3.36}$$

$$\text{Re}\,[\lambda(z)w(z)] = r(z),\ z \in L_3' \cup L_4',\ u(0) = b_0,\ \text{Im}\,[\lambda(z_3)w(z_3)] = b_1, \tag{3.37}$$

in which $z_3 = l + j\gamma_1(l)$, $\lambda(z)$, $r(z)$, b_1 on $L_3' \cup L_4'$ satisfy the condition (2.2), and $u(z)$ and $w(z)$ satisfy the relation

$$u(z) = 2\text{Re} \int_0^z w(z)dz + b_0 \text{ in } D'. \tag{3.38}$$

Suppose equation (1.16) in $\bar{D}'$ satisfies Condition C, through the transformation (3.33), we have $\xi_{\tilde{\nu}} = [2R_1 - 2\tau(\mu) + \mu]\xi_\nu/2R_1$, $\eta_{\tilde{\mu}} = \eta_\mu$, system (3.36) is reduced to

$$\xi_{\tilde{\nu}} = \frac{[2R_1 - 2\tau(\mu) + \mu][A\xi + B\eta + Cu + D]}{2R_1}, \quad \eta_{\tilde{\mu}} = A\xi + B\eta + Cu + D. \tag{3.39}$$

Moreover through the transformation (3.35), the boundary condition (3.37) on $L_3' \cup L_4'$ is reduced to

$$\text{Re}\,[\lambda(f^{-1}(\tilde{z}))w(f^{-1}(\tilde{z}))] = r(f^{-1}(\tilde{z})), \tilde{z} \in L_3 \cup L_4,$$

$$\text{Im}\,[\lambda(f^{-1}(\tilde{z}_3))w(f^{-1}(\tilde{z}_3))] = b_1,$$

$$(3.40)$$

in which $\tilde{z}_3 = f(z_3)$. Therefore the boundary value problem (3.36),(3.37) is transformed into the boundary value problem (3.39),(3.40). On the basis of Theorems 2.2 and 2.3, we see that the boundary value problem (3.39),(3.40) has a unique solution $w(\tilde{z})$, and then $w[f(z)]$ is just a solution of the boundary value problem (3.36),(3.37).

Theorem 3.4 *If equation (1.16) satisfies Condition C in the domain D' with the boundary $L_1 \cup L_2 \cup L_3' \cup L_4'$, then Problem P with the boundary condition (3.37)($w = u_z$) has a unique solution $u(z)$ as stated in (3.38).*

2. The boundaries L_3, L_4 of the domain D are replaced by two curves L_3'', L_4'' respectively, and the boundary of the domain D'' is $L_1 \cup L_2 \cup L_3'' \cup L_4''$, where the parameter equations of the curves L_3'', L_4'' are as follows:

$$L_3'' = \{x - y = \nu, \ y = \gamma_2(x), \quad l \le x \le R_2\},$$
$$L_4'' = \{x + y = \mu, \ y = \gamma_1(x), \quad 0 \le x \le l\}, \tag{3.41}$$

in which $\gamma_1(0) = 0$, $\gamma_2(R_2) = R_2 - 2R_1$; $\gamma_1(x) > 0$, $0 \le x \le l$; $\gamma_2(x) > 0$, $l \le x \le R_2$; $\gamma_1(x)$ on $0 \le x \le l$, $\gamma_2(x)$ on $l \le x \le R_2$ are continuous, and $\gamma_1(x)$, $\gamma_2(x)$ possess the derivatives on $0 \le x \le l$, $l \le x \le R_2$ except isolated points respectively, and $1+\gamma_1'(x) > 0$, $1-\gamma_2'(x) > 0$, $z_3' = l+j\gamma_1(l) = l+j\gamma_2(l) \in L_3$ (or L_4). By the conditions, we can find the inverse functions $x = \tau(\mu)$, $x = \sigma(\nu)$ of $x + \gamma_1(x) = \mu$, $x - \gamma_2(x) = \nu$ respectively, namely

$$\nu = 2\tau(\mu) - \mu, \quad 0 \le \mu \le 2R, \quad \mu = 2\sigma(\nu) - \nu, \quad 0 \le \nu \le l - \gamma_2(l). \tag{3.42}$$

We first make a transformation

$$\tilde{\mu} = \frac{2R\mu}{2\sigma(\nu) - \nu}, \quad \tilde{\nu} = \nu, \quad 0 \le \mu \le 2\sigma(\nu) - \nu, \quad 0 \le \nu \le 2R_1, \tag{3.43}$$

its inverse transformation is

$$\mu = \frac{(2\sigma(\nu) - \nu)\tilde{\mu}}{2R}, \quad \nu = \tilde{\nu}, \quad 0 \le \mu \le 2R, \quad 0 \le \nu \le 2R_1. \tag{3.44}$$

The above transformation can be expressed by

$$\tilde{x} = \frac{1}{2}(\tilde{\mu} + \tilde{\nu}) = \frac{2R(x + y) + (x - y)[2\sigma(x - \gamma_2(x)) - x + \gamma_2(x)]}{2[2\sigma(x - \gamma_2(x)) - x + \gamma_2(x)]},$$
$$\tilde{y} = \frac{1}{2}(\tilde{\mu} - \tilde{\nu}) = \frac{2R(x + y) - (x - y)[2\sigma(x - \gamma_2(x)) - x + \gamma_2(x)]}{2[2\sigma(x - \gamma_2(x)) - x + \gamma_2(x)]}, \tag{3.45}$$

and its inverse transformation is

$$x = \frac{1}{2}(\mu + \nu) = \frac{[2\sigma(x - \gamma_2(x)) - x + \gamma_2(x)](\tilde{x} + \tilde{y}) + 2R(\tilde{x} - \tilde{y})}{4R},$$
$$y = \frac{1}{2}(\mu - \nu) = \frac{[2\sigma(x - \gamma_2(x)) - x + \gamma_2(x)](\tilde{x} + \tilde{y}) - 2R(\tilde{x} - \tilde{y})}{4R}. \tag{3.46}$$

Denote by $\tilde{z} = \tilde{x} + j\tilde{y} = g(z)$, $z = x + jy = g^{-1}(\tilde{z})$ the transformation (3.45) and the inverse transformation (3.46) respectively. Through the transformation (3.44), we have

$$(u + v)_{\tilde{\nu}} = (u + v)_\nu, \quad (u - v)_{\tilde{\mu}} = \frac{2\sigma(\nu) - \nu}{2R}(u - v)_\mu, \tag{3.47}$$

system (3.38) in D'' is reduced to

$$\xi_{\tilde{\nu}} = A\xi + B\eta + Cu + D, \quad \eta_{\tilde{\mu}} = \frac{2\sigma(\nu) - \nu}{2R}[A\xi + B\eta + Cu + D] \text{ in } D'. \tag{3.48}$$

Moreover through the transformation (3.46), the boundary condition (3.37) on $L_3'' \cup L_4''$ is reduced to

$$
\mathrm{Re}\,[\lambda(g^{-1}(\tilde{z}))w(g^{-1}(\tilde{z}))] = r(g^{-1}(\tilde{z})),\, \tilde{z} \in L_3' \cup L_4',
$$

$$
\mathrm{Im}\,[\lambda(g^{-1}(\tilde{z}_3))w(g^{-1}(\tilde{z}_3))] = b_1,
$$

(3.49)

in which $\tilde{z}_3 = g(z_3')$. Besides the relation (3.38) is valid. Therefore the boundary value problem (3.36),(3.37) in $\bar{D}''$ is transformed into the boundary value problem (3.48),(3.49). On the basis of Theorem 3.4, we see that the boundary value problem (3.48),(3.49) has a unique solution $w(\tilde{z})$, and then $w[g(z)]$ is just a solution of the boundary value problem (3.36),(3.37) in D'', but we mention that the conditions of curve L_3', L_4' through the transformation $z = g^{-1}(\tilde{z})$ must satisfy the conditions of the curves in (3.31). For instance, if $z_3 \in L_3$, $\gamma_1(x) \geq x + 2l - 2R$ on $2R - 2l \leq x \leq l$, then the above condition holds. If $z_3 \in L_4$, $\gamma_2(x) \geq 2l - x$ on $l \leq x \leq 2l$, then we can similarly discuss. For other case, it can be discussed by using the method as stated in Section 2, Chapter VI below.

Theorem 3.5 *If equation* (1.16) *satisfies Condition C in the domain D'' with the boundary $L_1 \cup L_2 \cup L_3'' \cup L_4''$, then Problem P with the boundary conditions*

$$
\mathrm{Re}[\lambda(z)u_z] = r(z),\quad z \in L_3'' \cup L_4'',\ u(0) = b_0,\quad \mathrm{Im}[\lambda(z_3'')u_z(z_3'')] = b_1 \qquad (3.50)
$$

has a unique solution $u(z)$ as stated in (3.38) *in D''.*

By using the above method, we can generalize the domain D to more general domain including the disk $D'' = \{||z - R_1|| < R_1\}$. For the domain D'', we choose $R_2 = 2R_1$, the boundary L'' of the domain D'' consists of $L_1'', L_2'', L_3'', L_4''$, namely

$$
L_1'' = \left\{ y = -\gamma_1(x) = -\sqrt{R_1^2 - (x - R_1)^2},\ \ 0 \leq x \leq R_1 \right\},
$$

$$
L_2'' = \left\{ y = -\gamma_2(x) = -\sqrt{R_1^2 - (x - R_1)^2},\ \ R_1 \leq x \leq 2R_1 \right\},
$$

$$
L_3'' = \left\{ y = \gamma_3(x) = \sqrt{R_1^2 - (x - R_1)^2},\ \ R_1 \leq x \leq 2R_1 \right\},
$$

$$
L_4'' = \left\{ y = \gamma_4(x) = \sqrt{R_1^2 - (x - R_1)^2},\ \ 0 \leq x \leq R_1 \right\},
$$

where $1 + \gamma_1'(x) > 0, 1 + \gamma_4'(x) > 0$ on $0 < x \leq R_1$, $1 - \gamma_2'(x) > 0, 1 - \gamma_3'(x) > 0$ on $R_1 \leq x < 2R_1$. The above curves can be rewritten as

$$
L_1'' = \left\{ x = \sigma_1(\nu) = \frac{R_1 + \nu - \sqrt{R_1^2 + 2R_1\nu - \nu^2}}{2} \right\},
$$

$$
L_2'' = \left\{ x = \tau_1(\mu) = \frac{R_1 + \mu - \sqrt{R_1^2 + 2R_1\mu - \mu^2}}{2} \right\},
$$

$$L_3'' = \left\{ x = \sigma_2(\nu) = \frac{R_1 + \nu - \sqrt{R_1^2 + 2R_1\nu - \nu^2}}{2} \right\},$$

$$L_4'' = \left\{ x = \tau_2(\mu) = \frac{R_1 + \mu - \sqrt{R_1^2 + 2R_1\mu - \mu^2}}{2} \right\},$$

where $\sigma_1(\nu), \sigma_2(\nu)$ are the inverse functions of $x + \gamma_1(x) = \nu$, $x - \gamma_3(x) = \nu$ on $0 \le \nu \le 2R_1$, and $x = \tau_1(\mu)$, $x = \tau_2(\mu)$ are the inverse functions of $x - \gamma_2(x) = \mu$, $x + \gamma_4(x) = \mu$ on $0 \le \mu \le 2R_1$ respectively. Through a translation, we can discuss the unique solvability of corresponding boundary value problem for equation (1.16) in any disk $||z - z_0|| < R$, where z_0 is a hyperbolic number and R is a positive number.

4 Other Oblique Derivative Problems for Quasilinear Hyperbolic Equations of Second Order

In this section, we discuss other oblique derivative problems for quasilinear hyperbolic equations. Firstly the representation theorem of solutions for the above boundary value problems is given, moreover the uniqueness and existence of solutions for the above problem are proved. The results obtained include the corresponding result of the Dirichlet boundary value problem or the Darboux problem([12]3) as a special case.

4.1 Formulation of other oblique derivative problems for quasilinear hyperbolic equations

We first state four other oblique derivative problems for equation (1.16), here the domain D is the same as that in Section 1, but $R_2 = 2R_1$.

Problem P_1 Find a continuously differentiable solution $u(z)$ of (1.16) in $\bar{D}$ satisfying the boundary conditions

$$\mathrm{Re}\,[\lambda(z)u_z] = r(z), \quad z \in L_1 \cup L_2,$$
$$u(0) = b_0, \quad \mathrm{Im}\,[\lambda(z)u_z]|_{z=z_1} = b_1, \tag{4.1}$$

where b_0, b_1 are real constants, $\lambda(z) = a(z) + jb(z)$, $z \in L_1 \cup L_2$, and $\lambda(z), r(z), b_0, b_1$ satisfy the conditions

$$C_\alpha[\lambda(z), L_1 \cup L_2] \le k_0, \quad C_\alpha[r(z), L_1 \cup L_2] \le k_2,$$
$$|b_0|, |b_1| \le k_2, \quad \max_{z \in L_1} \frac{1}{|a(z) - b(z)|}, \quad \max_{z \in L_2} \frac{1}{|a(z) + b(z)|} \le k_0, \tag{4.2}$$

in which $\alpha(0 < \alpha < 1), k_0, k_2$ are non-negative constants.

If the boundary condition (4.1) is replaced by

$$\text{Re}\,[\lambda(z)u_z] = r(z), \quad z \in L_1, \quad \text{Re}\,[\Lambda(x)u_z(x)] = R(x), \quad x \in L_0 = (0, R_2),$$
$$u(0) = b_0, \quad \text{Im}\,[\lambda(z)u_z]|_{z=z_1} = b_1, \tag{4.3}$$

where b_0, b_1 are real constants, $\lambda(z) = a(z) + jb(z), z \in L_1, \Lambda(x) = 1 + j, x \in L_0$, and $\lambda(z), r(z), R(x), b_0, b_1$ satisfy the conditions

$$C_\alpha[\lambda(z), L_1] \leq k_0, \quad C_\alpha[r(z), L_1] \leq k_2, \quad |b_0|, |b_1| \leq k_2,$$
$$C_\alpha[R(x), L_0] \leq k_2, \quad \max_{z \in L_1} \frac{1}{|a(z) - b(z)|} \leq k_0, \tag{4.4}$$

in which $\alpha(0 < \alpha < 1), k_0, k_2$ are non-negative constants, then the boundary value problem for (1.16) will be called Problem P_2.

If the boundary condition in (4.1) is replaced by

$$\text{Re}\,[\lambda(z)u_z] = r(z), \quad z \in L_2, \quad \text{Re}\,[\Lambda(x)u_z(x)] = R(x), \quad x \in L_0 = (0, R_2),$$
$$u(0) = b_0, \quad \text{Im}\,[\lambda(z)u_z]|_{z=z_1} = b_1, \tag{4.5}$$

where b_0, b_1 are real constants, $\lambda(z) = a(z) + jb(z), z \in L_2, \Lambda(x) = 1 - j, x \in L_0$, and $\lambda(z), r(z), R(x), b_0, b_1$ satisfy the conditions

$$C_\alpha[\lambda(z), L_2] \leq k_0, \quad C_\alpha[r(z), L_2] \leq k_2, \quad |b_0|, |b_1| \leq k_2,$$
$$C_\alpha[R(x), L_0] \leq k_2, \quad \max_{z \in L_2} \frac{1}{|a(z) + b(z)|} \leq k_0, \tag{4.6}$$

then the boundary value problem for (1.16) is called Problem P_3. For Problem P_2 and Problem P_3, there is no harm in assuming that $w(z_1) = 0$, otherwise through the transformation $W(z) = \{w(z) - [a(z_1) - jb(z_1)][r(z_1) + jb_1]\}/[a^2(z_1) - b^2(z_1)]$, the requirement can be realized.

If the boundary condition in (4.1) is replaced by

$$u(x) = s(x), \quad u_y = R(x), \quad x \in L_0 = (0, R_2), \tag{4.7}$$

where $s(z), R(x)$ satisfy the conditions

$$C_\alpha^1[s(x), L_0] \leq k_2, \quad C_\alpha[R(x), L_0] \leq k_2, \tag{4.8}$$

then the boundary value problem for (1.16) is called Problem P_4.

In the following, we first discuss Problem P_2 and Problem P_3 for equation (2.11).

4.2 Representations of solutions and unique solvability of Problem P_2 and Problem P_3 for quasilinear hyperbolic equations

Similarly to Theorem 2.1, we can give the representation of solutions of Problem P_2 and Problem P_3 for equation (2.11), namely

Theorem 4.1 *Any solution $u(z)$ of Problem P_2 for the hyperbolic equation (2.11) can be expressed as*

$$u(z) = 2\mathrm{Re} \int_0^z w(z)dz + b_0 \text{ in } D, \tag{4.9}$$

where $w(z)$ is as follows

$$w(z) = f(x+y)e_1 + g(x-y)e_2$$
$$= \frac{1}{2}\{f(x+y) + g(x-y) + j[f(x+y) - g(x-y)]\}, \tag{4.10}$$

here $f(x+y)$, $g(x-y)$ possess the forms

$$f(x+y) = \mathrm{Re}\,[(1+j)u_z(x+y)] = R(x+y), \ 0 \le x+y \le 2R,$$

$$g(x-y) = \frac{2r((1-j)(x-y)/2) - [a((1-j)(x-y)/2) + b((1-j)(x-y)/2)]f(0)}{a((1-j)(x-y)/2) - b((1-j)(x-y)/2)},$$

$$0 \le x - y \le 2R_1, \tag{4.11}$$

$$f(0) = u(z_1) + v(z_1) = \frac{r((1-j)R_1) + b_1}{a((1-j)R_1) + b((1-j)R_1)}.$$

As for Problem P_3 for (2.11), its solution can be expressed as the forms (4.9), (4.10), but where $f(x+y)$, $g(x-y)$ possess the forms

$$f(x+y) = \frac{2r((1+j)(x+y)/2 + (1-j)R_1)}{a((1+j)(x+y)/2 + (1-j)R_1) - b((1+j)(x+y)/2 + (1-j)R_1)}$$
$$- \frac{[a((1+j)(x+y)/2 + (1-j)R_1) - b((1+j)(x+y)/2 + (1-j)R_1)]g(2R_1)}{a((1+j)(x+y)/2 + (1-j)R_1) + b((1+j)(x+y)/2 + (1-j)R_1)},$$
$$0 \le x + y \le 2R, \tag{4.12}$$

$$g(2R_1) = u(z_1) - v(z_1) = \frac{r((1-j)R_1) - b_1}{a((1-j)R_1) - b((1-j)R_1)},$$

$$g(x-y) = \mathrm{Re}\,[(1-j)u_z(x-y)] = R(x-y), \quad 0 \le x-y \le 2R_1.$$

Moreover the solution $u(z)$ of Problem P_2 and Problem P_3 satisfies the estimate

$$C_\alpha^1[u(z), \bar{D}] \le M_{23}, \quad C_\alpha^1[u(z), \bar{D}] \le M_{24}k_2, \tag{4.13}$$

where $M_{23} = M_{23}(a, k_0, k_2, D)$, $M_{24} = M_{24}(\alpha, k_0, D)$ are two non-negative constants.

Next we give the representation of solutions of Problem P_2 and Problem P_3 for the quasilinear hyperbolic equation (1.16).

Theorem 4.2   *Under Condition C, any solution $u(z)$ of Problem P_2 for the hyperbolic equation (1.16) can be expressed as*

$$u(z) = 2\text{Re} \int_0^z w(z)dz + b_0, \quad w(z) = W(z) + \Phi(z) + \Psi(z) \ \text{in} \ D,$$

$$W(z) = f(\mu)e_1 + g(\nu)e_2, \quad \Phi(z) = \tilde{f}(\mu)e_1 + \tilde{g}(\nu)e_2, \tag{4.14}$$

$$\Psi(z) = \int_{2R_1}^{\nu} [A\xi + B\eta + Cu + D]e_1 d\nu + \int_0^{\mu} [A\xi + B\eta + Cu + D]e_2 d\mu,$$

where $f(\mu)$, $g(\nu)$ are as stated in (4.11) and $\tilde{f}(\mu)$, $\tilde{g}(\nu)$ are similar to $f(\mu)$, $g(\nu)$ in (4.11), but the functions $r(z), R(x), b_1$ are replaced by the corresponding functions $-\text{Re}\,[\lambda(z)\Psi(z)], -\text{Re}\,[\Lambda(x)\Psi(x)], -\text{Im}\,[\lambda(z_1)\Psi(z_1)]$, namely

$$\text{Re}\,[\lambda(z)\Phi(z)] = -\text{Re}\,[\lambda(z)\Psi(z)], \quad z \in L_1,$$

$$\text{Re}\,[\Lambda(x)\Phi(x)] = -\text{Re}\,[\Lambda(x)\Psi(x)], \quad x \in L_0, \tag{4.15}$$

$$\text{Im}\,[\lambda(z_1)\Phi(z_1)] = -\text{Im}\,[\lambda(z_1)\Psi(z_1)].$$

As to Problem P_3 for (1.16), its solution $u(z)$ possesses the expression (4.14), where $W(z)$ is a solution of Problem P_3 for (2.11) and $\Phi(z)$ is also a solution of (2.11) satisfying the boundary conditions

$$\text{Re}\,[\lambda(z)\Phi(z)] = -\text{Re}\,[\lambda(z)\Psi(z)], \quad z \in L_2,$$

$$\text{Re}\,[\Lambda(x)\Phi(x)] = -\text{Re}\,[\Lambda(x)\Psi(x)], \quad x \in L_0, \tag{4.16}$$

$$\text{Im}\,[\lambda(z_1)\Phi(z_1)] = -\text{Im}\,[\lambda(z_1)\Psi(z_1)].$$

Proof   Let the solution $u(z)$ of Problem P_2 be substituted into the coefficients of equation (1.16). Due to Problem P_2 is equivalent to the Problem A_2 for the complex equation (2.8) with the relation (2.10), where the boundary conditions are as follows

$$\text{Re}\,[\lambda(z)w(z)] = r(z) \ \text{on} \ L_1, \quad \text{Re}\,[\Lambda(x)w(x)] = R(x) \ \text{on} \ L_0,$$

$$\text{Im}\,[\lambda(z_1)w(z_1)] = b_1. \tag{4.17}$$

According to Theorem 3.2 in Chapter I, it is not difficult to see that the function $\Psi(z)$ satisfies the complex equation

$$[\Psi]_{\bar{z}} = [A\xi + B\eta + Cu + D]e_1 + [A\xi + B\eta + Cu + D]e_2 \ \text{in} \ D, \tag{4.18}$$

and noting that $W(z)$ is the solution of Problem A_2 for the complex equation (2.12), hence $\Phi(z) = w(z) - W(z) - \Psi(z) = \xi e_1 + \eta e_2$ satisfies the complex equation and the boundary conditions

$$\xi_\mu e_1 + \eta_\lambda e_2 = 0, \tag{4.19}$$

$$\text{Re}\,[\lambda(z)(\xi e_1 + \eta e_2)] = -\text{Re}\,[\lambda(z)\Psi(z)] \text{ on } L_1,$$

$$\text{Re}\,[\Lambda(x)(\xi e_1 + \eta e_2)] = -\text{Re}\,[\Lambda(x)\Psi(x)] \text{ on } L_0, \tag{4.20}$$

$$\text{Im}\,[\lambda(z)(\xi e_1 + \eta e_2)]|_{z=z_1} = -\text{Im}\,[\lambda(z_1)\Psi(z_1)].$$

The representation of solutions of Problem A_2 for (1.16) is similar to (3.17) in Chapter I, we can obtain the representation (4.14) of Problem P_2 for (1.16). Similarly, we can verify that the solution of Problem P_3 for (1.16) possesses the representation (4.14) with the boundary condition (4.16).

By using the same method as stated in the proofs of Theorems 2.3 and 2.4, we can prove the following theorem.

Theorem 4.3 *If the complex equation* (1.16) *satisfies Condition C, then Problem P_2 (Problem P_3) for* (1.16) *has a unique solution.*

4.3 Representations of solutions and unique solvability of Problem P_4 for quasilinear hyperbolic equations

It is clear that the boundary condition (4.7) is equivalent to the following boundary condition

$$u_x = s'(x), \quad u_y = R(x), \quad x \in L_0 = (0, R_2), \quad u(0) = s(0), \quad \text{i.e.}$$

$$\text{Re}\,[(1+j)u_z(x)] = \sigma(x), \quad \text{Re}\,[(1-j)u_z(x)] = \tau(x), \quad u(0) = r(0), \tag{4.21}$$

in which

$$\sigma(x) = \frac{s'(x) + R(x)}{2}, \quad \tau(x) = \frac{s'(x) - R(x)}{2}. \tag{4.22}$$

Similarly to Theorem 4.1, we can give the representation of solutions of Problem P_4 for equation (2.11).

Theorem 4.4 *Any solution $u(z)$ of Problem P_4 for the hyperbolic equation* (2.11) *can be expressed as* (4.9), (4.10), *where $b_0 = s(0)$, $f(x+y)$, $g(x-y)$ possess the forms*

$$f(x+y) = \sigma(x+y) = \frac{s'(x+y) + R(x+y)}{2}, \ 0 \le x+y \le 2R,$$

$$g(x-y) = \tau(x-y) = \frac{s'(x-y) - R(x-y)}{2}, \ 0 \le x-y \le 2R_1, \tag{4.23}$$

and

$$f(0) = \frac{s'(0) + R(0)}{2}, \quad g(2R_1) = \frac{s'(2R_1) - R(2R_1)}{2}. \tag{4.24}$$

Moreover $u(z)$ of Problem P_4 satisfies the estimate (4.13).

Next we give the representation of Problem P_4 for the quasilinear hyperbolic equation (1.16).

Theorem 4.5   *Under Condition C, any solution $u(z)$ of Problem P_4 for the hyperbolic equation (1.16) can be expressed as (4.14), where $b_0 = s(0)$, $W(z)$ is a solution of Problem A_4 for (2.12) satisfying the boundary condition (4.21) ($W = u_z$), and $\Phi(z)$ is also a solution of (2.12) satisfying the boundary conditions*

$$\mathrm{Re}\,[(1+j)\Phi(x)] = -\mathrm{Re}\,[(1+j)\Psi(x)], \quad \mathrm{Re}\,[(1-j)\Phi(x)] = -\mathrm{Re}\,[(1-j)\Psi(x)]. \quad (4.25)$$

Proof   Let the solution $u(z)$ of Problem P_4 be substituted into the coefficients of equation (1.16). Due to Problem P_4 is equivalent to the Problem A_4 for the complex equation (2.8) with the relation (2.10) and the boundary conditions

$$\mathrm{Re}\,[(1+j)w(x)] = \sigma(z), \quad \mathrm{Re}\,[(1-j)w(x)] = \tau(x), \quad x \in L_0, \quad (4.26)$$

according to Theorem 3.2 in Chapter I, it can be seen that the function $\Psi(z)$ satisfies the complex equation (4.18) and noting that $W(z)$ is the solutions of Problem A_4 for the complex equation (2.12), i.e. (4.19), hence $\Phi(z) = w(z) - W(z) - \Psi(z)$ satisfies equation (4.19) and boundary conditions

$$\begin{aligned} \mathrm{Re}\,[(1+j)(\xi e_1 + \eta e_2)] &= -\mathrm{Re}\,[(1+j)\Psi(x)], \\ \mathrm{Re}\,[(1-j)(\xi e_1 + \eta e_2)] &= -\mathrm{Re}\,[(1-j)\Psi(x)], \end{aligned} \quad x \in L_0. \quad (4.27)$$

Similarly to Theorem 4.2, the representation of solutions of Problem A_4 for (2.8) is similar to (3.17) in Chapter I, we can obtain the representation (4.14) of Problem P_4 for (1.16).

By using the same method as stated in the proofs of Theorems 2.3 and 2.4, we can prove the following theorem.

Theorem 4.6   *If the complex equation (1.16) satisfies Condition C, then Problem P_4 for (1.16) has a unique solution.*

Besides we can discuss the unique solvability of Problem P_2, Problem P_3 and Problem P_4 for general quasilinear hyperbolic equation (3.15), and generalize the above results to the general domains D' with the conditions (3.31) and (3.41) respectively.

Similarly to Problem P as in Section 2, we can discuss Problem P_1 for equation (2.11), here the solution $w(z)$ of equation (2.12) is as follows

$$\begin{aligned} w(z) &= f(x+y)e_1 + g(x-y)e_2 \\ &= \frac{1}{2}\{f(x+y) + g(x-y) + j[f(x+y) - g(x-y)]\}, \\ f(0) &= u(z_1) + v(z_1) = \frac{r((1-j)R_1) + b_1}{a((1-j)R_1) + b((1-j)R_1)}, \\ g(2R_1) &= u(z_1) - v(z_1) = \frac{r((1-j)R_1) - b_1}{a((1-j)R_1) - b((1-j)R_1)}, \end{aligned} \quad (4.28)$$

here $f(x + y)$, $g(x - y)$ possess the forms

$$g(x - y) = \frac{2r((1-j)(x-y)/2) - [a((1-j)(x-y)/2) + b((1-j)(x-y)/2)]f(0)}{a((1-j)(x-y)/2) - b((1-j)(x-y)/2)},$$
$$0 \le x - y \le 2R_1,$$

(4.29)

$$f(x+y) = \frac{2r((1+j)(x+y)/2 + (1-j)R_1)}{a((1+j)(x+y)/2 + (1-j)R_1) - b((1+j)(x+y)/2 + (1-j)R_1)}$$
$$- \frac{[a((1+j)(x+y)/2 + (1-j)R_1) - b((1+j)(x+y)/2 + (1-j)R_1)]g(2R_1)}{a((1+j)(x+y)/2 + (1-j)R_1) + b((1+j)(x+y)/2 + (1-j)R_1)},$$
$$0 \le x + y \le 2R_1.$$

Moreover when we prove that Problem P_1 of equation (1.16) has a unique solution $u(z)$, the integrals in (2.16),(2.22),(2.25),(2.28) and (2.33) possess the similar forms, and the integral path in (2.21) can be chosen, for instance the integral $\Psi(z)$ in (2.16) is replaced by

$$\Psi(z) = \int_{2R_1}^{\nu} [A\xi + B\eta + Cu + D]e_1 d\nu + \int_0^{\mu} [A\xi + B\eta + Cu + D]e_1 d\mu. \quad (4.30)$$

Now we explain that the Darboux problem is a special case of Problem P_1. The so-called Darboux problem is to find a solution $u(z)$ for (1.16), such that $u(z)$ satisfies the boundary conditions

$$u(x) = s(x), \quad x \in L_0, \quad u(z) = \phi(x), \quad z = x + jy \in L_1, \quad (4.31)$$

where $s(x), \phi(x)$ satisfy the conditions

$$C^1[s(x), L_0] \le k_2, \quad C^1[\phi(x), L_1] \le k_2, \quad (4.32)$$

herein k_2 is a non-negative constant [12]3). From (4.31), we have

$$u(x) = s(x), \quad x \in L_0, \quad \text{Re}\,[(1-j)u_z] = \phi_x(x) = \phi'(x), \quad z \in L_1,$$
$$u(0) = s(0), \quad \text{Im}\,[(1-j)u_z]|_{z=z_1-0} = -\phi'(1), \quad \text{i.e.}$$
$$u(x) = s(x), \quad x \in L_0, \quad \text{Re}\,[\lambda(z)u_z] = r(z), \quad z \in L_1,$$
$$u(0) = b_0, \quad \text{Im}\,[\lambda(z)u_z]|_{z=z_1} = b_1,$$

(4.33)

where we choose

$$\lambda(z) = (1-j)/\sqrt{2}, \quad r(z) = \phi'(x)/\sqrt{2}, \quad z \in L_1, \quad b_0 = s(0), \quad b_1 = -\phi'(1)/\sqrt{2}, \quad (4.34)$$

it is easy to see that the boundary conditions (4.33),(4.34) possess the form of the boundary condition (4.1). This shows that the above Darboux problem is a special case of Problem P_1.

For more general hyperbolic equations of second order, the corresponding boundary value problems remain to be discussed.

5   Oblique Derivative Problems for Degenerate Hyperbolic Equations of Second Order

This section deals with the oblique derivative problem for the degenerate hyperbolic equation in a simply connected domain. We first give the representation theorem of solutions of the oblique derivative problem for the hyperbolic equation, and then by using the method of successive iteration, the existence and uniqueness of solutions for the above oblique derivative problem are proved.

5.1   Formulation of the oblique derivative problem for degenerate hyperbolic equations

It is known that the Chaplygin equation in the hyperbolic domain D possesses the form

$$K(y)u_{xx} + u_{yy} = 0 \text{ in } D, \tag{5.1}$$

where $K(y)$ possesses the first order continuous derivative $K'(y)$, and $K'(y) > 0$ on $y_1 < y < 0$, $K(0) = 0$, and the domain D is a simply connected domain with the boundary $L = L_0 \cup L_1 \cup L_2$, herein $L_0 = (0, 2)$,

$$L_1 = \left\{ x + \int_0^y \sqrt{-K(t)}dt = 0, x \in (0,1) \right\}, \quad L_2 = \left\{ x - \int_0^y \sqrt{-K(t)}dt = 2, x \in (1,2) \right\}$$

are two characteristic lines, and $z_1 = x_1 + jy_1 = 1 + jy_1$ is the intersection point of L_1 and L_2. In particular, if $K(y) = -|y|^m$, m is a positive constant, then

$$\int_0^y \sqrt{-K(t)}dt = \int_0^y |t|^{m/2}dt = -\int_0^{|y|} d\frac{2}{m+2}|t|^{(m+2)/2} = -\frac{2}{m+2}|y|^{(m+2)/2}.$$

In this case, the two characteristic lines L_1, L_2 are as follows:

$$L_1 : x - \frac{2}{m+2}|y|^{(m+2)/2} = 0, \quad L_2 : x + \frac{2}{m+2}|y|^{(m+2)/2} = 2, \quad \text{i.e.}$$

$$L_1 : y = -\left(\frac{m+2}{2}x\right)^{2/(m+2)}, \quad L_2 : y = -\left[\frac{m+2}{2}(2-x)\right]^{2/(m+2)}.$$

In this section we mainly consider the general Chaplygin equation of second order

$$K(y)u_{xx} + u_{yy} = du_x + eu_y + fu + g \text{ in } D, \tag{5.2}$$

where $D, K(y)$ are as stated in (5.1), its complex form is the following equation of second order

$$u_{z\bar{z}} = \text{Re}\left[Qu_{zz} + A_1u_z\right] + A_2u + A_3, \quad z \in D, \tag{5.3}$$

where

$$Q = \frac{K(y) + 1}{K(y) - 1}, \quad A_1 = \frac{d + je}{K(y) - 1}, \quad A_2 = \frac{f}{2(K(y) - 1)}, \quad A_3 = \frac{g}{2(K(y) - 1)},$$

and assume that the coefficients $A_j(z)(j = 1, 2, 3)$ satisfy Condition C. It is clear that equation (5.2) is a degenerate hyperbolic equation.

The oblique derivative boundary value problem for equation (5.2) may be formulated as follows:

Problem P_1 Find a continuous solution $u(z)$ of (5.3) in $\bar{D}$, which satisfies the boundary conditions

$$\frac{1}{2}\frac{\partial u}{\partial l} = \text{Re}\,[\lambda(z)u_{\tilde{z}}] = r(z), \quad z \in L = L_1 \cup L_2,$$

$$u(0) = b_0, \quad \text{Im}\,[\lambda(z)u_{\tilde{z}}]|_{z=z_1} = b_1,$$

(5.4)

where l is a given vector at every point $z \in L$, $u_z = [\sqrt{-K(y)}u_x + ju_y]/2$, $u_{\tilde{z}} = [\sqrt{-K(y)}u_x - ju_y]/2$, b_0, b_1 are real constants, $\lambda(z) = a(x) + jb(x) = \cos(l, x) + j\cos(l, y)$, $z \in L$, and $\lambda(z), r(z), b_0, b_1$ satisfy the conditions

$$C_\alpha^1[\lambda(z), L_j] \le k_0, \quad C_\alpha^1[r(z), L_j] \le k_2, \quad j = 1, 2,$$

$$|b_0|, |b_1| \le k_2, \quad \max_{z \in L_1} \frac{1}{|a(z) - b(z)|}, \quad \max_{z \in L_2} \frac{1}{|a(z) + b(z)|} \le k_0,$$

(5.5)

in which $\alpha\,(1/2 < \alpha < 1), k_0, k_2$ are non-negative constants. For convenience, we can assume that $u_{\tilde{z}}(z_1) = 0$, i.e. $r(z_1) = 0$, $b_1 = 0$, otherwise we make a transformation $u_{\tilde{z}} - [a(z_1) - jb(z_1)][r(z_1) + jb_1]/[a^2(z_1) - b^2(z_1)]$, the requirement can be realized. Problem P_1 with the conditions $A_3(z) = 0$, $z \in \bar{D}$, $r(z) = 0$, $z \in L$ and $b_0 = b_1 = 0$ will be called Problem P_0.

For the Dirichlet problem (Tricomi problem D) with the boundary condition:

$$u(x) = \phi(x) \text{ on } L_1 = AC = \left\{x = -\int_0^y \sqrt{-K(t)}dt, 0 \le x \le 1\right\},$$

$$L_2 = BC = \left\{x = 2 + \int_0^y \sqrt{-K(t)}dt, 1 \le x \le 2\right\},$$

(5.6)

we find the derivative for (5.6) according to $s = x$ on $L = L_1 \cup L_2$, and obtain

$$u_s = u_x + u_y y_x = u_x - (-K(y))^{-1/2}u_y = \phi'(x) \text{ on } L_1,$$

$$u_s = u_x + u_y y_x = u_x + (-K(y))^{-1/2}u_y = \phi'(x) \text{ on } L_2, \text{ i.e.}$$

$$(-K(y))^{1/2}U + V = (-K(y))^{1/2}\phi'(x)/2 = r(x) \text{ on } L_1,$$

$$(-K(y))^{1/2}U - V = (-K(y))^{1/2}\phi'(x)/2 = r(x) \text{ on } L_2, \text{ i.e.}$$

$$\text{Re}\,[(1 - j)(\tilde{U} + j\tilde{V})] = \tilde{U} - \tilde{V} = r(x) \text{ on } L_1,$$

$$\text{Im}\,[(1 - j)(\tilde{U} + j\tilde{V})] = [-\tilde{U} + \tilde{V}]|_{z=z_1-0} = -r(1 - 0),$$

$$\text{Re}\,[(1 + j)(\tilde{U} + j\tilde{V})] = \tilde{U} + \tilde{V} = r(x) \text{ on } L_2,$$

$$\text{Im}\,[(1 + j)(\tilde{U} + j\tilde{V})] = [\tilde{U} + \tilde{V}]|_{z=z_1+0} = r(1 + 0),$$

where

$$U = \sqrt{-K(y)}u_x/2 = \tilde{U}, \quad V = -u_y/2 = -\tilde{V},$$

$$a + jb = 1 - j, \quad a = 1 \neq b = -1 \text{ on } L_1,$$

$$a + jb = 1 + j, \quad a = 1 \neq -b = -1 \text{ on } L_2.$$

From the above formulas, we can write the complex forms of boundary conditions of $\tilde{U} + j\tilde{V}$:

$$\text{Re}\,[\lambda(z)(\tilde{U} + j\tilde{V})] = r(z), \quad z \in L_j\,(j = 1, 2),$$

$$\lambda(z) = \begin{cases} 1 - j = a + jb, \\ 1 + j = a - jb, \end{cases} \quad r(x) = \begin{cases} (-K(y))^{1/2}\phi'(x)/2 \text{ on } L_1, \\ (-K(y))^{1/2}\phi'(x)/2 \text{ on } L_2, \end{cases}$$

and

$$u(z) = 2\text{Re}\int_0^z (U - jV)dz + \phi(0) \text{ in } D. \tag{5.7}$$

Hence Problem D is a special case of Problem P_1.

5.2 Unique solvability of Problem P for Chaplygin equation (5.1) in the hyperbolic domain D

In the subsection, we discuss the Chaplygin equation (5.1) in the hyperbolic domain D, where the arcs $L_1 = AC$, $L_2 = BC$ are the characteristics of (5.1), i.e.

$$x + \int_0^y \sqrt{-K(t)}dt = 0, \quad 0 \leq x \leq 1, \quad x - \int_0^y \sqrt{-K(t)}dt = 2, \quad 1 \leq x \leq 2. \tag{5.8}$$

Setting that

$$\mu = x + \int_0^y \sqrt{-K(t)}dt, \quad \nu = x - \int_0^y \sqrt{-K(t)}dt, \tag{5.9}$$

and then

$$\mu + \nu = 2x, \quad \mu - \nu = 2\int_0^y \sqrt{-K(t)}dt,$$

$$(\mu - \nu)_y = 2\sqrt{-K(y)}, \quad \sqrt{-K(y)} = (\mu - \nu)_y/2, \tag{5.10}$$

$$x_\mu = 1/2 = x_\nu, \quad y_\mu = 1/2\sqrt{-K(y)} = -y_\nu,$$

hence we have

$$U_x = U_\mu + U_\nu, \quad V_y = \sqrt{-K(y)}(V_\mu - V_\nu),$$

$$V_x = V_\mu + V_\nu, \quad U_y = \sqrt{-K(y)}(U_\mu - U_\nu), \tag{5.11}$$

and

$$K(y)U_x - V_y = K(y)(U_\mu + U_\nu) - \sqrt{-K(y)}(V_\mu - V_\nu) = 0,$$

$$V_x + U_y = V_\mu + V_\nu + \sqrt{-K(y)}(U_\mu - U_\nu) = 0 \text{ in } D, \tag{5.12}$$

where $U = u_x/2$, $V = -u_y/2$ and $\tilde{U} = \sqrt{-K(y)}U$, $\tilde{V} = -V$. Noting that

$$\left(\sqrt{-K(y)}\right)_\mu = -\frac{1}{2}(-K)^{-1/2}K'(y)y_\mu = \frac{K'}{4K}, \quad \left(\sqrt{-K(y)}\right)_\nu = -K'/4K, \quad (5.13)$$

we have

$$(\tilde{U} - \tilde{V})_\mu = \sqrt{-K(y)}U_\mu + V_\mu + \frac{K'U}{4K},$$

$$(\tilde{U} + \tilde{V})_\nu = \sqrt{-K(y)}U_\nu - V_\nu - \frac{K'U}{4K}. \tag{5.14}$$

Moreover, by (5.11) and (5.14) we obtain

$$-2\sqrt{-K(y)}(\tilde{U} + \tilde{V})_\nu$$

$$= \sqrt{-K(y)}[(\tilde{U} + \tilde{V})_\mu - (\tilde{U} + \tilde{V})_\nu] - \sqrt{-K(y)}[(\tilde{U} + \tilde{V})_\mu + (\tilde{U} + \tilde{V})_\nu]$$

$$= -\sqrt{-K(y)}(\tilde{U} + \tilde{V})_x + (\tilde{U} + \tilde{V})_y$$

$$= -\frac{1}{2}(-K(y))^{-1/2}K'(y)U = \frac{K'(y)}{2K(y)}\tilde{U}, \; 2\sqrt{-K(y)}(\tilde{U} - \tilde{V})_\mu$$

$$= \sqrt{-K(y)}[(\tilde{U} - \tilde{V})_\mu + (\tilde{U} - \tilde{V})_\nu]$$

$$\quad + \sqrt{-K(y)}[(\tilde{U} - \tilde{V})_\mu - (\tilde{U} - \tilde{V})_\nu)]$$

$$= \sqrt{-K(y)}(\tilde{U} - \tilde{V})_x + (\tilde{U} - \tilde{V})_y$$

$$= -\frac{1}{2}(-K(y))^{-1/2}K'(y)U = \frac{K'(y)}{2K(y)}\tilde{U}. \tag{5.15}$$

Thus equation (5.15) can be written as a system of equations

$$(\tilde{U} + \tilde{V})_\nu = -(-K)^{-1/2}\frac{K'(y)}{4K(y)}\tilde{U}, \quad (\tilde{U} - \tilde{V})_\mu = (-K)^{-1/2}\frac{K'(y)}{4K(y)}\tilde{U}, \quad \text{i.e.}$$

$$\tilde{W}_{\bar{z}} = \frac{1}{2}[(-K(y))^{1/2}\tilde{W}_x - j\tilde{W}_y] = A_1\tilde{W}(z) + A_2\overline{\tilde{W}(z)} \text{ in } \Delta, \tag{5.16}$$

in which $\tilde{W} = \tilde{U} + j\tilde{V}$, $A_1(z) = A_2(z) = -jK'/8K$ in $\Delta = \{0 \le \mu \le 2, 0 \le \nu \le 2\}$, and

$$u(z) = 2\text{Re}\int_0^z (U - jV)dz + b_0 \text{ in } D, \tag{5.17}$$

where $U - jV = (-K)^{-1/2}\tilde{U} + j\tilde{V}$.

In the following, we first give the representation of solutions for the oblique derivative problem (Problem P_1) for system (5.16) in D. For this, we first discuss the system of equations

$$(\tilde{U} + \tilde{V})_\nu = 0, \quad (\tilde{U} - \tilde{V})_\mu = 0 \text{ in } D \tag{5.18}$$

with the boundary condition

$$\frac{1}{2}\frac{\partial u}{\partial l} = \text{Re}\,[\lambda(z)(\tilde{U} + j\tilde{V})] = r(z), \quad z \in L,$$

$$u(0) = b_0, \quad \text{Im}\,[\lambda(z)(\tilde{U} + i\tilde{V})]|_{z=z_1} = b_1,$$

(5.19)

in which $\lambda(z) = a(z) + jb(z)$ on $L_1 \cup L_2$. Similarly to Chapter I, the solution of Problem P_1 for (5.18) can be expressed as

$$\xi = \tilde{U} + \tilde{V} = f(\mu), \quad \eta = \tilde{U} - \tilde{V} = g(\nu),$$

$$\tilde{U}(x,y) = \frac{f(\mu) + g(\nu)}{2}, \quad \tilde{V}(x,y) = \frac{f(\mu) - g(\nu)}{2},$$

(5.20)

$$\text{i.e. } \tilde{W}(z) = \frac{(1+j)f(\mu) + (1-j)g(\nu)}{2},$$

in which $f(t), g(t)$ are two arbitrary real continuous functions on $[0,2]$. For convenience, denote by the functions $a(x), b(x), r(x)$ of x the functions $a(z), b(z), r(z)$ of z in (5.19), thus the first formula in (5.19) can be rewritten as

$$a(x)\tilde{U}(x,y) + b(x)\tilde{V}(x,y) = r(x) \text{ on } L, \quad \text{i.e.}$$

$$[a(x)-b(x)]f(x+y)+[a(x)+b(x)]g(x-y)=2r(x) \text{ on } L, \quad \text{i.e.}$$

$$[a(t/2) + b(t/2)]f(0) + [a(t/2) - b(t/2)]g(t) = 2r(t/2), \quad t \in [0,2],$$

$$[a(t/2+1)-b(t/2+1)]f(t)+[a(t/2+1)+b(t/2+1)]g(2)=2r(t/2+1), \quad t \in [0,2],$$

where

$$f(0) = \tilde{U}(0) + \tilde{V}(0) = \frac{r(1) + b_1}{a(1) + b(1)}, \quad g(2) = \tilde{U}(2) - \tilde{V}(2) = \frac{r(1) - b_1}{a(1) - b(1)}.$$

Noting that the boundary conditions in (5.19), we can derive

$$\tilde{U} = \frac{1}{2}\left\{f(\mu) + \frac{2r(\nu/2) - (a(\nu/2) + b(\nu/2))f(0)}{a(\nu/2) - b(\nu/2)}\right\},$$

$$\tilde{V} = \frac{1}{2}\left\{f(\mu) - \frac{2r(\nu/2) - (a(\nu/2) + b(\nu/2))f(0)}{a(\nu/2) - b(\nu/2)}\right\}, \quad \text{or}$$

(5.21)

$$\tilde{U} = \frac{1}{2}\left\{g(\nu) + \frac{2r(\mu/2 + 1) - (a(\mu/2 + 1) - b(\mu/2 + 1))g(2)}{a(\mu/2 + 1) + b(\mu/2 + 1)}\right\},$$

$$\tilde{V} = \frac{1}{2}\left\{-g(\nu) + \frac{2r(\mu/2 + 1) - (a(\mu/2 + 1) - b(\mu/2 + 1))g(2)}{a(\mu/2 + 1) + b(\mu/2 + 1)}\right\},$$

if $a(x) - b(x) \neq 0$ on $[0,1]$ and $a(x) + b(x) \neq 0$ on $[1,2]$ respectively. From the above formulas, it follows that

$$\text{Re}[(1+j)\tilde{W}(x)]=\tilde{U}+\tilde{V} = \frac{2r(x/2+1)-(a(x/2+1)+b(x/2+1))f(0)}{a(x/2+1)-b(x/2+1)},$$

$$\text{Re}[(1-j)\tilde{W}(x)]=\tilde{U}-\tilde{V} = \frac{2r(x/2)-(a(x/2)-b(x/2))g(2)}{a(x/2)+b(x/2)}, \quad x \in [0,2],$$

(5.22)

if $a(x) - b(x) \neq 0$ on $[0,1]$ and $a(x) + b(x) \neq 0$ on $[1,2]$ respectively. From (5.22),

$$
\tilde{W}(z) = \begin{cases}
\dfrac{1}{2}\left[(1+j)\dfrac{2r(\mu/2+1) - (a(\mu/2+1) - b(\mu/2+1))g(2)}{a(\mu/2+1) + b(\mu/2+1)} \right. \\[3mm]
\left. + (1-j)\dfrac{2r(\nu/2) - (a(\nu/2) + b(\nu/2))f(0)}{a(\nu/2) - b(\nu/2)} \right]
\end{cases}
\tag{5.23}
$$

can be derived.

Next, we find the solution of Problem P_1 for system (5.16). From (5.16), we have

$$
\tilde{U} + \tilde{V} = -\int_2^\nu (-K)^{-1/2}\frac{K'(y)}{4K(y)}\tilde{U}\,d\nu, \quad \tilde{U} - \tilde{V} = \int_0^\mu (-K)^{-1/2}\frac{K'(y)}{4K(y)}\tilde{U}\,d\mu,
$$

$$
\tilde{W} = \tilde{U} + j\tilde{V} = -\frac{1+j}{2}\int_2^\nu (-K)^{-1/2}\frac{K'(y)}{4K(y)}\tilde{U}\,d\nu + \frac{1-j}{2}\int_0^\mu (-K)^{-1/2}\frac{K'(y)}{4K(y)}\tilde{U}\,d\mu
$$

$$
= -\frac{1+j}{2}\int_2^\nu \frac{K'(y)}{4K(y)}U\,d\nu + \frac{1-j}{2}\int_0^\mu \frac{K'(y)}{4K(y)}U\,d\mu,
$$

the above last two integrals are along two characteristic lines s_2 and s_1 respectively. But according to the method in [66]1), if we denote by s_1 the member of the family of characteristic lines $dx/dy = -\sqrt{-K(y)}$, and by s_2 the member of the family of characteristic lines $dx/dy = \sqrt{-K(y)}$ passing through the point $P \in D$, and

$$
ds_1 = \sqrt{(dx)^2 + (dy)^2} = \sqrt{1 + \left(\frac{dy}{dx}\right)^2}\,dx = \sqrt{\frac{1-K}{-K}}\,dx = -\sqrt{1-K}\,dy,
$$

$$
ds_2 = \sqrt{(dx)^2 + (dy)^2} = \sqrt{1 + \left(\frac{dy}{dx}\right)^2}\,dx = \sqrt{\frac{1-K}{-K}}\,dx = \sqrt{1-K}\,dy,
\tag{5.24}
$$

then system (5.1) can be rewritten in the form

$$
(\tilde{U} + \tilde{V})_{s_1} = (\tilde{U} + \tilde{V})_x x_{s_1} + (\tilde{U} + \tilde{V})_y y_{s_1}
$$

$$
= \frac{1}{\sqrt{1-K}}\left[\sqrt{-K(y)}(\tilde{U} + \tilde{V})_x - (\tilde{U} + \tilde{V})_y\right]
$$

$$
= \frac{1}{2\sqrt{1-K}}(-K(y))^{-1/2}K'(y)U = -\frac{1}{2\sqrt{1-K}}\frac{K'(y)}{K(y)}\tilde{U},
$$

$$
(\tilde{U} - \tilde{V})_{s_2} = (\tilde{U} - \tilde{V})_x x_{s_2} + (\tilde{U} - \tilde{V})_y y_{s_2}
\tag{5.25}
$$

$$
= \frac{1}{\sqrt{1-K}}\left[\sqrt{-K(y)}(\tilde{U} - \tilde{V})_x + (\tilde{U} - \tilde{V})_y\right]
$$

$$
= -\frac{1}{2\sqrt{1-K}}(-K(y))^{-1/2}K'(y)U = \frac{1}{2\sqrt{1-K}}\frac{K'(y)}{K(y)}\tilde{U},
$$

thus we obtain the system of integral equations

$$\xi = \tilde{U} + \tilde{V} = -\int_0^{s_1} \frac{1}{2\sqrt{(1-K)}} \frac{K'(y)}{K(y)} \tilde{U} ds_1 = \int_0^{s_1} \frac{1}{2} \frac{K'(y)}{K(y)} \tilde{U} dy,$$

$$\eta = \tilde{U} - \tilde{V} = \int_0^{s_2} \frac{1}{2\sqrt{(1-K)}} \frac{K'(y)}{K(y)} \tilde{U} ds_2 = \int_0^{s_2} \frac{1}{2} \frac{K'(y)}{K(y)} \tilde{U} dy,$$

(5.26)

where the integrals are along the two families of characteristics s_1 and s_2 respectively. Similarly to next subsection, the solution $\tilde{U}+\tilde{V}$, $\tilde{U}-\tilde{V}$ can be obtained by the method of successive iteration, which can be expressed as

$$\tilde{W}(z) = \tilde{U} + j\tilde{V} = W(z) + \Phi(z) + \Psi(z),$$

$$\tilde{U}+\tilde{V} = -\int_0^{s_1} \frac{1}{2\sqrt{(1-K)}} \frac{K'(y)}{K(y)} \tilde{U} ds_1, \quad \tilde{U}-\tilde{V} = \int_0^{s_2} \frac{1}{2\sqrt{(1-K)}} \frac{K'(y)}{K(y)} \tilde{U} ds_2,$$

(5.27)

where $W(z)$, $\Phi(z)$ are the solutions of system (5.18) satisfying the boundary condition (5.19) respectively, but the function $r(z)$, b_1 in the boundary condition of $\Phi(z)$ should be replaced by $-\mathrm{Re}\,[\lambda(z)\Psi(z)]$ on $L_1 \cup L_2$ and $-\mathrm{Im}\,[\lambda(z_1)\Psi(z_1)]$, and then the function

$$u(z) = 2\mathrm{Re}\int_0^z (U - jV)dz + b_0 = 2\mathrm{Re}\int_0^z [(-K(y))^{-1/2}\tilde{U} + j\tilde{V}]dz + b_0 \quad \text{in } D, \quad (5.28)$$

is just the solution of Problem P_1 for (5.16) and the solution is unique. Here we mention that firstly it suffices to consider the case of $y_1 \leq y \leq -\delta$, where δ is a sufficiently small positive number, and when we find out the solution of Problem P_1 for equation (5.1) with the condition $y_1 \leq y \leq -\delta$, and then let $\delta \to 0$.

Theorem 5.1 *If the Chaplygin equation (5.1) in D satisfies the above conditions, then Problem P_1 for (5.1) in D has a unique solution.*

Finally we mention that the boundary condition of Problem P_1:

$$\mathrm{Re}\,[\lambda(z)(\tilde{U} + j\tilde{V})] = r(z), \quad z \in L_1 \cup L_2$$

can be replaced by

$$2\tilde{U}(x) = S(x) = \sqrt{-K[y(x)]}u_x, \quad \text{i.e. } u(x) = \int_0^x \frac{S(x)dx}{\sqrt{-K[y(x)]}} + u(0) = s(x),$$

(5.29)

$$2\tilde{V}(x) = R(x) = u_y \quad \text{on } AB = L_0 = (0, 2),$$

where $C_\alpha[R(x), AB] \leq k_2 < \infty$, $C_\alpha^1[s(x), AB] \leq k_2$.

5.3 Unique solvability of the oblique derivative problem for degenerate hyperbolic equations

In this subsection, we prove the uniqueness and existence of solutions of Problem P_4 for the degenerate hyperbolic equation (5.2), the boundary condition of Problem P_4 is as follows

$$u(x) = s(x), \quad u_y(x) = R(x) \text{ on } L_0 = (0,2), \tag{5.30}$$

where $s(x), R(x)$ satisfy the condition $C_\alpha^2[s(x), L_0], C_\alpha^2[R(x), L_0] \leq k_2$, the above boundary value problem is also called the Cauchy problem for (5.2). Making a transformation of function

$$v(z) = u(z) - yR(x) - s(x) \text{ in } D, \tag{5.31}$$

equation (5.2) and boundary condition (5.30) are reduced to the form

$$K(y)v_{xx} + v_{yy} = dv_x + ev_y + fv + G,$$
$$G = g + f(yR + s) + eR + d[yR'(x) + s'(x)] \tag{5.32}$$
$$-K(y)[yR''(x) + s''(x)] \text{ in } D,$$

$$v(x) = 0, \quad v_y(x) = 0 \text{ on } L_0. \tag{5.33}$$

Hence we may only discuss Cauchy problem (5.32),(5.33) and denote it by Problem P_4 again. According to Subsection 4.3, Problem P_4 for (5.32) is equivalent to the boundary value problem A for the hyperbolic system of first order equations, the relation and the boundary conditions

$$\xi_{s_1} = \frac{2\sqrt{-K}}{\sqrt{1-K}}\xi_\nu = \frac{1}{2\sqrt{1-K}}\left[\left(\frac{-d}{\sqrt{-K}} - e - \frac{1}{2}\frac{K'(y)}{K(y)}\right)\xi\right.$$
$$\left. + \left(\frac{-d}{\sqrt{-K}} + e - \frac{1}{2}\frac{K'(y)}{K(y)}\right)\eta - fv - G\right],$$

$$\eta_{s_2} = \frac{2\sqrt{-K}}{\sqrt{1-K}}\eta_\mu = \frac{1}{2\sqrt{1-K}}\left[\left(\frac{-d}{\sqrt{-K}} - e + \frac{1}{2}\frac{K'(y)}{K(y)}\right)\xi\right. \tag{5.34}$$
$$\left. + \left(\frac{-d}{\sqrt{-K}} + e + \frac{1}{2}\frac{K'(y)}{K(y)}\right)\eta - fv - G\right],$$

$$\xi = \tilde{U} + \tilde{V}, \quad \eta = \tilde{U} - \tilde{V}, \quad v_y = \xi - \eta, \quad v(x) = 0.$$

In particular, if $K(y) = -|y|^m h(y)$, m is a positive constant, then

$$\frac{K'(y)}{K(y)} = \frac{m|y|^{m-1}h(y)}{K(y)} - \frac{|y|^m h_y}{K(y)} = \frac{m}{y} + \frac{h_y}{h}.$$

Integrating the hyperbolic system in (5.34) along the characteristics s_1, s_2, we obtain the system of integral equations as follows

$$v(z) = \int_0^y (\xi - \eta) dy \text{ in } D,$$

$$\xi(z) = - \int_0^y [A_1 \xi + B_1 \eta + C_1(\xi + \eta) + Dv + E] dy, \quad z \in s_1, \qquad (5.35)$$

$$\eta(z) = \int_0^y [A_2 \xi + B_2 \eta + C_2(\xi + \eta) + Dv + E] dy, \quad z \in s_2,$$

in this case

$$A_1 = -\frac{e}{2} - \frac{h_y}{4h}, \quad B_1 = \frac{e}{2} - \frac{h_y}{4h}, \quad A_2 = -\frac{e}{2} + \frac{h_y}{4h}, \quad B_2 = \frac{e}{2} + \frac{h_y}{4h},$$

$$C_1 = \frac{1}{2}\frac{d}{\sqrt{-K}} - \frac{m}{4y}, \quad C_2 = -\frac{1}{2}\frac{d}{\sqrt{-K}} + \frac{m}{4y}, \quad D = -\frac{f}{2}, \quad E = -\frac{G}{2}.$$

In the following we may only discuss the case of $K(y) = -|y|^m h(y)$, because otherwise it can be similarly discussed. In order to find a solution of the system of integral equations (5.35), we need to add the condition

$$\lim_{y \to 0} \frac{|y| d(x, y)}{|y|^{m/2}} = 0, \quad \text{i.e. } d(x, y) \approx \varepsilon(y) |y|^{m/2 - 1}, \qquad (5.36)$$

where $\varepsilon(y) \to 0$ as $y \to 0$. It is clear that for two characteristics $s_1^0 : x = x_1(y, z_0)$, $s_2^0 : x = x_2(y, z_0)$ passing through $P_0 = z_0 = x_0 + jy_0 \in D$, we have

$$|x_1 - x_2| \le 2 \left| \int_0^y \sqrt{-K} dy \right| \le M |y|^{m/2 + 1} \text{ for } y' < y < 0, \qquad (5.37)$$

for any $z_1 = x_1 + jy \in s_1^0, z_2 = x_2 + jy \in s_2^0$, where $M(> \max[4\sqrt{h(y)}/(m+2), 1])$ is a positive constant. Suppose that the coefficients of (5.35) possess continuously differentiable with respect to $x \in L_0$ and satisfy the condition

$$|A_j|, \quad |A_{jx}|, \quad |B_j|, \quad |B_{jx}|, \quad |yC_j|, \quad |yC_{jx}|, \quad |D|, \quad |D_x|, \quad |E|,$$
$$|E_x|, \quad |1/\sqrt{h}|, \quad |h_y/h| \le M, \quad z \in \bar{D}, \quad j = 1, 2. \qquad (5.38)$$

According to the proof of Theorem 5.1, it is sufficient to find a solution of Problem P_4 for arbitrary segment $-\delta \le y \le 0$, where δ is a sufficiently small positive number, and choose a positive constant $\gamma(< 1)$ close to 1, such that the following inequalities hold:

$$\frac{3M\delta}{2} + \frac{[\varepsilon(y)M + m/2]\delta^{m/2}}{m+2} < \gamma,$$

$$\frac{6\delta^2 M^2}{m+6} + \frac{8\delta M}{m+2} + \frac{2\varepsilon(y)M + m}{m+2} < \gamma. \qquad (5.39)$$

Similar to [66]1), a solution of Problem P_4 for (5.35) on $-\delta < y < 0$ can be found. Firstly, let $y_0 \in (-\delta, 0)$ and D_0 be a domain bounded by the boundary $y = 0, s_1^0, s_2^0$, we choose $v_0 = 0, \xi_0 = 0, \eta_0 = 0$ and substitute them into the corresponding positions of v, ξ, η in the right-hand sides of (5.35), and obtain

$$\xi_1(z) = -\int_0^y [A_1\xi_0 + B_1\eta_0 + C_1(\xi_0 + \eta_0) + Dv_0 + E]dy = -\int_0^y Edy, \quad z \in s_1^0,$$

$$\eta_1(z) = \int_0^y [A_2\xi_0 + B_2\eta_0 + C_2(\xi_0 + \eta_0) + Dv_0 + E]dy = \int_0^y Edy, \quad z \in s_2^0, \qquad (5.40)$$

$$v_1(z) = \mathrm{Re}\int_0^y (\xi_0 - \eta_0)dy = 0 \text{ in } D_0.$$

By the successive iteration, we find the sequences of functions $\{v_k\}, \{\xi_k\}, \{\eta_k\}$, which satisfy the relations

$$\xi_{k+1}(z) = -\int_0^y [A_1\xi_k + B_1\eta_k + C_1(\xi_k + \eta_k) + Dv_k + E]dy, \quad z \in s_1^0,$$

$$\eta_{k+1}(z) = \int_0^y [A_2\xi_k + B_2\eta_k + C_2(\xi_k + \eta_k) + Dv_k + E]dy, \quad z \in s_2^0, \qquad (5.41)$$

$$v_{k+1}(z) = \int_0^y (\xi_k - \eta_k)dy \text{ in } D_0, \quad k = 0, 1, 2, \ldots.$$

We can prove that $\{v_k, \}, \{\xi_k\}, \{\eta_k\}$ in $\bar{D}_0$ satisfy the estimates

$$|v_k(z)|, \quad |\xi_k(z)|, \quad |\eta_k(z)| \le M\sum_{j=0}^k \gamma^j|y|, \quad |\xi_k(z) + \eta_k(z)|,$$

$$|v_k(z_1) - v_k(z_2)|, \quad |\xi_k(z_1) - \xi_k(z_2)|, \quad |\eta_k(z_1) - \eta_k(z_2)| \le M\sum_{j=0}^k \gamma^j|y|^{m/2+1},$$

$$|v_{k+1}(z) - v_k(z)|, \quad |\xi_{k+1}(z) - \xi_k(z)|, \quad |\eta_{k+1}(z) - \eta_k(z)| \le M\gamma^k|y|, \qquad (5.42)$$

$$|\xi_{k+1}(z) + \eta_{k+1}(z) - \xi_k(z) - \eta_k(z)|, \quad |v_{k+1}(z_1) - v_{k+1}(z_2)$$

$$-v_k(z_1) - v_k(z_2)|, \quad |\xi_{k+1}(z_1) - \xi_{k+1}(z_2) - \xi_k(z_1) + \xi_k(z_2)|,$$

$$|\eta_{k+1}(z_1) - \eta_{k+1}(z_2) - \eta_k(z_1) + \eta_k(z_2)| \le M\gamma^k|y|^{m/2+1}.$$

In fact, from (5.40), it follows that the first formula with $k = 1$ holds, namely

$$|v_1(z)| = 0 \le M|y|, \quad |\xi_1(z)| \le M|y|, \quad |\eta_1(z)| \le M|y| = M\gamma^0|y| \le M\sum_{j=0}^1 \gamma^j|y|.$$

Moreover we get

$$|v_1(z_1) - v_1(z_2)| = 0,$$

$$|\xi_1(z) + \eta_1(z)| \le |\int_0^y [E(z_1) - E(z_2)]dy| \le 2|\int_0^y E_\tau[x_1 - x_2]dy|$$

$$\le \frac{4}{m+4}M^2|y|^{m/2+2} \le M\gamma|y|^{m/2+1} \le M\sum_{j=0}^1 \gamma^j|y|^{m/2+1},$$

$$|\xi_1(z_1) - \xi_1(z_2)| \leq |\int_0^y [E(x_j(t,z_1) + jt) - E(x_j(t,z_2) + jt)]dt|$$

$$\leq |\int_0^y |E_x||x_j(t,z_1) - x_j(t,z_2)|dy| \leq M|\int_0^y |x_1 - x_2|dy|$$

$$\leq M|\int_0^y M|y|^{m/2+1}dy| \leq M\gamma|y|^{m/2+1} \leq M\sum_{j=0}^1 \gamma^j|y|^{m/2+1},$$

$$|\eta_1(z_1) - \eta_1(z_2)| = |\int_0^y [E(x_j(t,z_1) + jt) - E(x_j(t,z_2) + jt)]dt| \leq M\sum_{j=0}^1 \gamma^j|y|^{m/2+1},$$

$$|v_1(z) - v_0(z)| = |v_1(z)| \leq M\gamma|y|, \quad |\xi_1(z) - \xi_0(z)| = |\xi_1(z)| \leq M\gamma|y|,$$

$$|\eta_1(z) - \eta_0(z)| = |\eta_1(z)| \leq M\gamma|y|, \quad |v_1(z_1) - v_1(z_2) - v_0(z_1) - v_0(z_2)| \leq M\gamma|y|^{m/2+1},$$

$$|\xi_1(z) + \eta_1(z) - \xi_0(z) - \eta_0(z)| = |\xi_1(z) + \eta_1(z)| \leq \gamma|y|^{m/2+1} \leq M\gamma|y|^{m/2+1},$$

$$|\xi_1(z_1) - \xi_1(z_2) - \xi_0(z_1) + \xi_0(z_2)| = |\xi_1(z_1) - \xi_1(z_2)| \leq M\gamma|y|^{m/2+1},$$

$$|\eta_1(z_1) - \eta_1(z_2) - \eta_0(z_1) + \eta_0(z_2)| = |\eta_1(z_1) - \eta_1(z_2)| \leq M\gamma|y|^{m/2+1}.$$

In addition, we use the inductive method, namely suppose the estimates in (5.42) for $k = n$ are valid, then they are also valid for $k = n + 1$. In the following, we only give the estimates of $|v_{n+1}(z)|, |\xi_{n+1}(z)|, |\xi_{n+1} + \eta_{n+1}(z)|$, the other estimates can be similarly given. From (5.41), we have

$$|v_{n+1}(z)| \leq |\int_0^y [\xi_n - \eta_n]dy| \leq 2M|\int_0^y \sum_{j=0}^n \gamma^j y dy| \leq M\sum_{j=0}^n \gamma^j|y|^2 \leq M\sum_{j=0}^{n+1} \gamma^j|y|,$$

$$|\xi_{n+1}(z)| \leq |\int_0^y \left[(|A_1| + |B_1| + |D|)M\sum_{j=0}^n \gamma^j|y| + |C_1|M\sum_{j=0}^n \gamma^j|y|^{m/2+1} + |E|\right]dy|$$

$$\leq M|\int_0^y \left[3M\sum_{j=0}^n \gamma^j|y| + \left(\frac{\varepsilon(y)}{2|y|\sqrt{h}} + \frac{m}{4|y|}\right)\sum_{j=0}^n \gamma^j|y|^{m/2+1} + 1\right]dy|$$

$$\leq M|y|\left\{\left[\frac{3}{2}M|y| + \left(|\varepsilon(y)|M + \frac{m}{2}\right)\frac{|y|^{m/2}}{m+2}\right]\sum_{j=0}^n \gamma^j + 1\right\} \leq M\sum_{j=0}^{n+1} \gamma^j|y|,$$

and

$$|\xi_{n+1}(z) + \eta_{n+1}(z)| \leq |\int_0^y \left\{\sum_{j=1}^2 [A_2(z_2)\xi_n(z_2) - A_1(z_1)\xi_n(z_1) + B_2(z_2)\eta_n(z_2)\right.$$

$$- B_1(z_1)\eta_n(z_1)] + C_2(z_2)(\xi_n(z_2) + \eta_n(z_2))$$

$$- C_1(z_1)(\xi_n(z_1) + \eta_n(z_1))$$

$$\left.+ D(z_2)v_n(z_2) - D(z_1)v_n(z_1) + E(z_2) - E(z_1)]\right\}dy|.$$

Noting that

$$|D(z_2)v_n(z_2) - D(z_1)v_n(z_1)| = |[D(z_2) - D(z_1)]v_n(z_2) + D(z_1)$$
$$\times [v_n(z_2) - v_n(z_1)]|$$
$$\leq \sum_{j=0}^{n} \gamma^j |y||D(z_2) - D(z_1)| + M^2 \sum_{j=0}^{n} \gamma^j |y|^{m/2+1}$$
$$\leq M^2 \sum_{j=0}^{n} \gamma^j |y||x_2 - x_1| + M^2 \sum_{j=0}^{n} \gamma^j |y|^{m/2+1}$$
$$\leq (M|y| + 1)M^2 \sum_{j=0}^{n} \gamma^j |y|^{m/2+1},$$

and

$$|A_2(z_2)\xi_n(z_2) - A_1(z_1)\xi_n(z_1) + B_2(z_2)\eta_n(z_2) - B_1(z_1)\eta_n(z_1)|$$
$$\leq |[A_2(z_2) - A_2(z_1)]\xi_n(z_2) + [A_2(z_1) - A_1(z_1)]\xi_n(z_2) + A_1(z_1)[\xi_n(z_2) - \xi_n(z_1)]$$
$$+ [B_2(z_2) - B_2(z_1)]\eta_n(z_2) + B_1(z_1)[\eta_n(z_2) - \eta_n(z_1)]|$$
$$+ [B_2(z_1) - B_1(z_1)]\eta_n(z_2)$$
$$\leq 2M|y|[M|x_1 - x_2| + |y|^{m/2}] \sum_{j=0}^{n} \gamma^j + \left| \frac{h_y}{2h} \right| |\xi_n(z_2) - \eta_n(z_2)|$$
$$\leq (2M|y| + 3)M^2 \sum_{j=0}^{n} \gamma^j |y|^{m/2+1},$$

we get

$$|\xi_{n+1}(z) + \eta_{n+1}(z)| \leq \left| \int_0^y [(3M|y| + 4)M^2 \sum_{j=0}^{n} \gamma^j |y|^{m/2+1} \right.$$
$$+ (|C_1| + |C_2|)M \sum_{j=0}^{n} \gamma^j |y|^{m/2+1} + M^2 |y|^{m/2+1}] dy|$$
$$\leq M|y|^{m/2+1} \left[\frac{6M^2 y^2}{m+6} + \frac{8M|y|}{m+2} + \left(|\varepsilon(y)|M + \frac{m}{2} \right) \right.$$
$$\times \frac{2}{m+2} \sum_{j=0}^{n} \gamma^j + \frac{2M}{m+4}|y| \right]$$
$$\leq M \sum_{j=0}^{n+1} \gamma^j |y|^{m/2+1}.$$

On the basis of the estimates (5.42), we can derive that $\{v_k\}, \{\zeta_k\}, \{\eta_k\}$ in $\overline{D_0}$ uniformly converge to v_*, ξ_*, η_* satisfying the system of integral equations

$$\xi_*(z) = - \int_0^y [A_1\xi_* + B_1\eta_* + C_1(\xi_* + \eta_*) + Dv_* + E] dy, \quad z \in s_1,$$

$$\eta_*(z) = \int_0^y [A_2\xi_* + B_2\eta_* + C_2(\xi_* + \eta_*) + Dv_* + E]dy, \quad z \in s_2,$$

$$v_*(z) = \int_0^y (\xi_* - \eta_*)dy \text{ in } \overline{D_0},$$

and the function $v_*(z)$ satisfies equation (5.32) and boundary condition (5.33), hence $u^*(z) = v^*(z) + yR(x) + s(x)$ is a solution of Problem P_4 for (5.2). From the above discussion, we can see that the solution of Problem P_4 for (5.2) in D is unique.

Theorem 5.2 *If the equation (5.2) in D satisfies the above conditions, then Problem P_4 for (5.2) in D has a unique solution.*

Now we mention that if we denote

$$\tilde{W}(z) = \tilde{U} + j\tilde{V} = |y|^{m/2}U - jV = \frac{1}{2}[|y|^{m/2}u_x + ju_y],$$

$$\overline{\tilde{W}(z)} = \tilde{U} - j\tilde{V} = |y|^{m/2}U + jV = \frac{1}{2}[|y|^{m/2}u_x - ju_y],$$

then $\tilde{W}(z) = |y|^{m/2}U - jV$ is a solution of the first order hyperbolic complex equation

$$\tilde{W}_{\tilde{z}} = A_1(z)\tilde{W} + A_2(z)\overline{\tilde{W}} + A_3(z)u + A_4(z) \text{ in } D,$$

$$A_1 = -\frac{d}{4|y|^{m/2}} + j\left(\frac{m}{8|y|} - \frac{e}{4}\right), \quad A_3 = -\frac{f}{4}, \tag{5.43}$$

$$A_2 = -\frac{d}{4|y|^{m/2}} + j\left(\frac{m}{8|y|} + \frac{e}{4}\right), \quad A_4 = -\frac{g}{4},$$

and

$$u(z) = 2\text{Re}\int_0^z u_z dz = 2\text{Re}\int_0^z (U - jV)d(x+jy) = 2\text{Re}\int_0^z (U+jV)d(x-jy)$$

is a solution of equation (5.2) with $K(y) = -|y|^m$.

By using the similar method, we can prove the solvability of Problem P_1, Problem P_2 and Problem P_3 for equation (5.2). Moreover for general domain D' with non-characteristics boundary, we can also discuss the solvability of Problem P_1, Problem P_2, Problem P_3 and Problem P_4 for equation (5.2). Besides we can discuss the solvability of corresponding boundary value problems for the hyperbolic equation in the form

$$u_{xx} + K(y)u_{yy} = du_x + eu_y + fu + g \text{ in } D \tag{5.44}$$

under certain conditions, where $K(y)$ is as stated in (5.2).

The references for the chapter are [2],[7],[12],[13],[24],[25],[34],[41],[44],[47],[54], [60],[66],[70],[79],[85],[87],[89],[95].

CHAPTER III

NONLINEAR ELLIPTIC COMPLEX EQUATIONS OF FIRST AND SECOND ORDER

In this chapter, we discuss the representation and existence of solutions of discontinuous boundary value problems for nonlinear elliptic complex equations of first and second order, which will be used in latter chapters.

1   Generalizations of Keldych–Sedov Formula for Analytic Functions

It is known that the Keldych–Sedov formula gives the representation of solutions of the mixed boundary value problem for analytic functions in the upper half-plane (see [53]). But for many problems in mechanics and physics, one needs a more general formulas of solutions of the discontinuous Riemann–Hilbert boundary value problem for analytic functions in the upper half-plane and other special domains. In this section, we shall establish the representations of solutions of the general discontinuous boundary value problem for analytic functions in the upper half-plane and upper half-unit disk. In the following sections and chapters we shall give applications to some nonlinear elliptic complex equations and quasilinear equations of mixed type.

1.1   General discontinuous boundary value problem for analytic functions in the upper half-plane

Let D be the upper half-plane and $a(x)$, $b(x)$, $c(x)$ be known real functions on $L = \{-\infty < x < \infty, y = 0\}$, where $a(x)$, $b(x)$ possess discontinuities of first kind at m distinct points $x_j (j = 1, \ldots, m, -\infty < x_1 < \cdots < x_m < \infty)$, m is a positive integer, and $c(x) = O(|x - x_j|^{-\beta_j})$ in the neighborhood of $x_j (j = 1, 2, \ldots, m)$ on L, herein $\beta_j (< 1, j = 1, 2, \ldots, m)$ are non-negative constants, such that $\beta_j + \gamma_j < 1$, $\gamma_j (j = 1, \ldots, m)$ are as stated in (1.3) below. Denote $\lambda(x) = a(x) - ib(x)$ and $|a(x)| + |b(x)| \neq 0$, there is no harm in assuming that $|\lambda(x)| = 1$, $x \in L^* = L \setminus \{x_1, \ldots, x_m\}$. Suppose that $\lambda(x)$, $c(x)$ satisfy the conditions

$$\lambda(x) \in C_\alpha(L_j), \; |x - x_j|^{\beta_j} c(x) \in C_\alpha(L_j), \; j = 1, 2, \ldots, m, \tag{1.1}$$

where L_j is the line segment from the point x_{j-1} to x_j on L, $x_0 = x_m$, $L_j(j = 1, 2, \ldots, m)$ do not include the end points, $L_1 = \{x < x_1\} \cup \{x > x_m\}$, $\alpha(0 < \alpha < 1)$ is a constant, and the function $\lambda(x) \in C_\alpha(L_\infty)(L_\infty$ is a neighborhood of the point ∞) is indicated as $\lambda(1/x) \in C_\alpha(L_*)$, here $L_*(\subset L)$ is a neighborhood of the point $x = 0$.

The discontinuous Riemann–Hilbert boundary value problem for analytic functions in D may be formulated as follows.

Problem A   Find an analytic function $\Phi(z) = u(z) + iv(z)$ in D, which is continuous in $D^* = \bar{D} \backslash \{x_1, x_2, \ldots, x_m\}$ satisfying the boundary condition

$$\mathrm{Re}\,[\overline{\lambda(x)}\Phi(x)] = au - bv = c(x), \quad z \in L^*. \tag{1.2}$$

Problem A with the condition $c(x) = 0$ on L^* is called Problem A_0.

Denote by $\lambda(x_j - 0)$ and $\lambda(x_j + 0)$ the left limit and right limit of $\lambda(x)$ as $x \to x_j(j = 1, 2, \ldots, m)$ on L, and

$$e^{i\phi_j} = \frac{\lambda(x_j - 0)}{\lambda(x_j + 0)}, \quad \gamma_j = \frac{1}{\pi i}\ln\left[\frac{\lambda(x_j - 0)}{\lambda(x_j + 0)}\right] = \frac{\phi_j}{\pi} - K_j,$$

$$K_j = \left[\frac{\phi_j}{\pi}\right] + J_j, \quad J_j = 0 \text{ or } 1, \quad j = 1, \ldots, m, \tag{1.3}$$

in which $0 \leq \gamma_j < 1$ when $J_j = 0$ and $-1 < \gamma_j < 0$ when $J_j = 1$, $j = 1, \ldots, m$, and

$$K = \frac{1}{2}(K_1 + \cdots + K_m) = \frac{1}{2}\sum_{j=1}^m \left[\frac{\phi_j}{\pi} - \gamma_j\right] \tag{1.4}$$

is called the index of Problem A and Problem A_0. If $\lambda(x)$ on L is continuous, then $K = \Delta_\Gamma \arg \lambda(x)/2\pi$ is a unique integer. If the function $\lambda(x)$ on L is not continuous, we can choose $J_j = 0$ or 1, hence the index K is not unique. We can require that the solution $\Phi(z)$ satisfy the condition

$$\Phi(z) = O(|z - x_j|^{-\delta}), \quad \delta = \begin{cases} \beta_j + \tau, & \text{for } \gamma_j \geq 0, \text{ and } \gamma_j < 0, \beta_j > |\gamma_j|, \\ |\gamma_j| + \tau, & \text{for } \gamma_j < 0, \beta_j \leq |\gamma_j|, j = 1, \ldots, m, \end{cases} \tag{1.5}$$

in the neighborhood $(\subset D)$ of x_j, where $\tau (< \alpha)$ is an arbitrary small positive number.

In order to find the solution of Problem A for analytic functions, we first consider Problem A_0. Making a transformation

$$\Psi(z) = \frac{\Phi(z)}{\Pi(z)}, \quad \Pi(z) = \prod_{j=1}^m \left(\frac{z - x_j}{z + i}\right)^{\gamma_j}, \tag{1.6}$$

in which $\gamma_j (j = 1, \ldots, m)$ are as stated in (1.3), the boundary condition

$$\mathrm{Re}\,[\overline{\lambda(x)}\Phi_0(x)] = 0, \quad x \in L^*, \tag{1.7}$$

of Problem A_0 for analytic functions $\Phi_0(z)$ is reduced to the boundary condition

$$\operatorname{Re}\left[\overline{\Lambda(x)}\Psi_0(x)\right] = 0, \quad \Lambda(x) = \lambda(x)\frac{\overline{\Pi(x)}}{|\Pi(x)|}, \quad x \in L^*, \tag{1.8}$$

of Problem A_0^* for analytic functions $\Psi_0(z) = \Phi_0(z)/\Pi(z)$. Noting that

$$\frac{\Lambda(x_j - 0)}{\Lambda(x_j + 0)} = \frac{\lambda(x_j - 0)}{\lambda(x_j + 0)}\overline{\left(\frac{\Pi(x_j - 0)}{\Pi(x_j + 0)}\right)} = \frac{\lambda(x_j - 0)}{\lambda(x_j + 0)}e^{-i\pi\gamma_j} = \pm 1, \tag{1.9}$$

the index of $\Lambda(x)$ on L is

$$K = \frac{1}{2\pi}\Delta_L \arg \Lambda(x) = \frac{1}{2}\sum_{j=1}^m \left[\frac{\phi_j}{\pi} - \gamma_j\right] = \frac{1}{2}\sum_{j=1}^m K_j, \tag{1.10}$$

which is the same as the index of $\lambda(x)$ on L. If $2K$ is even, provided that we change the signs of $\Lambda(x)$ on some line segments of L_j $(j = 1, \ldots, m)$, then the new function $\Lambda^*(x)$ on L is continuous, its index is K too. When $2K$ is odd, we rewrite the boundary condition (1.8) in the form

$$\operatorname{Re}\left[\overline{\Lambda(x)}\frac{x - x_0}{x + i}\frac{x + i}{x - x_0}\Psi_0(x)\right] = 0, \quad x \in L^*, \tag{1.11}$$

where $x_0(\in L)$ is a real number and $x_0 \notin \{x_1, \ldots, x_m\}$, thus similarly to before we change the signs of $\Lambda(x)(x - x_0)|x + i|/(x + i)|x - x_0|$ on some line segments of L, such that the new function $\Lambda^*(x)$ on L is continuous, its index is $K^* = K - 1/2$. Next we find an analytic function

$$\Psi^*(z) = i\left(\frac{z - i}{z + i}\right)^{[K]}\left(\frac{z - x_0}{z + i}\right)e^{iS(z)} \text{ in } D, \tag{1.12}$$

which satisfies the homogeneous boundary condition

$$\operatorname{Re}\left[\overline{\Lambda^*(x)}\Psi^*(x)\right] = 0, \quad x \in L^*, \tag{1.13}$$

where $[K]$ is the integer part of K, $S(z)$ is an analytic function in D satisfying the boundary condition

$$\operatorname{Re}\left[S(x)\right] = \arg\left[\Lambda^*(x)\overline{\left(\frac{x - i}{x + i}\right)}^{[K]}\overline{\left(\frac{x - x_0}{x + i}\right)}\right], \quad x \in L, \quad \operatorname{Im}\left[S(i)\right] = 0. \tag{1.14}$$

Hence Problem A_0^* for analytic functions possesses the solution

$$\Psi_0(z) = \begin{cases} \Psi^*(z), & \text{when } 2K \text{ is even,} \\ (z - x_0)\Psi^*(z)/(z + i), & \text{when } 2K \text{ is odd,} \end{cases} \tag{1.15}$$

and then Problem A_0 for analytic functions has a non-trivial solution in the form

$$X(z) = \Pi(z)\Psi_0(z) = \begin{cases} i\left(\dfrac{z-i}{z+i}\right)^K \Pi(z)e^{iS(z)}, & \text{when } 2K \text{ is even,} \\[2ex] i\left(\dfrac{z-i}{z+i}\right)^{[K]} \dfrac{z-x_0}{z+i}\Pi(z)e^{iS(z)}, & \text{when } 2K \text{ is odd.} \end{cases} \tag{1.16}$$

Take into account that $X(z)$ has a zero of order $[K]$ at the point $z = i$ for $K \geq 0$ and a pole of order $|[K]|$ at the point $z = i$ for $K < 0$, and a zero of order 1 at the point $z = x_0$ when $2K$ is an odd integer, moreover, $X(z)$ satisfies the homogeneous boundary condition (1.7), it is clear that $i\overline{\lambda(x)}X(x)$ is a real-valued function on L. Let us divide the nonhomogeneous boundary condition (1.2) by $i\overline{\lambda(x)}X(x)$, and obtain

$$\mathrm{Re}\left[\frac{\Phi(x)}{iX(x)}\right] = \frac{c(x)}{i\overline{\lambda(x)}X(x)} = \frac{\lambda(x)c(x)}{iX(x)}, \quad x \in L^*. \tag{1.17}$$

By using the Schwarz formula, we get

$$\frac{\Phi(z)}{iX(z)} = \frac{1}{\pi i}\left[\int_{-\infty}^{\infty}\frac{\lambda(t)c(t)}{(t-z)iX(t)}dt + \frac{Q(z)}{i}\right],$$

$$\Phi(z) = \frac{X(z)}{\pi i}\left[\int_{-\infty}^{\infty}\frac{\lambda(t)c(t)}{(t-z)X(t)}dt + Q(z)\right]. \tag{1.18}$$

If $K \geq 0$, the function $Q(z)$ possesses the form

$$Q(z) = i\sum_{j=0}^{[K]}\left[c_j\left(\frac{z-i}{z+i}\right)^j + \overline{c_j}\left(\frac{z-i}{z+i}\right)^{-j}\right] + \begin{cases} 0, & \text{when } 2K \text{ is even,} \\[2ex] ic_*\dfrac{x_0+z}{x_0-z}, & \text{when } 2K \text{ is odd,} \end{cases} \tag{1.19}$$

where c_*, c_0 are arbitrary real constants, and c_j $(j=1,\ldots,[K])$ are arbitrary complex constants, from this we can see that the general solution $\Phi(z)$ includes $2K+1$ arbitrary real constants. If $2K$ is odd, we note $(z-x_0)/[(t-z)(t-x_0)] = 1/(t-z) - 1/(t-x_0)$, then the integral in (1.18) is understood as the difference of two integrals of Cauchy type. If $K < 0$, we have to take

$$Q(z) = \begin{cases} ic_* = 0, & \text{when } 2K \text{ is even,} \\[2ex] ic_*\dfrac{x_0+z}{x_0-z}, & \text{when } 2K \text{ is odd,} \end{cases} \tag{1.20}$$

and require that the function in the square bracket of (1.18) has at least a zero point of order $|[K]|$ at $z = i$. From

$$\int_{-\infty}^{\infty} \frac{\lambda(t)c(t)}{(t-z)X(t)} dt + ic_* \frac{x_0 + z}{x_0 - z}$$

$$= \int_{-\infty}^{\infty} \frac{\lambda(t)c(t)}{(1-(z-i)/(t-i))(t-i)X(t)} dt + ic_* \frac{1+(z-i)/(x_0+i)}{1-(z-i)/(x_0-i)} \frac{x_0+i}{x_0-i}$$

$$= \sum_{j=0}^{\infty} \int_{-\infty}^{\infty} \frac{\lambda(t)c(t)(z-i)^j}{(t-i)^{j+1} X(t)} dt + ic_* \left[1 + \frac{z-i}{x_0+i} \right] \frac{x_0+i}{x_0-i} \sum_{j=0}^{\infty} \left(\frac{z-i}{x_0-i} \right)^j \quad (1.21)$$

$$= \int_{-\infty}^{\infty} \frac{\lambda(t)c(t)}{(t-i)X(t)} dt + ic_* \frac{x_0+i}{x_0-i} + \sum_{j=1}^{\infty} \left[\int_{-\infty}^{\infty} \frac{\lambda(t)c(t)}{(t-i)^{j+1} X(t)} dt + \frac{2ic_* x_0}{(x_0-i)^{j+1}} \right] (z-i)^j$$

in the neighborhood of $z = i$, this shows that

$$\int_{-\infty}^{\infty} \frac{\lambda(t)c(t)}{X(t)(t-i)^j} dt = 0, \, j = 1, \ldots, -K(= |[K]|), \quad \text{when } 2K \text{ is even,}$$

$$\int_{-\infty}^{\infty} \frac{\lambda(t)c(t)}{X(t)(t-i)^j} dt + \frac{2ic_* x_0}{(x_0-i)^j} = 0, \, j = 2, \ldots, [-K]+1 (= |[K]|), \quad (1.22)$$

when $2K$ is odd,

then the function in the square bracket of (1.18) has a zero point of order $|[K]|$ at $z = i$, hence the function $\Phi(z)$ is analytic at $z = i$. Besides when $2K$ is odd,

$$c_* = i \frac{x_0 - i}{x_0 + i} \int_{-\infty}^{\infty} \frac{\lambda(t)c(t)}{X(t)(t-i)} dt \quad (1.23)$$

is a determined constant. Therefore when $K < 0$, Problem A has $-2K-1$ solvability conditions. Thus we have the following theorem.

Theorem 1.1 *Problem A for analytic functions in $D = \{\operatorname{Im} z > 0\}$ has the following solvability result.*

(1) *If the index $K \geq 0$, the general solution $\Phi(z)$ of Problem A possesses the form (1.18), (1.19), which includes $2K + 1$ arbitrary real constants.*

(2) *If the index $K < 0$, Problem A has $-2K - 1$ solvability conditions as stated in (1.22). When the conditions hold, the solution of Problem A is given by the second formula in (1.18), in particular*

$$\Phi(z) = \frac{X(z)(z-i)^{|[K]|}}{\pi i} \left[\int_{-\infty}^{\infty} \frac{\lambda(t)c(t)}{(t-z)(t-i)^{|[K]|}X(t)} dt + \frac{2ic_* x_0}{(x_0-z)(x_0-i)^{|[K]|}} \right] \quad (1.24)$$

in $|z - i| < 1$, where the constant c_* is determined as stated in (1.23).

Finally, we mention that if $x_1, \ldots, x_m$ are first kind of discontinuities of $c(x)$ and if $\gamma_j > 0$, $j = 1, 2, \ldots, m$, then the solution $\Phi(z)$ of Problem A is bounded in

$D^* = \overline{D}\backslash\{x_1, \ldots, x_m\}$. In general, if $\gamma_j \leq 0 (1 \leq j \leq m)$, the solution $\Phi(z)$ of Problem A may not be bounded in the neighborhood of x_j in $D^* = \overline{D}\backslash\{x_1, \ldots, x_m\}$. We have

$$\Phi(z) = \begin{cases} O(|z - x_j|^{-\gamma_j}), & \text{if } \gamma_j < 0, J_j = 1. \\ O(\ln|z - x_j|), & \text{if } \gamma_j = 0, J_j = 0, \end{cases} \tag{1.25}$$

in the neighborhood of x_j on D^*, but the integral

$$\int_i^z \Phi(z) dz \text{ in } D$$

is bounded. In particular, if $m = 2n$ and

$$\lambda(x) = \begin{cases} 1, & x \in (x_{2j-1}, x_{2j}), \\ i, & x \in (x_{2j}, x_{2j+1}), \end{cases} \quad j = 1, \ldots, n,$$

and $x_j(j = 1, \ldots, m = 2n)$ are first kind of discontinuous points of $c(x)$, we can choose $\gamma_{2j-1} = 1/2, K_{2j-1} = 0, \gamma_{2j} = -1/2, K_{2j} = 0, j = 1, \ldots, n$, and then the index of the mixed boundary value problem is $K = 0$. In this case one can choose $\Pi(z) = \sqrt{\Pi_{j=1}^n (z - x_{2j-1})/(z - x_{2j})}$. From the formula (1.18) with $K = 0$, the Keldych–Sedov formula of the mixed boundary value problem for analytic functions in the upper half-plane is derived [53]. If we chose $\gamma_{2j-1} = -1/2, K_{2j-1} = 1, \gamma_{2j} = -1/2, K_{2j} = 0, j = 1, \ldots, n$, and the index of the mixed boundary value problem is $K = n = m/2$, then the representation of solutions of the mixed boundary value problem for analytic functions can be written from (1.18) with $K = n$, which includes $2K + 1 = m + 1$ arbitrary real constants, where the function $\Pi(z) = 1/\sqrt{\Pi_{j=1}^n (z - x_{2j-1})(z - x_{2j})}$.

1.2 The general discontinuous boundary value problem for analytic functions in the upper half-disk

Now we first introduce the general discontinuous Riemann–Hilbert problem (Problem B) for analytic functions in the unit disk $D = \{|z| < 1\}$ with the boundary conditions

$$\text{Re}\left[\overline{\lambda(z)}\Phi(z)\right] = au - bv = c(z), \quad \Gamma = \{|z| = 1\}, \tag{1.26}$$

where $\lambda(z) = a(z) - ib(z), |\lambda(z)| = 1$ on Γ, and $Z = \{z_1, z_2, \ldots, z_m\}$ are first kind of discontinuous points of $\lambda(z)$ on Γ, and $\lambda(z), c(z)$ satisfy the conditions

$$\lambda(z) \in C_\alpha(\Gamma_j), \quad |z - z_j|^{\beta_j} c(z) \in C_\alpha(\Gamma_j), \quad j = 1, 2, \ldots, m, \tag{1.27}$$

herein Γ_j is the arc from the point z_{j-1} to z_j on Γ and $z_0 = z_m$, and $\Gamma_j(j = 1, 2, \ldots, m)$ does not include the end points, $\alpha(0 < \alpha < 1)$ is a constant.

Denote by $\lambda(z_j - 0)$ and $\lambda(z_j + 0)$ the left limit and right limit of $\lambda(z)$ as $z \to z_j(j = 1, 2, \ldots, m)$ on Γ, and

$$e^{i\phi_j} = \frac{\lambda(z_j - 0)}{\lambda(z_j + 0)}, \quad \gamma_j = \frac{1}{\pi i} \ln\left[\frac{\lambda(z_j - 0)}{\lambda(z_j + 0)}\right] = \frac{\phi_j}{\pi} - K_j,$$

$$K_j = \left[\frac{\phi_j}{\pi}\right] + J_j, \quad J_j = 0 \text{ or } 1, \quad j = 1, \ldots, m,$$

(1.28)

in which $0 \le \gamma_j < 1$ when $J_j = 0$ and $-1 < \gamma_j < 0$ when $J_j = 1$, $j = 1, \ldots, m$. The index K of Problem B is defined by (1.4). Let $\beta_j + \gamma_j < 1$, $j = 1, \ldots, m$, we require that the solution $\Phi(z)$ possesses the property

$$\Phi(z) = O(|z - z_j|^{-\delta}), \quad \delta = \begin{cases} \beta_j + \tau, & \text{for } \gamma_j \ge 0, \text{ and } \gamma_j < 0, \beta_j > |\gamma_j|, \\ |\gamma_j| + \tau, & \text{for } \gamma_j < 0, \beta_j \le |\gamma_j|, j = 1, \ldots, m, \end{cases}$$

(1.29)

in the neighborhood ($\subset D^*$) of z_j, where $\tau(< \alpha)$ is an arbitrary small positive number. By using a similar method as stated in Subsection 1, we can obtain the formula for solutions of the boundary value problem.

Theorem 1.2 *Problem B for analytic functions in $D = \{|z| < 1\}$ has the following solvability result.*

(1) *If the index $K \ge 0$, the general solution $\Phi(z)$ of Problem B possesses the form*

$$\Phi(z) = \frac{X(z)}{2\pi i}\left[\int_\Gamma \frac{(t + z)\lambda(t)c(t)}{(t - z)tX(t)}dt + Q(z)\right],$$

(1.30)

with

$$Q(z) = i\sum_{j=0}^{[K]}(c_j z^j + \overline{c_j}z^{-j}) + \begin{cases} 0, & \text{when } 2K \text{ is even}, \\ ic_*\dfrac{z_0 + z}{z_0 - z}, & \text{when } 2K \text{ is odd}, \end{cases}$$

(1.31)

where the constant c_*, c_0 are arbitrary real constants, and c_j $(j = 1, \ldots, [K])$ are arbitrary complex constants, which includes $2K + 1$ arbitrary real constants.

(2) *If the index $K < 0$, Problem B has $-2K - 1$ solvability conditions given by*

$$\int_\Gamma \frac{\lambda(t)c(t)}{X(t)t^j}dt = 0, \quad j = 1, \ldots, -K(= |[K]|), \quad \text{when } 2K \text{ is even},$$

$$\int_\Gamma \frac{\lambda(t)c(t)}{X(t)t^j}dt + ic_*z_0^{-j+1} = 0, \quad j = 1, \ldots, [-K]+1(= |[K]|), \quad \text{when } 2K \text{ is odd}.$$

(1.32)

When the conditions hold, the solution of Problem B possesses the form

$$\Phi(z) = \frac{X(z)z^{[K]}}{\pi i}\left[\int_\Gamma \frac{\lambda(t)c(t)}{(t - z)X(t)t^{|[K]|}}dt + \frac{ic_*}{(z_0 - z)z_0^{|[K]|-1}}\right],$$

(1.33)

where the constant c_* is determined via (1.32) as

$$c_* = i \int_\Gamma \frac{\lambda(t)c(t)}{X(t)t} dt.$$

In the above formula, $X(z)$ is a non-trivial solution of the homogeneous boundary value problem (Problem B_0) for analytic functions in the form

$$X(z) = \begin{cases} iz^K \Pi(z)e^{iS(z)}, & \text{when } 2K \text{ is even,} \\ iz^{[K]}(z-z_0)\Pi(z)e^{iS(z)}, & \text{when } 2K \text{ is odd,} \end{cases} \quad \Pi(z) = \prod_{j=1}^m (z-z_j)^{\gamma_j}, \quad (1.34)$$

in which $S(z)$ is an analytic function in D satisfying the boundary conditions

$$\text{Re}\,[S(z)] = \arg[\Lambda^*(z)\bar{z}^{[K]}], \quad z \in \Gamma, \quad \text{Im}\,[S(0)] = 0,$$

the function $\Lambda^*(z)$ is similar to that in (1.13) [85]11),[86]1).

In addition, through the conformal mapping from the upper half-unit disk $D = \{|z| < 1, \text{Im}\,z > 0\}$ onto the unit disk $G = \{|\zeta| < 1\}$, namely

$$\zeta(z) = -i\frac{z^2 + 2iz + 1}{z^2 - 2iz + 1}, \quad z(\zeta) = \frac{1}{\zeta + i}\left[1 + i\zeta - \sqrt{2(1-\zeta^2)}\right],$$

we can obtain the result of the general discontinuous Riemann–Hilbert problem (Problem C) for analytic functions in the upper half-unit disk $D = \{|z| < 1, \text{Im}\,z > 0\}$, namely

$$w(z) = \Phi[\zeta(z)] \text{ in } D = \{|z| < 1, \text{Im}\,z > 0\} \quad (1.35)$$

is the solution of Problem C for analytic functions.

In order to the requirement in latter chapters, we give a well posed version (Problem B') of Problem B for analytic functions in $D = \{|z| < 1\}$, namely we find an analytic function $\Phi(z)$, which is continuous in $\bar{D}\backslash Z$ and satisfies the boundary condition (1.26) and the point conditions

$$\text{Im}\,[\overline{\lambda(z'_j)}\Phi(z'_j)] = b_j, \quad j = 1, \ldots, m, \quad (1.36)$$

where $z'_1, \ldots, z'_m\,(\notin Z)$ are distinct points on Γ and $b_j\,(j = 1, \ldots, m)$ are real constants, and we choose the index $K = (m-1)/2$ of $\lambda(z)$ on $\Gamma = \{|z| = 1\}$. The homogeneous problem of Problem B' with the conditions $c(z) = 0$ on Γ and $b_j = 0\,(j = 1, \ldots, m)$ will be called Problem B'_0.

Theorem 1.3 *Problem B' for analytic functions in $D = \{|z| < 1\}$ has a unique solution.*

Proof First of all, we verify the uniqueness of solutions of Problem B'. Let $\Phi_1(z), \Phi_2(z)$ be two solutions of Problem B' for analytic functions. Then the function $\Phi(z) = \Phi_1(z) - \Phi_2(z)$ is a solution of Problem B'_0 with the homogeneous boundary conditions

$$\text{Re}\,[\overline{\lambda(z)}\Phi(z)] = 0 \text{ on } \Gamma = \{|z| = 1\}, \quad \text{Im}\,[\overline{\lambda(z'_j)}\Phi(z'_j)] = 0, \, j = 1, \ldots, m. \quad (1.37)$$

According to the method in the proof of Theorem 1.1 or [8]2),[80]1) and [85]11), we see that if $\Phi(z) \not\equiv 0$ in D, then

$$m = 2K + 1 \leq 2N_D + N_\Gamma \leq 2K, \tag{1.38}$$

where N_D, N_Γ are numbers of zero points in D and $\Gamma^* = \Gamma \backslash Z$ respectively. This contradiction proves that $\Phi(z) \equiv 0$, i.e. $\Phi_1(z) = \Phi_2(z)$ in D.

Now we prove the existence of solutions of Problem B' for analytic functions. By the representation (1.30) of the general solution of Problem B for analytic functions, it is easy to see that the general solution $\Phi(z)$ can be written as

$$\Phi(z) = \Phi_0(z) + \sum_{j=1}^{m} d_j \Phi_j(z), \tag{1.39}$$

where $\Phi_0(z)$ is a special solution of Problem B', and $\Phi_j(z)\,(j = 1,\ldots,m)$ are a complete system of linearly independent solutions of Problem B_0' and $d_j (j = 1,\ldots,m)$ are arbitrary real constants. In the following, we prove that there exists a unique system of real constants $d_j'(j = 1,\ldots,m)$ such that $|d_1'| + \cdots + |d_m'| \neq 0$ satisfying the equalities

$$\sum_{j=1}^{m} d_j' \Phi_j(z_j') = \lambda(z_j')[c(z_j') + ib_j] - \Phi_0(z_j'), \quad j = 1,\ldots,m. \tag{1.40}$$

Then the analytic function $\Phi(z) = \Phi_0(z) + \sum_{j=1}^{m} d_j' \Phi_j(z)$ satisfies the boundary conditions (1.26) and (1.36), and thus is a solution of Problem B'. According to the algebraic theory, it suffices to verify that the homogeneous system of algebraic system of equations: (1.40), i.e.

$$\Phi_*(z_j') = \sum_{j=1}^{m} d_j' \Phi_j(z_j') = 0, \quad j = 1,\ldots,m \tag{1.41}$$

has no non-trivial solution. Noting that the analytic function $\Phi_*(z) = \sum_{j=1}^{m} d_j' \Phi_j(z)$ is a solution of Problem B_0', from the uniqueness of solutions of Problem B', we see that $\Phi_*(z) = 0$. This proves the existence of solutions of Problem B' for analytic functions.

Next, we consider that D is the upper half-unit disk, $a(z), b(z)$ possess discontinuities of first kind at m distinct points $z_1, \ldots, z_m \in \Gamma \cup L_0 = \Gamma' = \partial D$, which are arranged according to the positive direction of ∂D. Here $\Gamma = \{|z| = 1, \operatorname{Im} z > 0\}$, $L_0 = \{-1 \leq x \leq 1, y = 0\}$ and $z_1,\ \ldots, z_{n-1} \in \Gamma = \{|z| = 1, \operatorname{Im} z > 0\}$, $x_n = -1,\ldots,x_m = x_0 = 1 \in L_0$, where $n\,(< m)$, m are positive integers, and $c(z) = O(|z - z_j|^{-\beta_j})$ in the neighborhood of $z_j\,(j = 1,2,\ldots,m)$ on Γ, in which $\beta_j(< 1, j = 1,2,\ldots,m)$ are non-negative constants, such that $\beta_j + \gamma_j < 1, \gamma_j (j = 1,\ldots,m)$ are as stated in (1.29). Denote $\lambda(z) = a(z) - ib(z)$ and $|a(z)| + |b(z)| \neq 0$, there is no harm in assuming that $|\lambda(z)| = 1, z \in \Gamma' = \Gamma \cup L_0$. Suppose that $\lambda(z), c(z)$ satisfy conditions again (1.27).

Problem C Find an analytic function $\Phi(z) = u(z) + iv(z)$ in D, which is continuous on $D^* = \bar{D} \backslash Z$ satisfying the boundary condition

$$\text{Re}\,[\overline{\lambda(z)}\Phi(z)] = au - bv = c(z), \quad z \in \Gamma^* = \Gamma' \backslash Z, \tag{1.42}$$

here $Z = \{z_1, \ldots, z_m\}$. Problem C with the condition $r(z) = 0$ on Γ^* is called Problem C_0.

The index K of Problem C is the same as stated in (1.4). We can require that the solution $\Phi(z)$ satisfies the condition (1.29).

In order to find the solution of Problem C for analytic functions, it suffices to choose a conformal mapping from the upper half-unit disk onto the upper half plane or the unit disk. In the following we shall use the other method, namely first find a solution of Problem A for analytic functions in $D^+ = \{\text{Im}\,z > 0\}$ with the boundary condition

$$\text{Re}\,[\overline{\lambda(x)}\Phi(x)] = r(x) \text{ on } L = (-\infty, \infty),$$

$$r(x) = \begin{cases} r(x) \text{ on } L_0 = (-1, 1), \\ c(x) \text{ on } L_1 = (-\infty < x < -1) \cup (1 < x < \infty), \end{cases} \tag{1.43}$$

in which $\lambda(x)$, $c(x)$ on $L_2 = (-\infty < x \le -1) \cup (1 \le x < \infty)$ are appropriate functions, such that $\lambda(x)$, $|x - x_j|^{\beta_j} c(x)$ are piecewise Hölder continuous functions and continuous at the points $x = -1, 1$, and the index of $\lambda(x)$ on L is $K = 0$. For instance, setting

$$\lambda(x) = \begin{cases} \lambda(-1 + 0) \text{ on } (-\infty, -1], \\ \lambda(x) \text{ on } (-1, 1), \\ \lambda(16/(x + 3) - 3) \text{ on } [1, \infty), \end{cases} \tag{1.44}$$

and denoting $x_{2m-j} = 16/(x_j + 3) - 3$, $j = n + 1, \ldots, m - 1$, we can determine that the index of above function $\lambda(x)$ on L is $K = 0$. On the basis of Theorem 1.1, the solution $\Psi(z)$ of Problem A can be expressed in the form (1.18),(1.19) with $K = 0$ and $\lambda(z), c(z)$ are as stated in (1.27). Thus the function $\tilde{\Phi}(z) = \Phi(z) - \Psi(z)$ is analytic in D and satisfies the boundary condition

$$\text{Re}\,[\overline{\lambda(z)}\tilde{\Phi}(z)] = \tilde{r}(z) = \begin{cases} r(z) - \text{Re}\,[\overline{\lambda(z)}\Psi(z)], \quad z \in \Gamma, \\ 0, \quad z \in L_0. \end{cases} \tag{1.45}$$

Next, similarly to Section 1, we make a transformation

$$\tilde{\Psi}(z) = \frac{\tilde{\Phi}(z)}{\Pi(z)}, \quad \Pi(z) = \prod_{j=n+1, j \ne m}^{2m-n-1} \left(\frac{z - x_j}{z + i}\right)^{\gamma_j}, \tag{1.46}$$

in which $\gamma_j(j = n + 1, \ldots, m - 1, m + 1, \ldots, 2m - n - 1)$ are similar to those in (1.28), the boundary condition

$$\text{Re}\,[\overline{\lambda(z)}\tilde{\Phi}(z)] = 0, \quad z \in L, \tag{1.47}$$

of Problem $\tilde{C}_0$ for the analytic function $\tilde{\Phi}(z)$ is reduced to the boundary condition

$$\text{Re}\,[\overline{\Lambda(z)}\tilde{\Psi}(z)] = 0, \quad \Lambda(z) = \lambda(z)\overline{\Pi(z)}/|\Pi(z)|, \quad z \in L, \tag{1.48}$$

for the analytic function $\tilde{\Psi}(z) = \tilde{\Phi}(z)/\Pi(z)$. Noting that

$$\frac{\Lambda(z_j - 0)}{\Lambda(z_j + 0)} = \frac{\lambda(z_j - 0)}{\lambda(z_j + 0)}\overline{\left(\frac{\Pi(z_j - 0)}{\Pi(z_j + 0)}\right)} = \frac{\lambda(z_j - 0)}{\lambda(z_j + 0)}e^{-i\pi\gamma_j} = \pm 1,$$

$$j = n + 1, \ldots, m - 1, m + 1, \ldots, 2m - n - 1, \tag{1.49}$$

the index of $\Lambda(z)$ on L is the same as the index of $\lambda(z)$ on L. Due to $2K = 0$ is even, provided that we change the sign of $\Lambda(z)$ on some arcs $L_j = (x_{j-1}, x_j)$ ($j = n + 1, \ldots, 2m - n - 1$), $L_{n+1} = (-\infty, x_{n+1}) \cup (x_{2m-n-1}, \infty)$, then the new function $\Lambda^*(z)$ on Γ is continuous, the index of $\lambda(z)$ on L has not been changed. Moreover we find a solution of Problem A for analytic functions in $\text{Im}\, z > 0$ with the boundary conditions

$$\text{Re}\,[S(z)] = \arg \Lambda^*(z) \text{ on } L = (-\infty, \infty), \quad \text{Im}\, S(i) = 0, \tag{1.50}$$

and denote $\hat{\Psi}(z) = \tilde{\Psi}(z)e^{-iS(z)}$.

Now we extend the analytic function $\hat{\Psi}(z)$ as follows

$$\hat{\Phi}(z) = \begin{cases} \hat{\Psi}(z) \text{ in } D = \{|z| < 1, \quad \text{Im}\, z > 0\}, \\ -\overline{\hat{\Psi}(\bar{z})} \text{ in } \tilde{D} = \{|z| < 1, \quad \text{Im}\, z < 0\}. \end{cases} \tag{1.51}$$

It can be seen that the analytic function $\hat{\Phi}(z)$ in $\{|z| < 1\}$ satisfies the boundary condition

$$\text{Re}\,[\overline{\Lambda(z)}\hat{\Phi}(z)] = R(z) \text{ on } |z| = 1, \tag{1.52}$$

where

$$\Lambda(z) = \begin{cases} \lambda(z), \\ \overline{\lambda(\bar{z})}, \end{cases} \quad R(z) = \begin{cases} \tilde{r}(z)e^{\text{Im}\, S(z)} \text{ on } \Gamma_0 = \{|z|=1, \quad \text{Im}\, z>0\}, \\ -\tilde{r}(\bar{z})e^{\text{Im}\, S(\bar{z})} \text{ on } \tilde{\Gamma}_0 = \{|z|=1, \quad \text{Im}\, z<0\}. \end{cases} \tag{1.53}$$

We can find the index K' of $\Lambda(z)$ on $|z| = 1$ and by Theorem 1.2, the analytic function $\hat{\Phi}(z)$ in D with the boundary condition (1.52) can be found, i.e.

$$\hat{\Phi}(z) = \hat{\Phi}\left(i\frac{1+\varsigma}{1-\varsigma}\right) \text{ in } D, \tag{1.54}$$

where $\hat{\Phi}(z)$ is an analytic function as the function $\Phi(z)$ in (1.18), but in which $\lambda(z), c(z), K$ are replaced by $\lambda[i(1+\varsigma)/(1-\varsigma)], R[i(1+\varsigma)/(1-\varsigma)], K'$ respectively, herein $\lambda(z), R(z)$ are as stated in (1.53). It is clear that $\hat{\Phi}(z)$ includes $2K' + 1$ arbitrary real constants when $K' \geq 0$ and $-2K' - 1$ solvability conditions when $K' < 0$.

Thus the solution of Problem C for analytic functions in the upper half-unit disk D is obtained, i.e.

$$w(z) = \Psi(z) + \hat{\Phi}(z)\Pi(z)e^{iS(z)} \text{ in } D. \tag{1.55}$$

Theorem 1.4 *When the index $K \geq 0$, Problem C for analytic functions in D has a solution in the form (1.55) including $2K + 1$ arbitrary real constants, and when $K < 0$, under $-2K - 1$ conditions, Problem C for analytic functions possesses the solution as stated in (1.55). Moreover, the above solution of Problem C for analytic functions can be expressed by (1.35).*

The Keldych–Sedov formula for analytic functions in the upper half-plane possesses important applications to the Tricomi problem for some equations of mixed type (see [12]1),3)). But more general boundary value problems for equations of mixed type cannot be solved by this formula. Due to we have Theorems 1.1–1.4, such that the above general problems can be solved. In addition, we can give the representation of solutions to the discontinuous Riemann–Hilbert boundary value problem for analytic functions in the zone domain $D = \{0 < \operatorname{Im} z < 1\}$, which can be used to solve some boundary value problems for nonlinear problems in mechanics.

2   Representation and Existence of Solutions for Elliptic Complex Equations of First Order

In this section, we shall establish the representations for solutions of the general discontinuous boundary value problem for elliptic complex equations of first order in the upper half-unit disk. Moreover, we shall prove the existence of solutions for nonlinear elliptic complex equations of first order.

2.1   Representation of solutions of the discontinuous Riemann–Hilbert problem for elliptic complex equations in the upper half-unit disk

Let D be an upper half-unit disk with the boundary $\Gamma' = \Gamma \cup L_0$ as stated in Section 1. We consider the nonlinear uniformly elliptic systems of first order equations

$$F_j(x, y, u, v, u_x, v_x, u_y, v_y) = 0 \text{ in } D, \quad j = 1, 2.$$

Under certain conditions, the system can be transformed into the complex form

$$w_{\bar{z}} = F(z, w, w_z), \quad F = Q_1 w_z + Q_2 \bar{w}_{\bar{z}} + A_1 w + A_2 \bar{w} + A_3. \quad z \in D \tag{2.1}$$

(see [86]1)), in which $F(z, w, U)$ satisfy the following conditions.

Condition C

1) $Q_j(z, w, U)$, $A_j(z, w)$ $(j = 1, 2)$. $A_3(z)$ are measurable in $z \in D$ for all continuous functions $w(z)$ in $D^* = \bar{D} \backslash Z$ and all measurable functions $U(z) \in L_{p_0}(D_*)$, and

satisfy

$$L_p[A_j, \bar{D}] \leq k_0, \quad j = 1, 2, \quad L_p[A_3, \bar{D}] \leq k_1, \tag{2.2}$$

where $Z = \{z_1, \ldots, z_m\}$, D_* is any closed subset in D, $p_0, p \, (2 < p_0 \leq p), k_0, k_1$ are non-negative constants.

2) The above functions are continuous in $w \in \mathbb{C}$ for almost every point $z \in D$, $U \in \mathbb{C}$, and $Q_j = 0 \, (j = 1, 2)$, $A_j = 0 \, (j = 1, 2, 3)$ for $z \notin D$.

3) The complex equation (2.1) satisfies the uniform ellipticity condition

$$|F(z, w, U_1) - F(z, w, U_2)| \leq q_0 |U_1 - U_2|, \tag{2.3}$$

for almost every point $z \in D$, in which $w, U_1, U_2 \in \mathbb{C}$ and $q_0(< 1)$ is a non-negative constant.

Problem A The discontinuous Riemann–Hilbert boundary value problem for (2.1) is to find a continuous solution $w(z)$ in D^* satisfying the boundary condition:

$$\text{Re}\,[\overline{\lambda(z)}w(z)] = c(z), \quad z \in \Gamma^* = \Gamma' \backslash Z, \tag{2.4}$$

where $\lambda(z), c(z)$ are as stated in Section 1 satisfying

$$C_\alpha[\lambda(z), \Gamma_j] \leq k_0, \quad C_\alpha[|z - z_j|^{\beta_j} c(z), \Gamma_j] \leq k_2, \quad j = 1, \ldots, m, \tag{2.5}$$

herein $\alpha \, (1/2 < \alpha < 1), k_0, k_2$ are non-negative constants. Assume that $(\beta_j + \gamma_j)/\beta < 1, \beta = \min(\alpha, 1 - 2/p_0)/2, \gamma_j, \beta_j (j = 1, \ldots, m)$ are as stated in (1.28), (1.29). Problem A with $A_3(z) = 0$ in D, $c(z) = 0$ on Γ^* is called Problem A_0. The index K of Problem A and Problem A_0 is defined as in (1.4).

In order to prove the solvability of Problem A for the complex equation (2.1), we need to give a representation theorem for Problem A.

Theorem 2.1 *Suppose that the complex equation (2.1) satisfies Condition C, and $w(z)$ is a solution of Problem A for (2.1). Then $w(z)$ is representable by*

$$w(z) = \Phi[\zeta(z)]e^{\phi(z)} + \psi(z), \tag{2.6}$$

where $\zeta(z)$ is a homeomorphism in $\bar{D}$, which quasiconformally maps D onto the unit disk $G = \{|\zeta| < 1\}$ with boundary $L = \{|\zeta| = 1\}$, where $\zeta(-1) = -1, \zeta(i) = i$, $\zeta(1) = 1$, $\Phi(\zeta)$ is an analytic function in G, $\psi(z), \phi(z), \zeta(z)$ and its inverse function $z(\zeta)$ satisfy the estimates

$$C_\beta[\psi, \bar{D}] \leq k_3, \quad C_\beta[\phi, \bar{D}] \leq k_3, \quad C_\beta[\zeta(z), \bar{D}] \leq k_3, \quad C_\beta[z(\zeta), \bar{G}] \leq k_3, \tag{2.7}$$

$$L_{p_0}[|\psi_{\bar{z}}| + |\psi_z|, \bar{D}] \leq k_3, \quad L_{p_0}[|\phi_{\bar{z}}| + |\phi_z|, \bar{D}] \leq k_3, \tag{2.8}$$

$$C_\beta[z(\zeta), \bar{G}] \leq k_3, \quad L_{p_0}[|\chi_{\bar{z}}| + |\chi_z|, D] \leq k_4, \tag{2.9}$$

in which $\chi(z)$ is as stated in (2.14) below, $\beta = \min(\alpha, 1 - 2/p_0)/2, p_0 \, (2 < p_0 \leq p)$, $k_j = k_j(q_0, p_0, k_0, k_1, D) \, (j = 3, 4)$ are non-negative constants. Moreover, if the *coefficients* $Q_j(z) = 0 \, (j = 1, 2)$ *of the complex equation* (2.1) *in* D, *then the representation* (2.6) *becomes the form*

$$w(z) = \Phi(z)e^{\phi(z)} + \psi(z), \tag{2.10}$$

and if $K < 0$, $\Phi(z)$ *satisfies the estimate*

$$C_\delta[X(z)\Phi(z), \bar{D}] \le M_1 = M_1(p_0, \beta, \delta, k, D), \tag{2.11}$$

in which

$$X(z) = \prod_{j=1, j\neq n,m}^{m} |z-z_j|^{\eta_j}|z-z_n|^{2\eta_n}|z-z_m|^{2\eta_m}, \quad \eta_j = \begin{cases} |\gamma_j|+\tau, \ \gamma_j < 0, \beta_j \le |\gamma_j|, \\ |\beta_j|+\tau, \ other \ case. \end{cases} \tag{2.12}$$

Here $\gamma_j \, (j = 1, \ldots, m)$ *are real constants as stated in* (1.28) *and* $\delta, \tau \, (0 < \delta < \min(\beta, \tau))$ *are sufficiently small positive constants, and* M_1 *is a non-negative constant.*

Proof We substitute the solution $w(z)$ of Problem A into the coefficients of equation (2.1) and consider the following system

$$\psi_{\bar{z}} = \quad Q\psi_z + A_1\psi + A_2\bar{\psi} + A_3, \quad Q = \begin{cases} Q_1 + Q_2\overline{w_z}/w_z \ \text{for} \ w_z \neq 0, \\ 0 \ \text{for} \ w_z = 0 \ \text{or} \ z \notin D, \end{cases}$$

$$\phi_{\bar{z}} = Q\phi_z + A, \quad A = \begin{cases} A_1 + A_2\bar{w}/w \ \text{for} \ w(z) \neq 0, \\ 0 \ \text{for} \ w(z) = 0 \ \text{or} \ z \notin D, \end{cases} \tag{2.13}$$

$$W_{\bar{z}} = QW_z, \quad W(z) = \Phi[\zeta(z)].$$

By using the continuity method and the principle of contracting mappings, we can find the solution

$$\psi(z) = Tf = -\frac{1}{\pi} \iint_D \frac{f(\zeta)}{\zeta - z} d\sigma_\zeta, \tag{2.14}$$

$$\phi(z) = Tg, \ \zeta(z) = \Psi[\chi(z)], \ \chi(z) = z + Th$$

of (2.13), where $f(z), g(z), h(z) \in L_{p_0,2}(\bar{D})$, $2 < p_0 \le p$, $\chi(z)$ is a homeomorphism in $\bar{D}$, $\Psi(\chi)$ is a univalent analytic function. which conformally maps $E = \chi(D)$ onto the unit disk G(see [85]11)), and $\Phi(\zeta)$ is an analytic function in G. We can verify that $\psi(z), \phi(z), \zeta(z)$ satisfy the estimates (2.7) and (2.8). It remains to prove that $z = z(\zeta)$ satisfies the estimate (2.9). In fact, we can find a homeomorphic solution of the last equation in (2.13) in the form $\chi(z) = z + Th$ such that $[\chi(z)]_z, [\chi(z)]_{\bar{z}} \in L_{p_0}(\bar{D})$ [80]1),[85]9). Next, we find a univalent analytic function $\zeta = \Psi(\chi)$, which maps $\chi(D)$ onto G, hence $\zeta = \zeta(z) = \Psi[\chi(z)]$. By the result on conformal mappings, applying the method of Lemma 2.1, Chapter II in [86]1), we can prove that (2.9) is true. When $Q_j(z) = 0$ in D, $j = 1, 2$, then we can choose $\chi(z) = z$ in (2.14), in this case $\Phi[\zeta(z)]$ can be replaced by the analytic function $\Phi(z)$, herein $\zeta(z), \Psi(z)$ are as stated in (2.14), it is clear that the representation (2.6) becomes the form (2.10). Thus the analytic function $\Phi(z)$ satisfies the boundary conditions

$$\text{Re}\,[\overline{\lambda(z)}e^{\phi(z)}\Phi(z)] = c(z) - \text{Re}\,[\overline{\lambda(z)}\psi(z)], \quad z \in \Gamma^*. \tag{2.15}$$

On the basis of Theorem 1.2 and the estimate (2.7), $\Phi(z)$ satisfies the estimate (2.11).

2.2 Existence of solutions of the discontinuous Riemann–Hilbert problem for nonlinear complex equations in the upper half-unit disk

Theorem 2.2 *Under the same conditions as in Theorem 2.1, the following statements hold.*

(1) *If the index $K \geq 0$, then Problem A for (2.1) is solvable, and the general solution includes $2K + 1$ arbitrary real constants.*

(2) *If $K < 0$, then Problem A has $-2K - 1$ solvability conditions.*

Proof Let us introduce a closed, convex and bounded subset B_1 in the Banach space $B = L_{p_0}(\bar{D}) \times L_{p_0}(\bar{D}) \times L_{p_0}(\bar{D})$, whose elements are systems of functions $q = [Q(z), f(z), g(z)]$ with norms $\| q \| = L_{p_0}(Q, \bar{D}) + L_{p_0}(f, \bar{D}) + L_{p_0}(g, \bar{D})$, which satisfy the condition

$$|Q(z)| \leq q_0 < 1 \, (z \in D), \quad L_{p_0}[f(z), \bar{D}] \leq k_3, \quad L_{p_0}[g(z), \bar{D}] \leq k_3, \qquad (2.16)$$

where q_0, k_3 are non-negative constants as stated in (2.3) and (2.7). Moreover introduce a closed and bounded subset B_2 in B, the elements of which are systems of functions $\omega = [f(z), g(z), h(z)]$ satisfying the condition

$$L_{p_0}[f(z), \bar{D}] \leq k_4, \quad L_{p_0}[g(z), \bar{D}] \leq k_4, \quad |h(z)| \leq q_0|1 + \Pi h|, \qquad (2.17)$$

where $\Pi h = -\frac{1}{\pi} \iint_D [h(\zeta)/(\zeta - z)^2] d\sigma_\zeta$.

We arbitrarily select $q = [Q(z), f(z), g(z)] \in B_1$, and using the principle of contracting mappings, a unique solution $h(z) \in L_{p_0}(\bar{D})$ of the integral equation

$$h(z) = Q(z)[1 + \Pi h] \qquad (2.18)$$

can be found, which satisfies the third inequality in (2.17). Moreover, $\chi(z) = z + Th$ is a homeomorphism in $\bar{D}$. Now, we find a univalent analytic function $\zeta = \Phi(\chi)$, which maps $\chi(D)$ onto the unit disk G as stated in Theorem 2.1. Moreover, we find an analytic function $\Psi(\zeta)$ in G satisfying the boundary condition in the form

$$\text{Re}\,[\overline{\Lambda(\zeta)}\Phi(\zeta)] = R(\zeta), \quad \zeta \in L = \zeta(\Gamma), \qquad (2.19)$$

in which $\zeta(z) = \Psi[\chi(z)]$, $z(\zeta)$ is its inverse function, $\psi(z) = Tf$, $\phi(z) = Tg$, $\Lambda(\zeta) = \lambda[z(\zeta)] \exp[\phi(z(\zeta))]$, $R(\zeta) = r[z(\zeta)] - \text{Re}\,[\overline{\lambda[z(\zeta)]}\psi(z(\zeta))]$, where $\Lambda(\zeta), R(\zeta)$ on L satisfy conditions similar to $\lambda(z), c(z)$ in (2.5) and the index of $\Lambda(\zeta)$ on L is K. In the following, we first consider the case of $K \geq 0$. On the basis of Theorem 1.2, we can find the analytic function $\Phi(\zeta)$ in the form (1.30), here $2K+1$ arbitrary real constants can be chosen. Thus the function $w(z) = \Phi[\zeta(z)]e^{\phi(z)} + \psi(z)$ is determined. Afterwards, we find out the solution $[f^*(z), g^*(z), h^*(z), Q^*(z)]$ of the system of integral equations

$$f^*(z) = F(z, w, \Pi f^*) - F(z, w, 0) + A_1(z, w)Tf^* + A_2(z, w)\overline{Tf^*} + A_3(z, w), \qquad (2.20)$$

$$Wg^*(z) = F(z, w, W\Pi g^* + \Pi f^*) - F(z, w, \Pi f^*) + A_1(z, w)W + A_2(z, w)\overline{W}, \qquad (2.21)$$

$$S'(\chi)h^*(z)e^{\phi(z)} = F[z, w, S'(\chi)(1 + \Pi h^*)e^{\phi(z)} + W\Pi g^* + \Pi f^*]$$
$$- F(z, w, W\Pi g^* + \Pi f^*),$$
(2.22)

$$Q^*(z) = \frac{h^*(z)}{[1 + \Pi h^*]}, \quad S'(\chi) = [\Phi(\Psi(\chi))]_\chi,$$
(2.23)

and denote by $q^* = E(q)$ the mapping from $q = (Q, f, g)$ to $q^* = (Q^*, f^*, g^*)$. According to Theorem 2.1 from Chapter IV in [86]1), we can prove that $q^* = E(q)$ continuously maps B_1 onto a compact subset in B_1. On the basis of the Schauder fixed-point theorem, there exists a system $q = (Q, f, g) \in B_1$, such that $q = E(q)$. Applying the above method, from $q = (Q, f, g)$, we can construct a function $w(z) = \Phi[\zeta(z)]e^{\phi(z)} + \psi(z)$, which is just a solution of Problem A for (2.1). As for the case of $K < 0$, it can be similarly discussed, but we first permit that the function $\Phi(\zeta)$ satisfying the boundary condition (2.15) has a pole of order $|[K]|$ at $\zeta = 0$, and find the solution of the nonlinear complex equation (2.1) in the form $w(z) = \Phi[\zeta(z)]e^{\phi(z)} + \psi(z)$. From the representation, we can derive the $-2K - 1$ solvability conditions of Problem A for (2.1).

Besides, we can discuss the solvability of the discontinuous Riemann–Hilbert boundary value problem for the complex equation (2.1) in the upper half-plane and the zone domain. For some problems in nonlinear mechanics as stated in [90], it can be solved by the results in Theorem 2.2.

2.3 The discontinuous Riemann–Hilbert problem for nonlinear complex equations in general domains

In this subsection, let D' be a general simply connected domain with the boundary $\Gamma' = \Gamma_1' \cup \Gamma_2'$, herein $\Gamma_1', \Gamma_2' \in C_\alpha^1$ ($0 < \alpha < 1$) and their intersection points z', z'' with the inner angles $\alpha_1\pi, \alpha_2\pi(0 < \alpha_1, \alpha_2 < 1)$ respectively. We discuss the nonlinear uniformly elliptic complex equation

$$w_{\bar{z}} = F(z, w, w_z), \quad F = Q_1 w_z + Q_2 \bar{w}_{\bar{z}} + A_1 w + A_2 \bar{w} + A_3, \quad z \in D',$$
(2.24)

in which $F(z, w, U)$ satisfies Condition C in D'. There exist m point $Z = \{z_1 = z', \ldots, z_{n-1}, z_n = z'', \ldots, z_m\}$ on Γ' arranged according to the positive direction successively. Denote by Γ_j' the curve on Γ' from z_{j-1} to z_j, $j = 1, 2, \ldots, m$, and $\Gamma_j'(j = 1, \ldots, m)$ does not include the end points.

Problem A' The discontinuous Riemann–Hilbert boundary value problem for (2.1) is to find a continuous solution $w(z)$ in $D^* = \overline{D'}\backslash Z$ satisfying the boundary condition:

$$\text{Re}\,[\overline{\lambda(z)}w(z)] = c(z), \quad x \in \Gamma^* = \Gamma'\backslash Z,$$
$$\text{Im}\,[\overline{\lambda(z_j')}w(z_j')] = b_j, \quad j = 1, \ldots, m,$$
(2.25)

where $z_j', b_j(j = 1, \ldots, m)$ are similar to those in (1.36), $\lambda(z), c(z), b_j(j = 1, \ldots, m)$ are given functions satisfying

$$C_\alpha[\lambda(z), \Gamma_j] \le k_0, \quad C_\alpha[|z - z_j|^{\beta_j}c(z), \Gamma_j] \le k_2, \quad |b_j| \le k_2, \quad j = 1, \ldots, m, \quad (2.26)$$

herein $\alpha\,(1/2 < \alpha < 1), k_0, k_2$ are non-negative constants, and assume that $(\beta_j + \gamma_j)/\beta < 1$, $\beta = \min(\alpha, 1 - 2/p_0)/\alpha_0$, $\beta_j (j = 1, \dots, m)$ are similar to those in (1.29), $\alpha_0 = \max(1/\alpha_1, 1/\alpha_2, 1)$. Problem A with $A_3(z) = 0$ in D, $r(z) = 0$ on Γ', and $b_j = 0\,(j = 1, \dots, m)$ is called Problem A_0'. The index $K = (m - 1)/2$ of Problem A and Problem A_0 is defined as in (1.4).

In order to give the unique result of solutions of Problem A' for equation (2.24), we need to add one condition: For any complex functions $w_j(z) \in C(D^*), U_j(z) \in L_{p_0}(D')(j = 1, 2, 2 < p_0 \le p)$, the following equality holds:

$$F(z, w_1, U_1) - F(z, w_1, U_2) = Q(U_1 - U_2) + A(w_1 - w_2) \text{ in } D', \qquad (2.27)$$

in which $|Q(z, w_1, w_2, U_1, U_2)| \le q_0$, $A(z, w_1, w_2, U_1, U_2) \in L_{p_0}(D)$. Especially, if (2.24) is a linear equation, then the condition (2.27) obviously is true.

Applying a similar method as before, we can prove the following theorem.

Theorem 2.3 *If the complex equation (2.24) in D' satisfies Condition C, then Problem A' for (2.24) is solvable. If Condition C and the condition (2.27) hold, then the solution of Problem A' is unique. Moreover the solution $w(z)$ can be expressed as (2.6)–(2.9), where $\beta = \min(\alpha, 1 - 2/p_0)/\alpha_0$. If $Q_j(z) = 0$ in D, $j = 1, 2$ in (2.24), then the representation (2.6) becomes the form*

$$w(z) = \Phi(z)e^{\phi(z)} + \psi(z), \qquad (2.28)$$

and $w(z)$ satisfies the estimate

$$C_\delta[X(z)w(z), \bar{D}] \le M_1 = M_1(p_0, \beta, \delta, k, D), \qquad (2.29)$$

in which

$$X(z) = \prod_{j=2, j \neq 1, n}^{m} |z - z_j|^{\eta_j} |z - z_1|^{\max(1/\alpha_1, 1)\eta_1} |z - z_n|^{\max(1/\alpha_2, 1)\eta_n},$$

$$\eta_j = \begin{cases} |\gamma_j| + \tau, & \text{if } \gamma_j < 0, \quad \beta_j \le |\gamma_j|, \\ |\beta_j| + \tau, & \text{if } \gamma_j \ge 0, \quad \text{and } \gamma_j < 0, \beta_j < |\gamma_j|, \end{cases} \qquad (2.30)$$

here $\gamma_j(j = 1, \dots, m)$ are real constants as stated in (1.28), $\delta, \tau\,(0 < \delta < \min(\beta, \tau))$ are sufficiently small positive constants, and M_1 is a non-negative constant.

3 Discontinuous Oblique Derivative Problems for Quasilinear Elliptic Equations of Second Order

This section deals with the oblique derivative boundary value problems for quasilinear elliptic equations of second order. We first give the extremum principle and representation of solutions for the above boundary value problem, and then obtain a priori estimates of solutions of the above problem, finally we prove the uniqueness and existence of solutions of the above problem.

3.1 Formulation of the discontinuous oblique derivative problem for elliptic equations of second order

Let D be the upper half-unit disk as stated in Section 1 and $\Gamma' = \Gamma \cup L_0$ of D be the boundary, where $\Gamma = \{|z| = 1, \operatorname{Im} z \geq 0\}$ and $L_0 = (-1,1)$. We consider the quasilinear uniformly elliptic equation of second order

$$au_{xx} + 2bu_{xy} + cu_{yy} + du_x + eu_y + fu = g \text{ in } D, \tag{3.1}$$

where a, b, c, d, e, f, g are given functions of $(x, y) \in \bar{D}$ and $u, u_x, u_y \in \mathbb{R}$. Under certain conditions, equation (3.1) can be reduced to the the complex form

$$u_{z\bar{z}} = F(z, u, u_z, u_{zz}), \quad F = \operatorname{Re}\left[Qu_{zz} + A_1 u_z\right] + A_2 u + A_3 \text{ in } D, \tag{3.2}$$

where $Q = Q(z, u, u_z)$, $A_j = A_j(z, u, u_z)$ and

$$z = x + iy, \quad u_z = \frac{1}{2}[u_x - iu_y], \quad u_{\bar{z}} = \frac{1}{2}[u_x + iu_y], \quad u_{z\bar{z}} = \frac{1}{4}[u_{xx} + u_{yy}],$$

$$Q(z) = \frac{-a + c - 2bi}{a + c}, \quad A_1(z) = \frac{-d - ei}{a + c}, \quad A_2(z) = \frac{-f}{2(a + c)}, \quad A_3(z) = \frac{g}{2(a + c)}.$$

Suppose that equation (3.2) satisfies the following conditions.

Condition C

1) $Q(z, u, w), A_j(z, u, w) \, (j = 1, 2, 3)$ are continuous in $u \in \mathbb{R}$, $w \in \mathbb{C}$ for almost every point $z \in D, u \in \mathbb{R}, w \in \mathbb{C}$, and $Q = 0$, $A_j = 0 \, (j = 1, 2, 3)$ for $z \notin D$.

2) The above functions are measurable in $z \in D$ for all continuous functions $u(z), w(z)$ on $D^* = \bar{D}\backslash Z$, and satisfy

$$L_p[A_j(z, u, w), \bar{D}] \leq k_0, j = 1, 2, \quad L_p[A_3(z, u, w), \bar{D}] \leq k_1, A_2(z, u, w) \geq 0 \text{ in } D, \tag{3.3}$$

in which $p_0, p \, (2 < p_0 \leq p), k_0, k_1$ are non-negative constants, $Z = \{-1, 1\}$.

3) Equation (3.2) satisfies the uniform ellipticity condition, namely for any number $u \in \mathbb{R}, w \in \mathbb{C}$, the inequality

$$|Q(z, u, w)| \leq q_0 < 1 \tag{3.4}$$

for almost every point $z \in D$ holds, where q_0 is a non-negative constant.

The discontinuous oblique derivative boundary value problem for equation (3.2) may be formulated as follows:

Problem P   Find a continuously differentiable solution $u(z)$ of (3.2) in $D^* = \bar{D}\backslash Z$, which is continuous in $\bar{D}$ and satisfies the boundary conditions

$$\frac{1}{2}\frac{\partial u}{\partial \nu} = \operatorname{Re}\left[\overline{\lambda(z)}u_z\right] = r(z), \quad z \in \Gamma^* = \Gamma'\backslash Z, \quad u(-1) = b_0, \quad u(1) = b_1, \tag{3.5}$$

where $Z = \{-1, 1\}$ is the set of discontinuous points of $\lambda(z)$ on Γ^*, ν is a given vector at every point on Γ^*, $\lambda(z) = a(x) + ib(x) = \cos(\nu, x) - i \cos(\nu, y)$, $\cos(\nu, n) \geq 0$ on Γ^*. If $\cos(\nu, n) \equiv 0$ on $\Gamma^* = \Gamma' \backslash Z$, then the condition $u(1) = b_1$ can be canceled. Here n is the outward normal vector at every point on Γ^*, $\delta_0 (< 1)$ is a constant, b_0, b_1 are real constants, and $\lambda(z)$, $r(z)$, b_0, b_1 satisfy the conditions

$$C_\alpha[\lambda(z), \Gamma_j] \leq k_0, \quad C_\alpha[|z - z_j|^{\beta_j} r(z), \Gamma_j] \leq k_2, \quad j = 1, 2, \quad |b_0|, |b_1| \leq k_2. \tag{3.6}$$

Herein $\alpha (1/2 < \alpha < 1)$, k_0, k_2 are non-negative constants. We assume that $(\beta_j + \gamma_j)/\beta < 1$, $\beta = \min(\alpha, 1 - 2/p_0)/2$, $\beta_j (j = 1, \ldots, m)$ are as stated in (1.27). Problem P with $A_3(z) = 0$ in D, $r(z) = 0$ on Γ, $b_0 = b_1 = 0$ is called Problem P_0. The index of Problem P is K, where K is defined as in (1.4), here we choose $K = 0$, and $K = -1/2$ if $\cos(\nu, n) \equiv 0$ on Γ^*. If $A_2(z) = 0$ in D, the last point condition in (3.5) can be replaced by

$$\text{Im} \, [\overline{\lambda(z)} u_z]|_{z=0} = b_2, \tag{3.7}$$

and we do not need the assumption $\cos(\nu, n) \geq 0$ on Γ, where b_2 is a real constant satisfying the condition $|b_2| \leq k_2$. Then the boundary value problem for (3.2) will be called Problem Q. In the following, we only discuss the case of $K = 0$, and the case of $K = -1/2$ can be similarly discussed.

3.2 The representation theorem of Problem P for equation (3.2)

We first introduce a theorem.

Theorem 3.1 *Suppose that equation (3.2) satisfies Condition C. Then there exist two solutions $\psi(z), \Psi(z)$ of the Dirichlet problem (Problem D) of (3.2) and its related homogeneous equation*

$$u_{z\bar{z}} - \text{Re} \, [Q(z, u, u_z)u_{zz} + A_1(z, u, u_z)u_z] - A_2(z, u, u_z)u = 0 \text{ in } D, \tag{3.8}$$

satisfying the boundary conditions

$$\psi(z) = 0, \quad \Psi(z) = 1 \text{ on } \Gamma \tag{3.9}$$

respectively, and $\psi(z), \Psi(z)$ satisfy the estimates

$$C_\beta^1[\psi(z), \bar{D}] \leq M_2, \, C_\beta^1[\Psi(z), \bar{D}] \leq M_2,$$
$$L_{p_0}[\psi_{z\bar{z}}, \bar{D}] \leq M_3, \quad L_{p_0}[\Psi_{z\bar{z}}, \bar{D}] \leq M_3, \quad \Psi \geq M_4 > 0 \text{ in } D, \tag{3.10}$$

where $\beta \, (0 < \beta \leq \alpha)$, $M_j = M_j(q_0, p_0, \beta, k_0, k_1, D) \, (j = 2, 3, 4)$ are non-negative constants.

Proof We first assume that the coefficients $Q = A_j = 0 \, (j = 1, 2, 3)$ of (3.2) in the ε-neighborhood of $z = -1, 1$, i.e. $D_\varepsilon = \{|z \pm 1| \leq \varepsilon, \text{Im} \, z \geq 0\}$, $\varepsilon > 0$, where $\varepsilon = 1/m$ (m is a positive integer). Introduce the transformation and its inversion

$$\zeta(z) = -i \frac{z^2 + 2iz + 1}{z^2 - 2iz + 1}, \quad z(\zeta) = \frac{1}{\zeta + i} \left[1 + i\zeta - \sqrt{2(1 - \zeta^2)} \right]. \tag{3.11}$$

The function $\zeta(z)$ maps D onto $G = \{|\zeta| < 1\}$, such that the boundary points $-1, 0, 1$ are mapped onto the points $-1, -i, 1$ respectively. Through the transformation, equation (3.2) is reduced to the equation

$$u_{\zeta\bar{\zeta}} = |z'(\zeta)|^2\{\operatorname{Re}\left[Qu_{\zeta\zeta}/(z'(\zeta))^2 + (A_1/z'(\zeta) - Qz''(\zeta)/(z'(\zeta))^3)u_\zeta\right] + A_2 u + A_3\} \quad (3.12)$$

in G. It is clear that equation (3.12) in $\bar{G}$ satisfies conditions similar to Condition C. Hence equation (3.12) and its related homogeneous equation

$$u_{\zeta\bar{\zeta}} = |z'(\zeta)|^2\{\operatorname{Re}\left[Qu_{\zeta\zeta}/(z'(\zeta))^2 + (A_1/z'(\zeta) - Qz''(\zeta)/(z'(\zeta))^3)u_\zeta\right] + A_2 u\} \text{ in } G \quad (3.13)$$

possess the solutions $\psi(\zeta)$, $\Psi(\zeta)$ satisfying the boundary conditions

$$\psi(\zeta) = 0, \quad \Psi(\zeta) = 1 \text{ on } L = \zeta(\Gamma),$$

and $\psi[\zeta(z)], \Psi[\zeta(z)]$ in D are the solutions of Problem D of (3.2),(3.8) satisfying the boundary condition (3.9) respectively, and $\psi(z), \Psi(z)$ satisfy the estimate (3.10), but the constants $M_j = M_j(q_0, p_0, \beta, k_0, k_1, D, \varepsilon)$ $(j = 2, 3, 4)$. Now we consider

$$\tilde{\psi}(z) = \begin{cases} \psi(z) \text{ in } D, \\ -\psi(\bar{z}) \text{ in } \tilde{D} = \{|z| < 1, \operatorname{Im} z < 0\}. \end{cases} \quad (3.14)$$

It is not difficult to see that $\tilde{\psi}(z)$ in $\Delta = \{|z| < 1\}$ is a solution of the elliptic equation

$$u_{z\bar{z}} - \operatorname{Re}\left[\tilde{Q}u_{zz} + \tilde{A}_1 u_z\right] - \tilde{A}_2 u = \tilde{A}_3 \text{ in } \Delta, \quad (3.15)$$

where the coefficients

$$\tilde{Q} = \begin{cases} Q(z), \\ \overline{Q(\bar{z})} \end{cases} \tilde{A}_1 = \begin{cases} A_1(z), \\ \overline{A_1(\bar{z})}, \end{cases} \tilde{A}_2 = \begin{cases} A_2(z), \\ A_2(\bar{z}), \end{cases} \tilde{A}_3 = \begin{cases} A_3(z) \\ -A_3(\bar{z}) \end{cases} \text{ in } \begin{cases} D \\ \tilde{D} \end{cases},$$

where $\tilde{D}$ is the symmetrical domian of D with respect to the real axis. It is clear that the coefficients in Δ satisfy conditions similar to those from Condition C. Obviously the solution $\tilde{\psi}(z)$ satisfies the boundary condition $\psi(z) = 0$ on $\partial\Delta = \{|z| = 1\}$. Denote by $\tilde{\psi}_m(z)$ the solution of equation (3.2) with $Q = A_j = 0(j = 1, 2, 3)$ in the $\varepsilon = 1/m$-neighborhood of $z = -1, 1$, we can derive that the function $\tilde{\psi}_m(z)$ in $\overline{\Delta}$ satisfies estimates similar to $\psi(z)$ in (3.10), where the constants $M_j(j = 2, 3)$ are independent of $\varepsilon = 1/m$. Thus we can choose a subsequence of $\{\tilde{\psi}_m(z)\}$, which uniformly converges to $\psi_*(z)$, and $\psi_*(z)$ is just a solution of Problem D for the original equation (3.2) in D. Noting that the solution $\Psi(z) = \psi(z) + 1$ of Problem D for equation (3.8) is equivalent to the solution $\psi(z)$ of Problem D for the equation

$$u_{z\bar{z}} - \operatorname{Re}\left[Qu_{zz} + A_1 u_z\right] - A_2 u = A_2 \text{ in } D \quad (3.16)$$

with the boundary condition $\psi(z) = 0$ on Γ, by using the same method, we can prove that there exists a solution $\Psi(z)$ of Problem D for (3.8) with the boundary condition $\Psi(z) = 1$ on Γ, and the solution satisfies the estimates in (3.10).

Theorem 3.2 *Suppose that equation (3.2) satisfies Condition C, and $u(z)$ is a solution of Problem P for (3.2). Then $u(z)$ can be expressed as*

$$u(z)=U(z)\Psi(z)+\psi(z), U(z)=2\mathrm{Re}\int_0^z w(z)dz+b_0, w(z)=\Phi[\zeta(z)]e^{\phi(z)}, \qquad (3.17)$$

where $\psi(z), \Psi(z)$ are as stated in Theorem 3.1 satisfying the estimate (3.10), $\zeta(z)$ is a homeomorphism in $\bar{D}$, which quasiconformally maps D onto the unit disk $G=\{|\zeta| < 1\}$ with boundary L, where $\zeta(-1) = -1, \zeta(1) = 1, \zeta(i) = i$, $\Phi(\zeta)$ is an analytic function in G, $\phi(z), \zeta(z)$ and its inverse function $z(\zeta)$ satisfy the estimates

$$C_\beta[\phi(z), \bar{D}] \leq k_3, \quad C_\beta[\zeta(z), \bar{D}] \leq k_3,$$

$$C_\beta[z(\zeta), \bar{G}] \leq k_3, \quad L_{p_0}[|\phi_{\bar{z}}| + |\phi_z|, \bar{D}] \leq k_3, \qquad (3.18)$$

$$C_\beta[z(\zeta), \bar{G}] \leq k_3, \quad L_{p_0}[|\chi_{\bar{z}}| + |\chi_z|, \bar{D}] \leq k_4,$$

in which $\chi(z)$ is as stated in (2.14), $\beta = \min(\alpha, 1 - 2/p_0)/2, p_0(2 < p_0 \leq p), k_j = k_j(q_0, p_0, k_0, k_1, D)(j = 3, 4)$ are non-negative constants.

Proof We substitute the solution $u(z)$ of Problem P into the coefficients of equation (3.2). It is clear that (3.2) in this case can be seen as a linear equation. Firstly, on the basis of Theorem 3.1 there exist two solutions $\psi(z), \Psi(z)$ of Problem D of (3.2) and its homogeneous equation (3.8) satisfying the estimate (3.10). Thus the function

$$U(z) = \frac{u(z) - \psi(z)}{\Psi(z)} \text{ in } D, \qquad (3.19)$$

is a solution of the equation

$$U_{z\bar{z}} - \mathrm{Re}\,[QU_{zz} + AU_z] = 0, \quad A = A_1 - 2(\ln\psi)_{\bar{z}} + 2Q(\ln\Psi)_z \text{ in } D, \qquad (3.20)$$

and $w(z) = U_z$ is a solution of the first order equation

$$w_{\bar{z}} = \frac{1}{2}[Qw_z + \bar{Q}\bar{w}_{\bar{z}} + Aw + A\bar{w}] \text{ in } D \qquad (3.21)$$

satisfying the boundary condition

$$\frac{1}{2}[\frac{\partial U}{\partial\nu} + (\ln\Psi)_\nu U] = r(z) - \mathrm{Re}\,[\overline{\lambda(z)}\psi_z] \text{ on } \Gamma^*, \quad \text{i.e.}$$

$$\mathrm{Re}\,[\overline{\lambda(z)}U_z + (\ln\Psi)_\nu U/2] = r(z) - \mathrm{Re}\,[\overline{\lambda(z)}\psi_z] \text{ on } \Gamma^*. \qquad (3.22)$$

By the following Lemma 3.3, we see that $(\ln\Psi)_\nu > 0$ on Γ^*, and similarly to Theorem 2.1, the last formula in (3.17) can be derived, and $\phi(z), \zeta(z)$ and its inverse function $z(\zeta), \chi(z)$ satisfy the estimates (2.7)–(2.9).

Now we consider the linear homogeneous equation

$$u_{z\bar{z}} - \mathrm{Re}\,[Qu_{zz} + A_1(z)u_z] - A_2(z)u = 0 \text{ in } D, \qquad (3.23)$$

and give a lemma.

Lemma 3.3   *Let the equation (3.23) in D satisfy Condition C, and $u(z)$ be a continuously differentiable solution of (3.23) in $\overline{D}$. If $M = \max_{z \in \overline{D}} u(z) \geq 0$, then there exists a point $z_0 \in \partial D$, such that $u(z_0) = M$. If $z_0 = x_0 \in (-1,1)$, and $u(z) < u(z_0)$ in $\overline{D} \backslash \{z_0\}$, then*

$$\frac{\partial u}{\partial l} = \lim_{z(\in l) \to z_0} \frac{u(z_0) - u(z)}{|z - z_0|} > 0, \tag{3.24}$$

where $z (\in D)$ approaches z_0 along a direction l, such that $\cos(l, y) > 0$.

Proof   From the result in Section 2, Chapter III, [86]1), we see that the solution $u(z)$ in $\overline{D}$ attains its non-negative maximum M at a point $z_0 \in \partial D$. There is no harm in assuming that z_0 is a boundary point of $\Delta = \{|z| < R\}$, because we can choose a subdomain$(\in \overline{D})$ with smooth boundary and the boundary point z_0, and then make a conformal mapping. Thus this requirement can be realized. By Theorem 3.1, we find a continuously differentiable solution $\Psi(z)$ of (3.23) in $\overline{\Delta}$ satisfying the boundary condition: $\Psi(z) = 1$, $z \in \partial \Delta = \{|z| = R\}$, and can derive that $0 < \Psi(z) \leq 1$, $z \in \overline{\Delta}$. Due to $V(z) = u(z)/\Psi(z)$ is a solution of the following equation

$$LV = V_{z\bar{z}} - \operatorname{Re}\left[A(z)V_z\right] = 0, \quad A(z) = -2(\ln \Psi)_z + A_1(z) \text{ in } \Delta, \tag{3.25}$$

it is clear that $V(z) < V(z_0)$, $z \in \Delta$, and $V(z)$ attains the maximum at the point z_0. Afterwards, we find a continuously differentiable solution $\tilde{V}(z)$ of (3.25) in $\tilde{\Delta} = \{R/2 \leq |z| \leq R\}$ satisfying the boundary condition

$$\tilde{V}(z) = 0, \quad z \in \partial \Delta; \quad \tilde{V}(z) = 1, \quad |z| = \frac{R}{2}.$$

It is easy to see that $\partial \tilde{V}/\partial s = 2\operatorname{Re}\left[iz\tilde{V}_z\right]$, $z \in \partial \tilde{\Delta}$, and

$$\frac{\partial \tilde{V}}{\partial n} = 2\operatorname{Re}\frac{z\tilde{V}_z}{R}, \quad z \in \partial \Delta, \quad \frac{\partial \tilde{V}}{\partial n} = -4\operatorname{Re}\frac{z\tilde{V}_z}{R}, \quad |z| = \frac{R}{2},$$

where s, n are the tangent vector and outward normal vector on the boundary $\partial \tilde{\Delta}$. Noting that $W(z) = \tilde{V}_z$ satisfies the equation

$$W_{\bar{z}} - \operatorname{Re}\left[A(z)W\right] = 0, \quad z \in \tilde{\Delta},$$

and the boundary condition $\operatorname{Re}\left[izW(z)\right] = 0$, $z \in \partial \tilde{\Delta}$, and the index of $\overline{iz}$ on the boundary $\partial \tilde{\Delta}$ equals to 0, hence $W(z)$ has no zero point on $\partial \Delta$, thus $\partial \tilde{V}/\partial n = 2\operatorname{Re}\left[zW(z)/R\right] < 0$, $z \in \partial \Delta$. The auxiliary function

$$\hat{V}(z) = V(z) - V(z_0) + \varepsilon \tilde{V}(z), \quad z \in \tilde{\Delta},$$

by selecting a sufficiently small positive number ε, such that $\hat{V}(z) < 0$ on $|z| = R/2$, obviously satisfies $\hat{V}(z) \leq 0$, $z \in \partial \Delta$. Due to $L\hat{V} = 0$, $z \in \tilde{\Delta}$, on the basis of the maximum principle, we have

$$\hat{V}(z) \leq 0, z \in \partial \Delta, \quad \text{i.e. } V(z_0) - V(z) \geq -\varepsilon[V(z_0) - V(z)], \quad z \in \tilde{\Delta}.$$

Thus at the point $z = z_0$ we have

$$\frac{\partial V}{\partial n} \geq -\varepsilon \frac{\partial \tilde{V}}{\partial n} > 0, \quad \frac{\partial u}{\partial n} = \Psi \frac{\partial V}{\partial n} + V \frac{\partial \Psi}{\partial n} \geq -\varepsilon \frac{\partial \tilde{V}}{\partial n} + V \frac{\partial \Psi}{\partial n} > 0.$$

Moreover, noting the condition $\cos(l, n) > 0, \cos(l, s) > 0, \partial U / \partial s = 0$ at the point z_0, where s is the tangent vector at z_0, it follows the inequality

$$\frac{\partial u}{\partial l} = \cos(l, n) \frac{\partial u}{\partial n} + \cos(l, s) \frac{\partial u}{\partial s} > 0. \tag{3.26}$$

Theorem 3.4 *If equation (3.2) satisfies Condition C and for any $u_j(z) \in C^1(D^*), j = 1, 2, u_{zz} \in \mathcal{C}$, the following equality holds:*

$$F(z, u_1, u_{1z}, u_{1zz}) - F(z, u_2, u_{2z}, u_{2zz}) = -\mathrm{Re}\,[\tilde{Q}u_{zz} + \tilde{A}_1 u_z] - \tilde{A}_2 u,$$

where $L_p[\tilde{A}_j, \bar{D}] < \infty, j = 1, 2$, then the solution $u(z)$ of Problem P is unique.

Proof Suppose that there exist two solutions $u_1(z), u_2(z)$ of Problem P for (3.2), it can be seen that $u(z) = u_1(z) - u_2(z)$ satisfies the homogeneous equation and boundary conditions

$$u_{z\bar{z}} = \mathrm{Re}\,[\tilde{Q}u_{zz} + \tilde{A}_1 u_z] + \tilde{A}_2 u \text{ in } D,$$

$$\frac{1}{2}\frac{\partial u}{\partial \nu} = 0, \quad z \in \Gamma^*, \quad u(-1) = 0, \quad u(1) = 0. \tag{3.27}$$

If the maximum $M = \max_{\bar{D}} u(z) > 0$, it is clear that the maximum point $z^* \neq -1$ and 1. On the basis of Lemma 3.3, the maximum of $u(z)$ cannot attain on $(-1, 1)$, hence its maximum M attains at a point $z^* \in \Gamma^*$. If $\cos(\nu, n) > 0$ at z^*, from Lemma 3.3, we get $\partial u / \partial \nu > 0$ at z^*, this contradicts the boundary condition in (3.27); if $\cos(\nu, n) = 0$ at z^*, denote by Γ' the longest curve of Γ including the point z^*, so that $\cos(\nu, n) = 0$ and $u(z) = M$ on Γ', then there exists a point $z' \in \Gamma \backslash \Gamma'$, such that at z', $\cos(\nu, n) > 0, \partial u / \partial n > 0, \cos(\nu, s) > 0 (< 0), \partial u / \partial s \geq 0 (\leq 0)$, hence (3.26) at z' holds, it is impossible. This shows $z^* \notin \Gamma$. Hence $\max_{\overline{D^+}} u(z) = 0$. By the similar method, we can prove $\min_{\overline{D^+}} u(z) = 0$. Therefore $u(z) = 0, u_1(z) = u_2(z)$ in $\bar{D}$.

Theorem 3.5 *Suppose that equation (3.2) satisfies Condition C, then the solution $u(z)$ of Problem P for (3.2) satisfies the estimates*

$$\tilde{C}_\delta^1[u(z), \bar{D}] = C_\beta[u(z), \bar{D}] + C_\delta[|X(z)|^{1/\beta} u_z, \bar{D}] \leq M_5, \tag{3.28}$$

$$\tilde{C}_\delta^1[u(z), \bar{D}] \leq M_6(k_1 + k_2),$$

in which $\beta = \min(\alpha, 1 - 2/p_0), X(z) = |z + 1|^{2\eta_1} |z - 1|^{2\eta_2}, M_5 = M_5(p_0, \beta, \delta, k, D), M_6 = M_6(p_0, \beta, \delta, k_0, D)$ are two non-negative constants.

Proof We first verify that any solution $u(z)$ of Problem P for (3.2) satisfies the estimate

$$S(u) = C[u(z), \bar{D}] + C[|X(z)|^{1/\beta} u_z, \bar{D}] \leq M_7 = M_7(p_0, \alpha, k, D). \tag{3.29}$$

Otherwise, if the above inequality is not true, there exist sequences of coefficients: $\{Q^m\}$, $\{A_j^m\}(j = 1, 2, 3)$, $\{\lambda^m\}$, $\{r^m\}$, $\{b_j^m\}(j = 1, 2)$ satisfying the same conditions of $Q, A_j(j = 1, 2, 3)$, $\lambda, r, b_j(j = 0, 1)$, and $\{Q^m\}, \{A_j^m\}(j = 1, 2, 3)$ weakly converge in D to $Q^0, A_j^0(j = 1, 2, 3)$, and $\{\lambda^m\}, \{r^m\}, \{b_j^m\}(j = 0, 1)$ uniformly converge on Γ^* to $\lambda^0, r^0, b_j^0(j = 0, 1)$ respectively. Let u^m is a solution of Problem P for (3.2) corresponding to $\{Q^m\}, \{A_j^m\} \ (j = 1, 2, 3)$, $\{\lambda^m\}, \{r^m\}$, $\{b_j^m\}(j = 0, 1)$, but $\max_{\bar{D}} |u^m(z)| = H_m \to \infty$ as $m \to \infty$. There is no harm in assuming that $H_m \geq 1$. Let $U^m = u^m/H_m$. It is clear that $U^m(z)$ is a solution of the boundary value problem

$$U_{z\bar{z}}^m - \text{Re}\,[Q^m U_{zz}^m + A_1^m U_z^m] - A_2^m U^m = \frac{A_3^m}{H_m},$$

$$\frac{\partial U^m}{\partial \nu_m} = \frac{r^m(z)}{H_m}, \ z \in \Gamma^*, \quad U^m(-1) = \frac{b_0^m}{H_m}, U^m(1) = \frac{b_1^m}{H_m}.$$

From the conditions in the theorem, we have

$$L_p[A_2^m U^m + \frac{A_3^m}{H_m}, \bar{D}] \leq M_7, \, C[\lambda_j, \Gamma_j] \leq M_7,$$

$$C_\alpha \left[|z - z_j|^{\beta_j} \frac{r^m(z)}{H_m}, \Gamma^* \right] \leq M_7, j = 1, \ldots, m, \quad \left| \frac{b_j^m}{H_m} \right| \leq M_7, j = 0, 1,$$

where $M_7 = M_7(q_0, p_0, \alpha, K, D)$ is a non-negative constant. According to the method in the proof of Theorem 2.3, we denote

$$w_m = U_z^m, \, U^m(z) = 2\text{Re} \int_{-1}^z w_m(z)dz + \frac{b_0^m}{H_m},$$

and can obtain that $U_m(z)$ satisfies the estimate

$$C_\beta[U^m(z), \bar{D}] + C_\delta[|X(z)|^{1/\beta}U_z^m, \bar{D}] \leq M_8, \qquad (3.30)$$

in which $M_8 = M_8(q_0, p_0, \delta, \alpha, K, D)$, $\delta\,(> 0)$ are non-negative constants. Hence from $\{U^m(z)\}$ and $\{|X(z)|^{1/\beta}U_z^m\}$, we can choose subsequences $\{U^{m_k}(z)\}$ and $\{|X(z)|^{1/\beta}U_z^{m_k}\}$, which uniformly converge to $U^0(z)$ and $|X(z)|^{1/\beta}U_z^0$ in $\bar{D}$ respectively, and $U^0(z)$ is a solution of the following boundary value problem

$$U_{z\bar{z}}^0 = \text{Re}\,[Q^0 U_{zz}^0 + A_1^0 u_z] + A_2^0 U^0 = 0 \ \text{in} \ D,$$

$$\frac{\partial U^0}{\partial \nu} = 0 \ \text{on} \ \Gamma^*, \quad U^0(-1) = 0, \quad U^0(1) = 0.$$

By the result as stated before, we see that the solution $U^0(z) = 0$. However, from $S(U^m) = 1$, the inequality $S(U^0) > 0$ can be derived. Hence the estimate (3.29) is true. Moreover, by using the method from $S(U^m) = 1$ to (3.30), we can prove the first estimate in (3.28). The second estimate in (3.28) can be derived from the first one.

3.3 Existence of solutions of the discontinuous oblique derivative problem for elliptic equations in the upper half-unit disk

Theorem 3.6 *If equation (3.2) satisfies Condition C, then Problem P for (3.2) is solvable.*

Proof Noting that the index $K = 0$, we introduce the boundary value problem P_t for the linear elliptic equation with a parameter $t(0 \le t \le 1)$:

$$Lu = u_{z\bar{z}} - \mathrm{Re}\,[Qu_{zz} + A_1(z)u_z] = G(z, u), \quad G = tA_2(z)u + A(z) \qquad (3.31)$$

for any $A(z) \in L_{p_0}(\bar{D})$ and the boundary condition (3.5). It is evident that when $t = 1$, $A(z) = A_3(z)$, Problem P_t is just Problem P. When $t = 0$, the equation in (3.31) is

$$Lu = u_{z\bar{z}} - \mathrm{Re}\,[Qu_{zz} + A_1u_z] = A(z), \quad \text{i.e. } w_{\bar{z}} - \mathrm{Re}\,[Qw_z + A_1w] = A(z), \qquad (3.32)$$

where $w = u_z$. By Theorem 3.7 below, we see that Problem P for the first equation in (3.32) has a unique solution $u_0(z)$, which is just a solution of Problem P for equation (3.31) with $t = 0$. Suppose that when $t = t_0\ (0 \le t_0 < 1)$, Problem P_{t_0} is solvable, i.e. Problem P_t for (3.31) has a unique solution $u(z)$ such that $|X(z)|^{1/\beta}u_z \in C_\delta(\bar{D})$. We can find a neighborhood $T_\varepsilon = \{|t - t_0| < \varepsilon, 0 \le t \le 1, \varepsilon > 0\}$ of t_0, such that for every $t \in T_\varepsilon$, Problem P_t is solvable. In fact, Problem P_t can be written in the form

$$Lu - t_0[G(z, u) - G(z, 0)] = (t - t_0)[G(z, u) - G(z, 0)] + A(z), \quad z \in D \qquad (3.33)$$

and (3.5). Replacing $u(z)$ in the right-hand side of (3.33) by a function $u_0(z)$ with the condition $|X(z)|^{1/\beta}u_{0z} \in C_\delta(\bar{D})$, especially, by $u_0(z) = 0$, it is obvious that the boundary value problem (3.33),(3.5) then has a unique solution $u_1(z)$ satisfying the conditions $|X(z)|^{1/\beta}u_{1z} \in C_\delta(\bar{D})$. Using successive iteration, we obtain a sequence of solutions: $\{u_n(z)\}$ satisfying the conditions $|X(z)|^{1/\beta}u_{nz} \in C_\delta(\bar{D})(n = 1, 2, \dots)$ and

$$Lu_{n+1} - t_0[G(z, u_{n+1}) - G(z, 0)] = (t - t_0)[G(z, u_n) - G(z, 0)] + A(z), \quad z \in D,$$

$$\mathrm{Re}\,[\overline{\lambda(z)}u_{n+1z}] = r(z), \quad z \in \Gamma, \quad u_{n+1}(-1) = b_0, \quad u_{n+1}(1) = b_1, \quad n = 1, 2, \dots.$$

From the above formulas, it follows that

$$L(u_{n+1} - u_n)_{\bar{z}} - t_0[G(z, u_{n+1}) - G(z, u_n)]$$

$$= (t - t_0)[G(z, u_n) - G(z, u_{n-1})], \quad z \in D,$$

$$\mathrm{Re}\,[\overline{\lambda(z)}(u_{n+1z} - u_{nz})] = 0, \quad z \in \Gamma, \qquad (3.34)$$

$$u_{n+1}(-1) - u_n(-1) = 0, \quad u_{n+1}(1) - u_n(1) = 0.$$

Noting that

$$L_p[(t - t_0)(G(z, u_n) - G(z, u_{n-1})), \bar{D}] \le |t - t_0|k_0\tilde{C}_\delta^1[u_n - u_{n-1}, \bar{D}], \qquad (3.35)$$

where $\tilde{C}_\delta^1[u_n - u_{n-1}, \bar{D}] = C_\beta[u_n - u_{n-1}, \bar{D}] + C_\delta[|X(z)|^{1/\beta}(u_{nz} - u_{n-1z}), \bar{D}]$, and applying Theorem 3.5, we get

$$\tilde{C}_\delta^1[u_{n+1} - u_n, \bar{D}] \leq |t - t_0|M_6\tilde{C}_\delta^1[u_n - u_{n-1}, \bar{D}]. \tag{3.36}$$

Choosing the constant ε so small that $2\varepsilon M_6 < 1$, it follows that

$$\tilde{C}_\delta^1[u_{n+1} - u_n, \bar{D}] \leq \tilde{C}_\delta^1 \frac{u_n - u_{n-1}, \bar{D}}{2}, \tag{3.37}$$

and when $n, m \geq N_0 + 1$ (N_0 is a positive integer),

$$\tilde{C}_\delta^1[u_{n+1} - u_n, \bar{D}] \leq 2^{-N_0}\sum_{j=0}^{\infty} 2^{-j}\tilde{C}_\delta^1[u_1 - u_0, \bar{D}] \leq 2^{-N_0+1}\tilde{C}_\delta^1[u_1 - u_0, \bar{D}]. \tag{3.38}$$

Hence $\{u_n(z)\}$ is a Cauchy sequence. According to the completeness of the Banach space $\tilde{C}_\delta^1(\bar{D})$, there exists a function $u_*(z) \in \tilde{C}_\delta^1(\bar{D})$, so that $\tilde{C}_\delta^1[u_n - u_*, \bar{D}] \to 0$ for $n \to \infty$. From (3.38), we can see that $u_*(z)$ is a solution of Problem P_t for every $t \in T_\varepsilon = \{|t - t_0| \leq \varepsilon\}$. Because the constant ε is independent of t_0 ($0 \leq t_0 < 1$), therefore from the solvability of Problem P_t when $t = 0$, we can derive the solvability of Problem P_t when $t = \varepsilon, 2\varepsilon, \ldots, [1/\varepsilon]\varepsilon, 1$. In particular, when $t = 1$ and $A(z) = A_3(z)$, Problem P_1 for the linear case of equation (3.2) is solvable.

Next, we discuss the quasilinear equation (3.2) satisfying Condition C, but we first assume that the coefficients $Q = 0$, $A_j(j = 1, 2, 3) = 0$ in $D_m = \{z \in \bar{D}, \text{dist}(z, \Gamma) < 1/m\}$, here $m(\geq 2)$ is a positive integer, namely consider

$$u_{z\bar{z}} = \text{Re}\,[Q^m u_{zz} + A_1^m u_z] + A_2^m u + A_3^m \text{ in } D, \tag{3.39}$$

where

$$Q^m = \begin{cases} Q(z, u, u_z), \\ 0, \end{cases} \quad A_j^m = \begin{cases} A_j(z, u, u_z) \\ 0 \end{cases} \text{ in } \begin{cases} \tilde{D}_m = D\backslash D_m \\ D_m \end{cases}, j = 1, 2, 3.$$

Now, we introduce a bounded, closed and convex set B_M in the Banach space $B = \tilde{C}_\delta^1(\bar{D})$, any element of which satisfies the inequality

$$\tilde{C}_\delta^1[u(z), \bar{D}] \leq M_5, \tag{3.40}$$

where M_5 is a non-negative constant as stated in (3.28). We are free to choose an arbitrary function $U(z) \in B_M$ and insert it into the coefficients of equation (3.39). It is clear that the equation can be seen as a linear equation, hence there exists a unique solution $u(z)$ of Problem P, and by Theorem 3.5, we see $u(z) \in B_M$. Denote by $u(z) = S[U(z)]$ the mapping from $U(z) \in B_M$ to $u(z)$, obviously $u(z) = S[U(z)]$ maps B_M onto a compact subset of itself. It remains to verify that $u(z) = S[U(z)]$ continuously maps the set B_M onto a compact subset. In fact, we arbitrarily select a sequence of functions: $\{U_n(z)\}$, such that $\tilde{C}_\delta^1[U_n(z) - U_0(z), \bar{D}] \to 0$ as $n \to \infty$.

Setting $u_n(z) = S[U_n(z)]$, and subtracting $u_0(z) = S[U_0(z)]$ from $u_n(z) = S[U_n(z)]$, we obtain the equation for $\tilde{u}_n = u_n(z) - u_0(z)$:

$$\tilde{u}_{n\bar{z}} - \text{Re}\left[Q^m(z, U_n, U_{nz})\tilde{u}_{nzz} + A_1^m(z, U_n, U_{nz})\tilde{u}_{nz}\right] - A_2^m(z, U_n, U_{nz})\tilde{u}_n = C_n,$$
$$C_n = C_n(z, U_n, U_0, u_0) = \tilde{A}_3^m - \text{Re}\left[\tilde{Q}^m u_{0zz} + \tilde{A}_1^m u_{0z}\right] - \tilde{A}_2^m u_0, \tag{3.41}$$

in which $\tilde{Q}^m = Q^m(z, U_n, U_{nz}) - Q^m(z, U_0, U_{0z})$, $\tilde{A}_j^m = A_j^m(z, U_n, U_{nz}) - A_j^m(z, U_0, U_{0z})$, $j = 1, 2, 3$, and the solution $\tilde{u}_n(z)$ satisfies the homogeneous boundary conditions

$$\text{Re}\left[\overline{\lambda(z)}u_z\right] = 0, \quad z \in \Gamma^* = \Gamma \backslash Z, \quad u(-1) = 0, \quad u(1) = 0. \tag{3.42}$$

Noting that the function $C_n = 0$ in D_m, according to the method in the formula (2.43), Chapter II, [86]1), we can prove that

$$L_p[C_n, \bar{D}] \to 0 \text{ as } n \to \infty.$$

On the basis of the second estimate in (3.28), we obtain

$$\tilde{C}_\delta^1[u_n(z) - u_0(z), \bar{D}] \leq M_6 L_p[C_n, \bar{D}], \tag{3.43}$$

thus $\tilde{C}_\delta^1[u_n(z) - u_0(z), \bar{D}] \to 0$ as $n \to \infty$. This shows that $u(z) = S[U(z)]$ in the set B_M is a continuous mapping. Hence by the Schauder fixed-point theorem, there exists a function $u(z) \in B_M$, such that $u(z) = S[u(z)]$, and the function $u(z)$ is just a solution of Problem P for the quasilinear equation (3.39).

Finally we cancel the conditions: the coefficients $Q = 0$, A_j $(j = 1, 2, 3) = 0$ in $D_m = \{z, \text{dist}(z, \Gamma) < 1/m\}$. Denote by $u^m(z)$ a solution of Problem P for equation (3.39). By Theorem 3.5, we see that the solution satisfies the estimate (3.28). Hence from the sequence of solutions: $u^m(z), m = 2, 3, \ldots$, we can choose a subsequence $\{u^{m_k}(z)\}$, for convenience denote $\{u^{m_k}(z)\}$ by $\{u^m(z)\}$ again, which uniformly converges to a function $u_0(z)$ in $\bar{D}$, and $u_0(z)$ satisfies the boundary condition (3.5) of Problem P. At last, we need to verify that the function $u_0(z)$ is a solution of equation (3.2). Construct a twice continuously differentiable function $g_n(z)$ as follows

$$g_n(z) = \begin{cases} 1, & z \in \tilde{D}_n = \bar{D}\backslash D_n, \\ 0, & z \in D_{2n}, \end{cases} \quad 0 \leq g_n(z) \leq 1 \text{ in } D_n\backslash D_{2n}, \tag{3.44}$$

where $n(\geq 2)$ is a positive integer. It is not difficult to see that the function $u_n^m(z) = g_n(z)u^m(z)$ is a solution of the following Dirichlet boundary value problem

$$u_{nz\bar{z}}^m - \text{Re}\left[Q^m u_{nzz}^m\right] = C_n^m \text{ in } D, \tag{3.45}$$

$$u_n^m(z) = 0 \text{ on } \Gamma, \tag{3.46}$$

where

$$C_n^m = g_n[\text{Re}\,(A_1^m u_z^m) + A_2^m u^m] + u^m[g_{nz\bar{z}} - \text{Re}\,(Q^m g_{nzz})] + 2\text{Re}\,[g_{nz}u_{\bar{z}}^m - Q^m g_{nz}u_z^m]. \tag{3.47}$$

By using the method from the proof of Theorem 3.5, we can obtain the estimates of $u_n^m(z,t) = u^m(z,t)$ in $\tilde{D}_n$, namely

$$C_\beta^1[u_n^m, \tilde{D}_n] \le M_9, \quad \| u_n^m \|_{W_{p_0}^2(\tilde{D}_n)} \le M_{10}, \tag{3.48}$$

where $\beta = \min(\alpha, 1 - 2/p_0), 2 < p_0 \le p$, $M_j = M_j(q_0, p_0, \alpha, k_0, k_1, M_n', g_n, D_n)$, $j = 9, 10$, here $M_n' = \max_{1 \le m < \infty} C^{1,0}[u^m, \tilde{D}_{2n}]$. Hence from $\{u_n^m(z)\}$, we can choose a subsequence $\{u_{nm}(z)\}$, such that $\{u_{nm}(z)\}, \{u_{nmz}(z)\}$ uniformly converge to $u_0(z), u_{0z}(z)$ and $\{u_{nmzz}(z)\}, \{u_{nmz\bar{z}}(z)\}$ weakly converge to $u_{0zz}(z), u_{0z\bar{z}}(z)$ in $\tilde{D}_n$, respectively. For instance, we take $n = 2$, $u_2^m(z) = u^m(z)$ in $\tilde{D}_2$, $\{u_2^m(z)\}$ has a subsequence $\{u_{m2}(z)\}$ in $\tilde{D}_2$, the limit function of which is $u_0(z)$ in $\tilde{D}_2$. Next, we take $n = 3$, from $\{u_3^m(z)\}$ we can select a subsequence $\{u_{m3}(z)\}$ in $\tilde{D}_3$, the limit function is $u_0(z)$ in $\tilde{D}_3$. Similarly, from $\{u_n^m(z)\}(n > 3)$, we can choose a subsequence $\{u_{mn}(z)\}$ in $\tilde{D}_n$ and the limit of which is $u_0(z)$ in $\tilde{D}_n$. Finally from $\{u_{mn}(z)\}$ in $\tilde{D}_n$, we choose the diagonal sequence $\{u_{mm}(z)\}$ ($m = 2, 3, 4, \ldots$), such that $\{u_{mm}(z)\}, \{u_{mmz}(z)\}$ uniformly converge to $u_0(z), u_{0z}(z)$ and $\{u_{mmzz}(z)\}, \{u_{mmz\bar{z}}(z)\}$ weakly converge to $u_{0zz}(z), u_{0z\bar{z}}(z)$ in any closed subset of D respectively, the limit function $u(z) = u_0(z)$ is just a solution of equation (3.2) in D. This completes the proof.

Theorem 3.7   *If equation (3.2) with $A_2(z) = 0$ satisfies Condition C, then Problem Q for (3.2) has a unique solution.*

Proof   By Theorem 2.3, we choose $D' = D, n = m = 2, z_1 = -1, z_2 = 1$ and $K = 0$, the second linear equation in (3.32) with $A(z) = A_3(z)$ has a unique solution $w_0(z)$, and the function

$$u_0(z) = 2\,\mathrm{Re} \int_{-1}^z w_0(z)dz + b_0 \tag{3.49}$$

is a solution of Problem Q for the first linear equation in (3.32). If $u_0(1) = b' = b_1$, then the solution is just a solution of Problem P for the linear equation (3.2) with $A_2(z) = 0$. Otherwise, $u_0(1) = b' \ne b_1$, we find a solution $u_1(z)$ of Problem Q with the boundary conditions

$$\mathrm{Re}\,[\overline{\lambda(z)}u_{1z}] = 0 \text{ on } \Gamma, \quad \mathrm{Im}\,[\overline{\lambda(z)}u_{1z}]|_{z=0} = 1, \quad u_1(-1) = 0.$$

On the basis of Theorem 3.4, it is clear that $u_1(1) \ne 0$, hence there exists a real constant $d \ne 0$, such that $b_1 = b' + du_1(1)$, thus $u(z) = u_0(z) + du_1(z)$ is just a solution of Problem P for the linear equation (3.2) with $A_2(z) = 0$. As for the quasilinear equation (3.2) with $A_2 = 0$, the existence of solutions of Problem Q and Problem P can be proved by the method as stated in the proof of last theorem.

3.4   The discontinuous oblique derivative problem for elliptic equations in general domains

In this subsection, let D' be a general simply connected domain, whose boundary $\Gamma' = \Gamma_1' \cup \Gamma_2'$, herein $\Gamma_1', \Gamma_2' \in C_\alpha^2(1/2 < \alpha < 1)$ have two intersection points z', z'' with the inner angles $\alpha'\pi, \alpha''\pi \, (0 < \alpha', \alpha'' < 1)$, respectively. We discuss the quasilinear uniformly elliptic equation

$$u_{z\bar{z}} = F(z, u, u_z, u_{zz}), \quad F = \mathrm{Re}\,[Qu_{zz} + A_1 u_z] + A_1 u + A_3, \quad z \in D', \qquad (3.50)$$

in which $F(z, u, u_z, u_{zz})$ satisfy Condition C in D'. There are m points $Z = \{z_1 = z'$ $\ldots, z_{n-1}, z_n = z'', \ldots, z_m\}$ on Γ' arranged according to the positive direction successively. Denote by Γ'_j the curve on Γ' from z_{j-1} to z_j, $j = 1, 2, \ldots, m$, $z_0 = z_m$, and $\Gamma'_j(j = 1, 2, \ldots, m)$ does not include the end points.

Problem P' The discontinuous oblique derivative boundary value problem for (3.50) is to find a continuous solution $w(z)$ in $D^* = \overline{D'}\backslash Z$ satisfying the boundary condition:

$$\frac{1}{2}\frac{\partial u}{\partial \nu} = \mathrm{Re}\,[\overline{\lambda(z)}u_z] = c(z), \quad z \in \Gamma^* = \Gamma'\backslash Z, \quad u(z_j) = b_j, \quad j = 1, \ldots, m, \qquad (3.51)$$

where $\cos(\nu, n) \geq 0$, $\lambda(z), c(z)$ are given functions satisfying

$$C_\alpha[\lambda(z), \Gamma_j] \leq k_0, \quad C_\alpha[|z - z_j|^{\beta_j}c(z), \Gamma_j] \leq k_2, \quad |b_j| \leq k_2, \quad j = 1, \ldots, m, \qquad (3.52)$$

herein $\alpha \, (1/2 < \alpha < 1), k_0, k_2$ are non-negative constants, and assume that $(\beta_j + \gamma_j)/\beta < 1$, $\beta = \min(\alpha, 1 - 2/p_0)/\alpha_0$, $\beta_j(j = 1, \ldots, m)$ are as stated in (1.27), $\alpha_0 = \max(1/\alpha_1, 1/\alpha_2, 1)$. Problem P' with $A_3(z) = 0$ in D, $r(z) = 0$ on Γ' is called Problem P'_0. If $\cos(\nu, n) \not\equiv 0$ on each of $\Gamma_j(j = 1, \ldots, m)$, we choose the index $K = m/2 - 1$ of Problem P', which is defined as that in Subsection 2.3. If $A_2 = 0$ in D, the last point conditions in (3.51) can be replaced by

$$u(z_n) = b_n, \quad \mathrm{Im}\,[\overline{\lambda(z)}u_z]|_{z'_j} = b_j, \quad j = 1, \ldots, n-1, n+1, \ldots, m. \qquad (3.53)$$

Here $z'_j(\not\in Z, j = 1, \ldots, n-1, n+1, \ldots, m) \in \Gamma'$ are distinct points and the condition $\cos(\nu, n) \geq 0$ on Γ' can be canceled. This boundary value problem is called Problem Q'.

Applying a similar method as before, we can prove the following theorem.

Theorem 3.8 *Let equation (3.50) in D' satisfy Condition C similar to before. Then Problem P' and Problem Q' for (3.50) are solvable, and the solution $u(z)$ can be expressed by (3.17), but where $\beta = \min(\alpha, 1 - 2/p_0)/\alpha_0$. Moreover, if $Q(z) = 0$ in D, then the solution $u(z)$ of equation (3.50) possesses the form in (3.17), where $w(z) = \Phi(z)e^{\phi(z)} + \psi(z)$ and $u(z)$ satisfies the estimate*

$$\tilde{C}^1_\delta[u, \bar{D}] = C_\delta[u(z), \bar{D}] + C_\delta[X(z)w(z), \bar{D}] \leq M_{11} = M_{11}(p_0, \beta, \delta, k, D), \qquad (3.54)$$

in which $X(z)$ is given as

$$X(z) = \prod_{j=2, j\neq n}^{m} |z - z_j|^{\eta_j}|z - z_1|^{\eta_1/\alpha'}|z - z_n|^{\eta_n/\alpha''}, \qquad (3.55)$$

*where $\eta_j (j = 1, \ldots , m)$ are as stated in (2.30). Besides the solution of Problem P'
and Problem Q' for (3.50) are unique, if the following condition holds: For any real
functions $u_j(z) \in C^1(D^*), V_j(z) \in L_{p_0}(D)(j = 1, 2)$, the equality*

$$F(z, u_1, u_{1z}, V_1) - F(z, u_1, u_{1z}, V_2) = \mathrm{Re}\,[\tilde{Q}(V_1 - V_2) + \tilde{A}_1(u_1 - u_2)_z] + \tilde{A}_2(u_1 - u_2) \text{ in } D',$$

holds, where $|\tilde{Q}| \leq q_0$ in D', $\tilde{A}_1, \tilde{A}_2 \in L_{p_0}(D')$.

Finally we mention that the above results can be generalized to the case of second
order nonlinear elliptic equation in the form

$$u_{z\bar{z}} = F(z, u, u_z, u_{zz}),$$

$$F = \mathrm{Re}\,[Q(z, u, u_z, u_{zz}) + A_1(z, u, u_z)u_z] + A_2(z, u, u_z)u + A_3(z, u, u_z) \text{ in } D$$

satisfying the conditions similar to Condition C, which is the complex form of the
second order nonlinear elliptic equation

$$\Phi(x, y, u, u_x, u_y, u_{xx}, u_{xy}, u_{yy}) = 0 \text{ in } D$$

with certain conditions (see [86]1)).

4   Boundary Value Problems for Degenerate Elliptic Equations of Second Order in a Simply Connected Domain

This section deals with the oblique derivative problem for the degenerate elliptic
equation of second order in a simply connected domain. We first give a boundedness
estimate of solutions of the oblique derivative problem for the equation, and then by
using the principle of compactness, the existence of solutions for the above oblique
derivative problem is proved.

4.1   Formulation of boundary value problems for degenerate elliptic equations

Let D be a simply connected domain with the boundary $\Gamma' = \Gamma \cup L_0$, where $\Gamma \in
C_\alpha^2 (0 < \alpha < 1)$ in the upper half plane with the end points $-1, 1$ and $L_0 = [-1, 1]$,
and the inner angles of D at $-1, 1$ equal $\alpha_1 \pi, \alpha_2 \pi$, herein $0 < \alpha_1, \alpha_2 < 1$. We consider
the elliptic equation of second order

$$Lu = y^m u_{xx} + u_{yy} + a(x, y)u_x + b(x, y)u_y + c(x, y)u = d(x, y) \text{ in } D, \tag{4.1}$$

here m is a positive number. Its complex form is as follows

$$u_{z\bar{z}} - \mathrm{Re}\,[Q(z)u_{zz} + A_1(z)u_z] + A_2 u = A_3 \text{ in } D, \tag{4.2}$$

where

$$Q(z) = \frac{1 - y^m}{1 + y^m}, \quad A_1(z) = -\frac{a + bi}{1 + y^m}, \quad A_2(z) = -\frac{c}{2(1 + y^m)}, \quad A_3(z) = \frac{d}{2(1 + y^m)}.$$

Suppose that equation (4.2) satisfies the following conditions.

Condition C

The coefficients $A_j(z)(j = 1, 2)$ are continuously differentiable in $\bar{D}$ and satisfy

$$C_\alpha^1[A_j(z), \bar{D}] \leq k_0, \quad j = 1, 2, 3, \quad A_2 = -\frac{c}{2(1 + y^m)} \geq -\frac{c}{4} \geq 0 \text{ on } D, \qquad (4.3)$$

in which $\alpha(0 < \alpha < 1), k_0$ are non-negative constants.

The oblique derivative boundary value problem is as follows.

Problem P In the domain D, find a solution $u(z)$ of equation (4.1), which is continuously differentiable in $\bar{D}$, and satisfies the boundary condition

$$lu = \frac{1}{2}\frac{\partial u}{\partial \nu} + \sigma(z)u = \phi(z), \quad z \in \Gamma, \quad \frac{\partial u}{\partial y} = \psi(x), x \in L_0,$$

$$u(-1) = b_0, u(1) = b_1,$$
$$(4.4)$$

where ν is any unit vector at every point on $\Gamma, \cos(\nu, n) \geq 0, \sigma(z) \geq \sigma_0 > 0, n$ is the unit outer normal at every point on $\Gamma, \lambda(z) = \cos(\nu, x) - i\cos(\nu, y)$, and $\lambda(z), \phi(z)$, $\psi(x)$ are known functions and b_0, b_1 are known constants satisfying the conditions

$$C_\alpha^1[\eta, \Gamma] \leq k_0, \eta = \lambda, \sigma, \quad C_\alpha^1[\phi, L_0] \leq k_0, \quad C_\alpha^1[\psi, L_0] \leq k_0, \quad |b_0|, |b_1| \leq k_0, \quad (4.5)$$

in which $\alpha(1/2 < \alpha < 1), k_0, \sigma_0$ are non-negative constants. Problem P with the conditions: $A_3(z) = 0$ in $\bar{D}$, $\phi(z) = 0$ on Γ, $\psi(z) = 0$ on L_0 and $b_0 = b_1 = 0$ is called Problem P_0. If $\cos(\nu, n) = 1$, here n is a outward normal vector on Γ, then Problem P is the Neumann boundary value problem (Problem N), and if $\cos(\nu, n) > 0, \sigma(z) = 0$ on Γ, then Problem P is the regular oblique derivative problem, i.e. third boundary value problem (Problem O), in this case we choose $\sigma(z) > 0$ on Γ. If $\cos(\nu, n) = 0$ and $\sigma(z) = 0$ on Γ, then from (4.4), we can derive

$$u(z) = 2\,\text{Re}\int_{-1}^{z} u_z dz + b_0 = r(z) \text{ on } \Gamma, \quad u(1) = b_1 = 2\,\text{Re}\int_{-1}^{1} u_z dz + b_0. \qquad (4.6)$$

In this case, Problem P is called Problem D. In the following, there is no harm in assuming $d(z) = 0$ in (4.1).

4.2 A priori estimates of solutions for Problem P for (4.1)

First of all, we give a lemma and then give a priori estimate of boundedness of solutions of Problem P for (4.1).

Lemma 4.1 *Suppose that equation (4.1) or (4.2) satisfies Condition C and $Lu \geq 0$ (or $Lu \leq 0$) in D, if the solution $u(z) \in C^2(D) \cap C(\bar{D})$ of (4.1) attains its positive maximum (or negative minimum) at a point $x_0 \in (-1,1)$, and $\max_\Gamma u(z) < u(x_0)$ (or $\min_\Gamma u(z) > u(x_0)$) on Γ, then*

$$\lim_{y \to 0} \frac{\partial u(x_0, y)}{\partial y} < 0 \ (\text{or} \ \lim_{y \to 0} \frac{\partial u(x_0, y)}{\partial y} > 0), \qquad (4.7)$$

if the limit exists.

Proof Assume that the first inequality is not true, namely

$$\lim_{y \to 0} \frac{\partial u(x_0, y)}{\partial y} = M' \geq 0. \qquad (4.8)$$

Obviously $M' = 0$. Denote $M = u(x_0)$, $B = \max_{\bar{D}} |b(z)|$ and by d the diameter of D. Thus there exists a small positive constant $\varepsilon < M$ such that $\max_\Gamma u(z) \leq M - \varepsilon$. Making a function

$$v(z) = \frac{\varepsilon u(z)}{(Me^{Bd} - \varepsilon e^{By})},$$

we have

$$v(z) \leq \frac{\varepsilon(M - \varepsilon)}{Me^{Bd} - \varepsilon e^{Bd}} < \frac{\varepsilon M}{Me^{Bd} - \varepsilon} \ \text{on} \ \Gamma, \quad v(x) \leq v(x_0) = \frac{\varepsilon M}{Me^{Bd} - \varepsilon} \ \text{on} \ L_0. \qquad (4.9)$$

Noting that $Lu \geq 0$, the function $v(x, y)$ satisfies the inequality

$$y^m v_{xx} + v_{yy} + a(x, y)v_x + \tilde{b}(x, y)v_y + \tilde{c}(x, y)v \geq 0 \ \text{in} \ D,$$

where $\tilde{b} = b - 2\varepsilon Be^{By}/(Me^{Bd} - \varepsilon e^{By})$, $\tilde{c}(x, y) = c - \varepsilon(B+b)Be^{By}/(Me^{Bd} - \varepsilon e^{By}) \leq 0$ in D. According to the above assumption, we get

$$\lim_{y \to 0} \frac{\partial v(x_0, y)}{\partial y} = \frac{\varepsilon^2 BM}{(Me^{Bd} - \varepsilon)^2} > 0.$$

Hence $v(x, y)$ attains its maximum in D, but from (4.9), it is impossible. This proves the first inequality in (4.7). Similarly we can prove the second inequality in (4.7).

Now we choose a positive constant $\eta < 1$, and consider the equation

$$L_\eta u = (y+\eta)^m u_{xx} + u_{yy} + a(x, y)u_x + b(x, y)u_y + c(x, y)u = d \ \text{in} \ D. \qquad (4.10)$$

It is easy to see that (4.10) is a uniformly elliptic equation in D. From Theorem 3.6, we can derive that for every one of $\eta = 1/n > 0 \, (n = 2, 3, \ldots)$, there exists a solution $u_n(z)$ of Problem D for equation (4.10). In the following, we shall give some estimates of the solution $u_n(z)$.

Lemma 4.2 *If Condition C holds, then any solution $u_n(z)$ of Problem P for (4.10) with $d = 0$ satisfies the estimate*

$$C[u_n(z), \bar{D}] \leq M_{12} = M_{12}(\alpha, k_0, D), \qquad (4.11)$$

where M_{12} is a non-negative constant.

Proof We first discuss Problem D and choose two positive constants c_1, c_2 such that

$$c_1 \geq c_2 + \max_\Gamma |r(z)| + \max_{\bar{D}} e^{c_2 y}, \quad c_2 > \max_{L_0} |\psi(x)| + \max_{\bar{D}} |b| + 2 \max_{\bar{D}} |d| + 1,$$

and make a transformation of function $v(z) = c_1 - e^{c_2 y} \pm u_n(z)$, thus we have

$$L_\eta v \leq -c_2(c_2 + b)e^{c_2 y} + c(c_1 - e^{c_2 y}) + 2 \max_{\bar{D}} |d| < 0, \quad z = x + iy \in D,$$

$$v > 0 \text{ on } \Gamma, \quad v_y = \frac{\partial v}{\partial y} = -c_2 e^{c_2 y} \pm \psi(z) < 0 \text{ on } L_0, \tag{4.12}$$

by the extremum principle for elliptic equations, the function $v(z)$ cannot take the negative minimum in D, hence

$$v(z) = c_1 - e^{c_2 y} \pm u_n(z) \geq 0, \quad \text{i.e. } c_1 \geq e^{c_2 y} \mp u_n(z) \text{ in } \bar{D}, \tag{4.13}$$

hence $|u_n(z)| \leq c_1 - e^{c_2 y} \leq c_1 = M_{12}$.

For other case, we introduce an auxiliary function $v(z) = c_1 - e^{c_2 y} \pm u_n(z)$, where c_1, c_2 are two positive constants satisfying the conditions

$$c_2 > \max_{\bar{D}} |b(z)| + \max_{L_0} |\psi(x)| + \max_{\bar{D}} e^{c_2 y} + 2 \max_{\bar{D}} |d|,$$

$$c_1 > c_2 + \max_{\bar{D}} e^{c_2 y} \left(1 + \frac{c_2}{\sigma_0}\right) + \max_\Gamma \frac{|\phi(z)|}{\sigma_0}. \tag{4.14}$$

We can verify that the function $v(z)$ satisfies the conditions

$$L_\eta v < 0 \text{ in } D, \quad lv > 0 \text{ on } \Gamma, \quad v_y < 0 \text{ on } L_0, \tag{4.15}$$

hence $v(z)$ cannot attain the negative minimum in $\bar{D}$. Thus $|u_n(z)| \leq c_1 - e^{c_2 y} \leq c_1 = M_{12}$. This completes the proof.

Secondly from the sequence of solutions: $\{u_n(z)\}$ of Problem P for equation (4.10), we can choose a subsequence $\{u_{n_k}(z)\}$, which uniformly converges to a solution $u_*(z)$ of (4.1) in any closed subset of $\bar{D} \backslash L$. In fact by Condition C and the estimate (4.11), we can derive the estimate of the solution $u_n(z)$ as follows

$$C_\beta^1[u_n(z), \bar{D}] \leq M_{13} = M_{13}(\beta, k_0, D, \eta), \tag{4.16}$$

where $\eta = 1/n > 0$ and $\beta \, (0 < \beta \leq \alpha)$ is a constant.

Lemma 4.3 *If Condition C holds, then any solution $u_n(z)$ of Problem P for (4.10) satisfies the estimate (4.16).*

From the above lemma, we can derive that the limit function $u_*(z)$ of $\{u_{n_k}(z)\}$ satisfies the first boundary condition in (4.4). In order to prove that $u_*(z)$ satisfies the second boundary condition in (4.4), we write the similar results in [24]1) as a lemma.

Lemma 4.4   *Suppose that Condition C holds and $0 < m < 2$, or $m \geq 2$,*

$$a(x,y) = O(y^{m/2-1+\varepsilon}), \quad a_y = O(y^{m/2-2+\varepsilon}), \qquad (4.17)$$

where ε is sufficiently small positive number. Then any solution $u_m(z)$ of Problem P for (4.10) with $d = 0$ satisfies the estimate

$$|u_{ny}|, |((y+\eta)^{m+\varepsilon} - \eta^{m+\varepsilon})u_{nx}^2| \leq M_{14} = M_{14}(\alpha, k_0, D) \text{ in } \overline{R_{n,2\delta_0}}, \qquad (4.18)$$

where $R_{n,\delta} = \{|x - x_0| < \rho - \delta, 0 < y < \delta_l\}(\overline{R_{n,\delta}} \subset D), x_0 \in (-1,1), \delta_0, \delta_1, \delta, \rho$ $(0 < \delta \leq 2\delta_0 < \rho, \delta_1 < 1/n)$ are small positive constants, and M_{14} is a non-negative constant.

Proof   (1) First of all, we prove the estimate

$$(u_y)^2 \leq [(y+\eta)^{m+\varepsilon_1} - \eta^{m+\varepsilon_1}](u_x)^2 + M_{15} \text{ in } \overline{R_{n,\delta_0}} = D^*, \qquad (4.19)$$

in which $\varepsilon_1(< \varepsilon)$, M_{15} are non-negative constants. $f = f(x) = X^4 = [\rho^2 - (x - x_0)^2]^4$, $g = g(x) = X^2 = [\rho^2 - (x - x_0)^2]^2$ are functions of x and $F = \eta^{m+\varepsilon_1} - Y^{m+\varepsilon_1}$, $G = 1 - Y^{\varepsilon_1}$, $H = -Y^{\varepsilon_1}$ are functions of $Y = y + \eta$, and introducing an auxiliary function

$$v(z) = f[F(u_x)^2 + G(u_y)^2] + gu^2 + H \text{ in } D^*, \qquad (4.20)$$

if $v(z)$ attains a positive maximum value at a point $z^* \in D^*$, then

$$v(z) > 0, \ v_x = v_y = 0, \ \tilde{L}_\eta(v) = L_\eta(v) + cv \leq 0 \text{ at } z^*. \qquad (4.21)$$

From (4.20), we get

$$v_x = 2f[Fu_xu_{xx} + Gu_yu_{xy}] + 2guu_x + f'[F(u_x)^2 + G(u_y)^2] + g'u^2 = 0,$$

$$v_y = 2f[Fu_xu_{xy} + Gu_yu_{yy}] + 2guu_y + f[F'(u_x)^2 + G'(u_y)^2] + H' = 0,$$

$$\tilde{L}_\eta(v) = 2f[Fu_xL_\eta(u_x) + Gu_yL_\eta(u_y)] + 2guL_\eta(u) + 2fF[Y^m(u_{xx})^2 + (u_{xy})^2] \qquad (4.22)$$

$$+ 2fG[Y^m(u_{xy})^2 + (u_{yy})^2] + 2g[Y^m(u_x)^2 + (u_y)^2] + 4Y^m f'[Fu_xu_{xx} + Gu_yu_{xy}]$$

$$+ 4f[F'u_xu_{xy} + G'u_yu_{yy}] + Y^m f''[F(u_x)^2 + G(u_y)^2] + f[F''(u_x)^2 + G''(u_y)^2]$$

$$+ 4Y^m g'uu_x + Y^m g''u^2 + H'' + af'[F(u_x)^2 + G(u_y)^2]$$

$$+ ag'u^2 + bf[F'(u_x)^2 + G'(u_y)^2] + bH' + 2cH,$$

in which $Y = y + \eta$, and from (4.10), we obtain

$$L_\eta(u_x) = -(a_xu_x + b_xu_y + c_xu),$$

$$L_\eta(u_y) = -(mY^{m-1}u_{xx} + a_yu_x + b_yu_y + c_yu), \qquad (4.23)$$

$$2fFY^m(u_{xx})^2 = 2fFY^{-m}(u_{yy} + au_x + bu_y + cu)^2,$$

and then we have

$$2fGu_yL_\eta(u_y) = -\frac{m}{Y}[2guu_y + f[F'(u_x)^2 + G'(u_y)^2] + H'] - \frac{2mfF}{Y}$$

$$\times u_xu_{xy} + \frac{2mfG}{Y}u_y(au_x + bu_y + cu) - 2fGu_y(a_yu_x + b_yu_y + c_yu). \tag{4.24}$$

Substituting (4.23),(4.24) into $\tilde{L}_\eta(v)$, it is not difficult to derive

$$\frac{1}{f}\tilde{L}_\eta(v) = 2(FY^{-m} + G)(Y^m(u_{xy})^2 + (u_{yy})^2) + 2(2F' - \frac{mF}{Y})u_xu_{xy}$$

$$+4[G'u_y + FY^{-m}(au_x + bu_y + cu)]u_{yy} - 2Fu_x(a_xu_x + b_xu_y + c_xu)$$

$$-2Gu_y(a_yu_x + b_yu_y + c_yu) + \frac{2mG}{Y}u_y(au_x + bu_y + cu)$$

$$+2FY^{-m}(au_x + bu_y + cu)^2 + \left[Y^m\left(\frac{f''}{f} - 2\frac{f'^2}{f^2}\right) + a\frac{f'}{f}\right] \tag{4.25}$$

$$\times[F(u_x)^2 + G(u_y)^2] + \left(-\frac{m}{Y} + b\right)[F'(u_x)^2 + G'(u_y)^2] + F''(u_x)^2$$

$$+G''(u_y)^2 + 2\frac{g}{f}[Y^m(u_x)^2 + (u_y)^2] + 4Y^m\left(\frac{g'}{f} - \frac{f'g}{f^2}\right)uu_x - 2\frac{mg}{fY}uu_y$$

$$+\left[Y^m\left(\frac{g''}{f} - 2\frac{f'g'}{f^2}\right) + a\frac{g'}{f}\right]u^2 + \frac{H'' + (-m/Y + b)H' + 2cH}{f}.$$

Moreover by (4.11),(4.17),(4.20) and (4.25), we obtain

$$\frac{1}{f}\tilde{L}_\eta(v) = (2+o(1))[Y^m(u_{xy})^2 + (u_{yy})^2] + O\Big\{Y^{m+\varepsilon_1-1}|u_xu_{xy}|$$

$$+Y^{m-2+2\varepsilon+\varepsilon_1}|u_x|^2 + (Y^{m/2-1+\varepsilon+\varepsilon_1}|u_x| + Y^{\varepsilon_1-1}|u_y| + Y^{\varepsilon_1})|u_{yy}|$$

$$+Y^{m/2-2+\varepsilon}|u_xu_y| + Y^{-1}(u_y)^2 + Y^{m/2-1+\varepsilon+\varepsilon_1}|u_x| + Y^{-1}|u_y| + Y^{\varepsilon_1}$$

$$+\left(\frac{Y^m}{X^2} + \frac{Y^{m/2-1+\varepsilon}}{X}\right)[Y^{m+\varepsilon_1}(u_x)^2 + (u_y)^2]\Big\}$$

$$+(1-\varepsilon_1)(m+\varepsilon_1)^2(1+o(1))Y^{m-2+\varepsilon_1}(u_x)^2 \tag{4.26}$$

$$+\varepsilon_1(m+1-\varepsilon_1)(1+o(1))Y^{\varepsilon_1-2}(u_y)^2$$

$$+2\frac{Y^m}{X^2}(u_x)^2 + 2\frac{(u_y)^2}{X^2} + O\left(\frac{Y^m}{X^3}|u_x| + \frac{Y^{-1}}{X^2}|u_y|\right)$$

$$+\frac{\varepsilon_1(m+1-\varepsilon_1)(1+o(1))}{X^4}Y^{\varepsilon_1-2}.$$

When $0 < \varepsilon_1 < \min(\varepsilon,1)$, it is easy to see that the right-hand side of (4.26) is positive, which contradicts (4.21), hence $v(z)$ cannot have a positive maximum in D^*.

On the basis of the estimate (4.11), we see that $v(s)$ on the upper boundary $\{|x - x_0| < \rho, y = 1/n\}$ of D^* is bounded, and $v(x) \leq fG\psi^2 + gM_{12}^2$ on the lower boundary $\{|x - x_0| < \rho, y = 0\}$ of D^*, moreover $v(x) < 0$ on the left-hand side and right-hand side $\{|x - x_0| = \rho, 0 < y < 1/n\}$ of D^*. Thus the estimate (4.19) is derived.

(2) Now we give the estimate

$$|((y+\eta)^{m+\varepsilon}-\eta^{m+\varepsilon})u_{nx}^2| \le M_{16} \in \overline{R_{n,\delta_0}} = D^*, \tag{4.27}$$

in which M_{16} is independent of η. In fact, we introduce the auxiliary function (4.20), where we choose that $f=f(x)=X^4=[(\rho-\delta_0)^2-(x-x_0)^2]^4, g=g(x)=X^2=[(\rho-\delta_0)^2-(x-x_0)^2]^2$, and $F=Y^{m+\varepsilon_1}-\eta^{m+\varepsilon_1}, G=Y^{\varepsilon_2}, H=Y^{\varepsilon_3}$, herein $Y=y+\eta, \varepsilon_2, \varepsilon_3$ are positive constants satisfying $0<2\varepsilon_3<\varepsilon_2\le\varepsilon_1/2$. If $v(z)$ attains a positive maximum value at a point $z^* \in D^*$, then we have (4.21). Substituting (4.10),(4.20),(4.23) into (4.22), we get

$$\frac{\tilde{L}_\eta(v)}{f} = 2F[Y^m(u_{xx})^2+(u_{xy})^2]+2G[Y^m(u_{xy})^2 \\ +(u_{yy})^2]+2\frac{g}{f}[Y^m(u_x)^2+(u_y)^2]+\Sigma, \tag{4.28}$$

in which

$$\Sigma = 4F'u_xu_{xy}+2\left(2G'+\frac{mG}{Y}\right)u_yu_{yy}-2Fu_x(a_xu_x+b_xu_y+c_xu)$$

$$-2Gu_y(a_yu_x+b_yu_y+c_yu)+2\frac{mG}{Y}u_y(au_x+bu_y+cu)$$

$$+\left[Y^m\left(\frac{f''}{f}-2\frac{f'^2}{f^2}\right)+a\frac{f'}{f}\right][F(u_x)^2+G(u_y)^2]+b[F'(u_x)^2 \tag{4.29}$$

$$+G'(u_y)^2]+F''(u_x)^2+G''(u_y)^2+4Y^m\left(\frac{g'}{f}-\frac{f'g}{f^2}\right)uu_x$$

$$+\left[Y^m\left(\frac{g''}{f}-2\frac{f'g'}{f^2}\right)+a\frac{g'}{f}\right]u^2+\frac{H''+bH'+2cH}{f}.$$

From (4.11),(4.17),(4.20) and (4.29), it follows that

$$\Sigma = O\left\{Y^{m+\varepsilon_1-1}|u_xu_{xy}|+Y^{\varepsilon_2-1}|u_yu_{yy}|+Y^{m+\varepsilon_1-2}|u_x|^2+Y^{m/2-2+\varepsilon+\varepsilon_2}\right.$$

$$\times|u_xu_y|+Y^{\varepsilon_2-2}|u_y|^2+Y^{m+\varepsilon_1}|u_x|+Y^{\varepsilon_2-1}|u_y|+\left(\frac{Y^m}{X^2}+\frac{Y^{m/2-1+\varepsilon}}{X}\right)$$

$$\times(Y^{m+\varepsilon_1}|u_x|^2+Y^{\varepsilon_2}|u_y|^2)+\frac{Y^m}{X^3}|u_x|+\frac{Y^{\varepsilon_3-2}}{X^4}\right\}\ge-G(Y^m|u_{xy}|^2+|u_{yy}|^2) \tag{4.30}$$

$$-2\frac{g}{f}(Y^m|u_x|^2+|u_y|^2)+O\left(Y^{m+\varepsilon_1-2}|u_x|^2+Y^{\varepsilon_2-2}|u_y|^2+\frac{Y^{\varepsilon_3-2}}{X^4}\right).$$

By (4.28),(4.30), if we can verify the following inequality

$$G[Y^m(u_{xy})^2+(u_{yy})^2]+O\left(Y^{m+\varepsilon_1-2}|u_x|^2+Y^{\varepsilon_2-2}|u_y|^2+\frac{Y^{\varepsilon_3-2}}{X^4}\right)>0, \tag{4.31}$$

then the inequality $\tilde{L}_\eta(v) > 0$. Noting that F', G', H' are positive, and from (4.11), (4.20), (4.22), we have

$$2|Fu_x u_{xy} + Gu_y u_{yy}| \geq F'|u_x|^2 + G'|u_y|^2 - 2\frac{g}{f}|uu_y| + \frac{H'}{f}$$

$$\geq F'|u_x|^2 + \frac{1}{f}(H' - \frac{u^2}{G'^2}) \geq F'|u_x|^2 + \frac{1+o(1)}{f}H'.$$

Hence

$$(F^2|u_x|^2 + Y^m G^2|u_y|^2)(Y^m|u_{xy}|^2 + |u_{yy}|^2)$$

$$= Y^m(Fu_x u_{xy} + Gu_y u_{yy})^2 + (Y^m Gu_y u_{xy} - Fu_x u_{yy})^2 \tag{4.32}$$

$$\geq \frac{Y^m}{4}(F'|u_x|^2 + \frac{1+o(1)}{f}H')^2.$$

By (4.19),(4.20),(4.32), we obtain

$$(F^2|u_x|^2 + Y^m G^2|u_y|^2)\left[Y^m|u_{xy}|^2 + |u_{yy}|^2\right.$$

$$\left. + \frac{1}{G}O\left(Y^{m+\varepsilon_1-2}|u_x|^2 + Y^{\varepsilon_2-2}|u_y|^2 + \frac{Y^{\varepsilon_3-2}}{X^4}\right)\right]$$

$$\geq \frac{Y^m}{4}(mY^{m+\varepsilon_1-1}|u_x|^2 + \frac{\varepsilon_3+o(1)}{X^4}Y^{\varepsilon_3-1})^2 \tag{4.33}$$

$$+ Y^{m+\varepsilon_2}(Y^{m+\varepsilon_1+\varepsilon_2}|u_x|^2 + Y^{\varepsilon_2}M_{15})Y^{-\varepsilon_2}O\left(Y^{m+\varepsilon_1-2}|u_x|^2 + \frac{Y^{\varepsilon_2-2}}{X^4}\right) > 0.$$

From (4.22), we see that u_x, u_y cannot simultaneously be zero. By (4.33), we have (4.31), such that $\tilde{L}_\eta(v) > 0$ holds. This contradicts (4.21). Therefore $v(z)$ cannot attain a positive maximum in D^*.

On the basis of (4.11),(4.20) and the boundary condition (4.4), we see that $v \leq fG\psi^2 + gM_{12}^2 + H$ on the lower boundary of D^* is uniformly bounded. Moreover the function $v(z)$ is uniformly bounded on the upper, left-hand and right-hand boundaries of D^*. Thus the estimates in (4.18) are derived.

Now we prove a theorem as follows.

Theorem 4.5 *Suppose that Condition C, (4.17) hold and $a_x + c \leq 0$ in $\bar{D}$. Then any solution $u_n(z)$ of Problem P for (4.10) satisfies the estimate*

$$|u_x| \leq M_{17} = M_{17}(\alpha, k_0, D) \text{ in } D, \tag{4.34}$$

where we assume that the inner angles $\alpha_j \pi (j=1,2)$ of D at $z=-1,1$ satisfy the conditions $0 < \alpha_j (\neq 1/2) < 1, j=1,2$ and M_{17} is a non-negative constant.

Proof   We find the derivative with respect to x to equation (4.10), and obtain

$$(y+\eta)^m u_{xxx} + u_{xyy} + a(x,y)u_{xx} + b(x,y)u_{xy} + [a_x + c(x,y)]u_x = F(x,y),$$
$$F = f(x,y) - b_x u_y - c_x u \text{ in } D. \tag{4.35}$$

On the basis of Lemmas 4.1 and 4.4, we have

$$|F(x,y)| = |f(x,y) - b_x u_y - c_x u| \le M_{18} < \infty \text{ in } D,$$

and equation (4.35) can be seen as a elliptic equation of u_x, and the solution u_x satisfies the boundary conditions

$$\frac{\partial u}{\partial s} = \cos(s,x)u_x + \cos(s,y)u_y = \frac{\partial r(z)}{\partial s} \text{ on } \Gamma, \quad (u_x)_y = \psi'(x) \text{ on } L_0, \tag{4.36}$$

in which s is the tangent vector at every point Γ. Noting that the angles $\alpha_j \pi (j=1,2)$ satisfy the conditions $0 < \alpha_j (\ne 1/2) < 1, j = 1,2$, it is easy to see that $\cos(s,x) \ne 0$ at $z = -1, 1$. Thus the first boundary condition in (4.6) can be rewritten in the form

$$u_x = R(z) = -\frac{\cos(s,y)}{\cos(s,x)}u_y + \frac{1}{\cos(s,x)}\frac{\partial r(z)}{\partial s} \text{ on } \Gamma, \tag{4.37}$$

here $R(z)$ is a bounded function in the neighborhood $(\subset \Gamma)$ of $z = -1, 1$, hence by the method in the proof of Lemma 4.2, we can prove that the estimate (4.34) holds. As for $\cos(s,x) = 0$ at $z = -1$ or $z = 1$, the problem remains to be solved?

Theorem 4.6   *Suppose that Condition C and (4.17) hold. Then Problem P for (4.1) or (4.2) has a unique solution.*

Proof   As stated before, for a sequence of positive numbers: $\eta = 1/n, n = 2, 3, \ldots$, we have a sequence of solutions: $\{u_n(z)\}$ of the corresponding equations (4.10) with $\eta = 1/n (n = 2, 3, \ldots)$, which satisfy the estimate (4.16), hence from $\{u_n(z)\}$, we can choose a subsequence $\{u_{n_k}(z)\}$, which converges to a solution $u_0(z)$ of (4.2) in $D \cup \Gamma$ satisfying the first boundary condition in (4.4). It remains to prove that $u_0(z)$ satisfies the other boundary condition in (4.4). For convenience, we denote $u_{n_k}(z)$ by $u(z)$, x_0 is any point in $-1 < x_0 < 1$, and give a small positive number β, there exists a sufficiently small positive number δ, such that $|\psi(x) - \psi(x_0)| < \beta$ when $|x - x_0| < \delta$. Moreover we consider an auxiliary function

$$v(z) = F(u_x)^2 \pm u_y + G + f, \quad G = -Cy^{\varepsilon_2} - \tau \mp \psi(x_0), \quad f = -C(x-x_0)^2,$$

$$F = \begin{cases} Y^{m+1+\varepsilon_2} - \eta^{m+1+\varepsilon_2}, & 0 < m < 1, \\ Y^{m+\varepsilon_1} - (m+\varepsilon_1)\eta^{m-1+\varepsilon_1}Y + (m+\varepsilon_1-1)\eta^{m+\varepsilon_1}, & m \ge 1, \end{cases} \tag{4.38}$$

where $Y = y + \eta, \eta = 1/n, \varepsilon_2 (0 < \varepsilon_2 \le \varepsilon_1/3)$ are positive constants and C is an undetermined positive constants. We first prove that $v(z)$ cannot attain its positive maximum in $D^* = \{|x-x_0|^2 + y^2 < \sigma^2, y > 0\}$. Otherwise there exists a point z^*, such that

$v(z^*) = \max_{\overline{D^*}} v(z) > 0$, and then

$$v_x = 2F u_x u_{xx} \pm u_{xy} + f' = 0, \quad v_y = 2F u_x u_{xy} \pm u_{yy} + F'(u_x)^2 + G' = 0,$$

$$\tilde{L}_\eta(v) = 2F u_x L_\eta(u_x) \pm L_\eta(u_y) + 2F[Y^m(u_{xx})^2 + (u_{xy})^2] + 4F' u_x u_{xy}$$

$$+ [F'' + bF' + cF](u_x)^2 \pm c u_y + Y^m f''$$

$$+ a f' + 2cf + G'' + bG' + 2cG,$$

(4.39)

and from (4.23) and (4.39), we obtain

$$\pm L_\eta(u_y) = -\frac{m}{Y}[2F u_x u_{xy} + F'(u_x)^2 + G']$$

$$\pm \frac{m}{Y}(a u_x + b u_y + cu) \mp (a_y u_x + b_y u_y + c_y u).$$

(4.40)

Moreover by (4.23) and (4.39)–(4.40), we have

$$\tilde{L}_\eta(v) = -2F u_x(a_x u_x + b_x u_y + c_x u) - \frac{m}{Y}[2F u_x u_{xy} + F'(u_x)^2 + G'$$

$$\mp (a u_x + b u_y + cu)] \mp (a_y u_x + b_y u_y + c_y u)$$

$$+ 2F[Y^m(u_{xx})^2 + (u_{xy})^2] \mp 4F' u_x(2F u_x u_{xx} + f')$$

$$+ [F'' + bF' + cF](u_x)^2 \pm c u_y + Y^m f'' + a f' + 2cf + G'' + bG' + 2cG$$

$$= 2FY^m[u_{xx} \mp 2F' Y^{-m}(u_x)^2]^2 + 2F\left(u_{xy} - \frac{m}{2Y} u_x\right)^2$$

(4.41)

$$+ [F'' - \frac{m}{Y}F' + bF' + cF - 8F'^2 Y^{-m} F(u_x)^2](u_x)^2$$

$$+ \left[-2F(b_x u_y + c_x u) \pm \frac{ma}{Y} \mp a_y \mp 4F' f'\right] u_x + G'' - \frac{m}{Y}G' + bG'$$

$$+ 2cG - \left(\frac{m^2}{2Y^2} + 2a_x\right)F(u_x)^2 \pm \frac{m}{Y}(b u_y + cu) \mp (b_y u_y + c_y u)$$

$$\pm c u_y + Y^m f'' + a f' + 2cf.$$

Choosing a sufficiently small positive number σ, such that the domain $D^* = \{|x - x_0|^2 + y^2 < \sigma^2\} \cap \{y > 0\} \subset D, \sigma < \min[\rho - 2\delta_0, \delta_1]$, where δ_0, δ_1 are constants as stated in Lemma 4.4, and $|\psi(z) - \psi(x_0)| < \tau$, we can obtain

$$\tilde{L}_\eta(v) \geq \varepsilon_2(m+1+\varepsilon_2)(1+o(1))Y^{m+\varepsilon_2-1}(u_x)^2 + O\left(\frac{1}{Y}\right)|u_x|$$

$$+ C\varepsilon_2(m+1-\varepsilon_2)(1+o(1))y^{\varepsilon_2-2} > 0 \text{ if } 0 < m < 1,$$

(4.42)

and

$$\tilde{L}_\eta(v) \geq (m+\varepsilon_1)(m+\varepsilon_1-1)Y^{m+\varepsilon_1-2}(u_x)^2 + O(Y^{m/2+\varepsilon-2})|u_x|$$

$$+ C\varepsilon_2(m+1-\varepsilon_2)(1+o(1))Y^{\varepsilon_2-2} > 0 \text{ if } m \geq 1, \quad \varepsilon - \varepsilon_1,$$

(4.43)

in which we use Lemmas 4.2 and 4.4, the conditions (4.17), (4.38), (4.41) and

$$F(u_x)^2 = O(Y^{2\varepsilon_2}), F'(u_x)^2 = O(Y^{-1+2\varepsilon_2}) \text{ if } m \geq 1.$$

It is clear that (4.42),(4.43) contradict (4.21), hence $v(z)$ cannot attain a positive maximum in D^*. From (4.38), we get

$$v(z) = F(u_x)^2 \pm [u_y - \psi(x_0)] - \tau - C[(x - x_0)^2 + y^{\varepsilon_2}]. \tag{4.44}$$

Moreover it is easy to see that $v(z) < 0$ on the boundary of D^*, provided that the constant C is large enough. Therefore $v(z) \leq 0$ in D^*. From $F \geq 0$ and (4.42)–(4.44), the inequality

$$\pm [u_y - \psi(x_0)] - \tau - C[(x - x_0)^2 + Y^{\varepsilon_2}] \leq 0, \text{ i.e.}$$

$$|u_y - \psi(x_0)| \leq \tau + C[(x - x_0)^2 + Y^{\varepsilon_2}] \text{ in } D^* \tag{4.45}$$

is derived. Firstly let $\eta \to 0$ and then let $z \to x_0, \tau \to 0$, we obtain $\lim \partial u / \partial y \to \psi(x_0)$. Similarly we can verify $\lim_{z(\in D^*) \to x_0} u_y = \psi(x_0)$ when $x_0 = -1, 1$. Besides we can also prove that $u(z) \to u(x_0)$ as $z(\in D^*) \to x_0$, when $x_0 = -1, 1$. This shows that the limit function $u(z)$ of $\{u_n(z)\}$ is a solution of Problem P for (4.1).

Now we prove the uniqueness of solutions of Problem P for (4.1), it suffices to verify that Problem P_0 has no non-trivial solution. Let $u(z)$ be a solution of Problem P_0 for (4.1) with $d = 0$ and $u(z) \not\equiv 0$ in D. Similarly to the proof of Theorem 3.4, we see that its maximum and minimum cannot attain in $D \cup \Gamma$. Moreover by using Lemma 4.1, we can prove that the maximum and minimum cannot attain at a point in $(-1, 1)$. Hence $u(z) \equiv 0$ in D.

Finally we mention that for the degenerate elliptic equation

$$K(y)u_{xx} + u_{yy} = 0, \quad K(0) = 0, \quad K'(y) > 0 \text{ in } \bar{D}, \tag{4.46}$$

which is similar to equation (4.1) satisfying Condition C and other conditions as before, hence any solution $u(z)$ of Problem P_0 for (4.46) satisfies the estimates (4.11),(4.18) and (4.34) in D, provided that the inner angles $\alpha_j \pi (j = 1, 2)$ of D at $z = -1, 1$ satisfy the conditions $0 < \alpha_j (\neq 1/2) < 1, j = 1, 2$. Equation (4.46) is the Chaplygin equation in elliptic domain. Besides, oblique derivative problems for the degenerate elliptic equations of second order

$$u_{xx} + y^m u_{yy} + a(x, y)u_x + b(x, y)u_y + c(x, y)u = f(x, y) \text{ in } D,$$

$$y^{m_1} u_{xx} + y^{m_2} u_{yy} + a(x, y)u_x + b(x, y)u_y + c(x, y)u = f(x, y) \text{ in } D$$

needs to be considered, where m, m_1, m_2 are non-negative constants.

The references for this chapter are [3],[6],[11],[15],[18],[23],[24],[30],[33],[38],[39], [40],[46],[48],[50],[53],[58],[60],[65],[67],[76],[78],[80],[81],[82], [85],[86],[94],[96],[99].

CHAPTER IV

FIRST ORDER COMPLEX EQUATIONS OF MIXED TYPE

In this chapter, we introduce the Riemann–Hilbert boundary value problem for first order complex equation of mixed (elliptic-hyperbolic) type in a simply connected domain. We first prove uniqueness and existence of solutions for the above boundary value problem, and then give a priori estimates of solutions for the problem, finally discuss the solvability of the above problem in general domains. The results in this chapter will be used in the following chapters.

1   The Riemann–Hilbert Problem for Simplest First Order Complex Equation of Mixed Type

In this section, we discuss the Riemann–Hilbert boundary value problem for the simplest mixed complex equation of first order in a simply connected domain. Firstly, we verify a unique theorem of solutions for the above boundary value problem. Moreover, the existence of solutions for the above problem is proved.

1.1   Formulation of the Riemann–Hilbert problem for the simplest complex equation of mixed type

Let D be a simply connected bounded domain in the complex plane $\mathbb{C}$ with the boundary $\partial D = \Gamma \cup L$, where $\Gamma(\subset \{y > 0\}) \in C_\alpha(0 < \alpha < 1)$ with the end points $z = 0, 2$ and $L = L_1 \cup L_2, L_1 = \{x = -y, 0 \le x \le 1\}$, $L_2 = \{x = y + 2, 1 \le x \le 2\}$. Denote $D^+ = D \cap \{y > 0\}$, $D^- = D \cap \{y < 0\}$ and $z_1 = 1 - i$. Without loss of generality, we may assume that $\Gamma = \{|z - 1| = 1, y \ge 0\}$, otherwise through a conformal mapping, this requirement can be realized.

We discuss the mixed system of first order equations

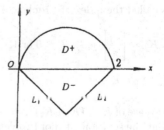

Figure 1.1

$$u_x - v_y = 0, \quad v_x + \mathrm{sgn}\, y\, u_y = 0 \quad \text{in} \quad D. \qquad (1.1)$$

Its complex form is the following complex equation of first order

$$\left\{ \begin{matrix} w_{\bar{z}} \\ w_{\overline{z^*}} \end{matrix} \right\} = 0 \text{ in } \left\{ \begin{matrix} D^+ \\ D^- \end{matrix} \right\}, \tag{1.2}$$

where

$$w = u + iv, \quad z = x + iy, \quad w_{\bar{z}} = \frac{1}{2}[w_x + iw_y], \quad w_{\overline{z^*}} = \frac{1}{2}[w_x - i\bar{w}_y].$$

The Riemann–Hilbert boundary value problem for the complex equation (1.2) may be formulated as follows:

Problem A Find a continuous solution $w(z)$ of (1.2) in $D^* = \bar{D}\backslash(\{0, 2\} \cup \{x \pm y = 2, \text{Im } z \leq 0\})$ or $D^* = \bar{D}\backslash(\{0, 2\} \cup \{x \pm y = 0, \text{Im } z \leq 0\})$ satisfying the boundary conditions

$$\text{Re}\,[\overline{\lambda(z)}w(z)] = r(z), \quad z \in \Gamma, \tag{1.3}$$

$$\text{Re}\,[\overline{\lambda(z)}w(z)] = r(z), \quad z \in L_j (j = 1 \text{ or } 2), \quad \text{Im}\,[\overline{\lambda(z_1)}w(z_1)] = b_1, \tag{1.4}$$

where $\lambda(z) = a(z) + ib(z), |\lambda(z)| = 1, z \in \Gamma \cup L_j (j = 1 \text{ or } 2), b_1$ is a real constant, and $\lambda(z), r(z), b_1$ satisfy the conditions

$$C_\alpha[\lambda(z), \Gamma] \leq k_0, \quad C_\alpha[r(z), \Gamma] \leq k_2, \quad |b_1| \leq k_2,$$

$$C_\alpha[\lambda(z), L_j] \leq k_0, \quad C_\alpha[r(z), L_j] \leq k_2, \quad j = 1 \text{ or } 2, \tag{1.5}$$

$$a(z) - b(z) \neq 0 \text{ on } L_1, \quad \text{or } a(z) + b(z) \neq 0 \text{ on } L_2,$$

in which $\alpha(0 < \alpha < 1), k_0, k_2$ are non-negative constants. For convenience, we may assume that $w(z_1) = 0$, otherwise through a transformation of the function $w(z) - \lambda(z_1)[r(z_1) + ib_1]$, the requirement can be realized.

This Riemann–Hilbert problem (Problem A) for (1.2) with $r(z) = 0, z \in \Gamma \cup L_1$ (or L_2) and $b_1 = 0$ will be called Problem A_0. The number

$$K = \frac{1}{2}(K_1 + K_2), \tag{1.6}$$

is called the index of Problem A and Problem A_0, where

$$K_j = \left[\frac{\phi_j}{\pi}\right] + J_j, J_j = 0 \text{ or } 1, \quad e^{i\phi_j} = \frac{\lambda(t_j - 0)}{\lambda(t_j + 0)}, \quad \gamma_j = \frac{\phi_j}{\pi} - K_j, \quad j = 1, 2, \tag{1.7}$$

in which $t_1 = 2, t_2 = 0, \lambda(t) = (1 - i)/\sqrt{2}$ on $L_0 = [0, 2]$ or $\lambda(t) = (1 + i)/\sqrt{2}$ on $L_0 = [0, 2]$ and $\lambda(t_1 - 0) = \lambda(t_2 + 0) = \exp(7\pi i/4)$ or $\exp(\pi i/4)$. Here we only discuss the case of $K = (K_1 + K_2)/2 = -1/2$ on the boundary ∂D^+ of D^+. In order to ensure that the solution $w(z)$ of Problem A is continuous in the neighborhood($\subset \overline{D^-}$) of the point $z = 0$ or $z = 2$, we need to choose $\gamma_1 > 0$ or $\gamma_2 > 0$ respectively.

1.2   Uniqueness of solutions of the Riemann–Hilbert problem for the simplest complex equation of mixed type

Theorem 1.1   *Problem A for* (1.2) *has at most one solution.*

Proof   Let $w_1(z), w_2(z)$ be any two solutions of Problem A for (1.2). It is clear that $w(z) = w_1(z) - w_2(z)$ is a solution of Problem A_0 for (1.2) with boundary conditions

$$\mathrm{Re}\,[\overline{\lambda(z)}w(z)] = 0, \quad z \in \Gamma, \tag{1.8}$$

$$\mathrm{Re}\,[\overline{\lambda(z)}w(z)] = 0, \quad z \in L_j(j = 1 \text{ or } 2), \quad \mathrm{Im}\,[\overline{\lambda(z_1)}w(z_1)] = 0. \tag{1.9}$$

Due to the complex equation (1.2) in D^- can be reduced to the form

$$\xi_\nu = 0, \quad \eta_\mu = 0 \text{ in } D^-, \tag{1.10}$$

where $\mu = x + y, \nu = x - y, \xi = u + v, \eta = u - v$, the general solution of system (1.10) can be expressed as

$$\xi = u + v = f(\mu) = f(x + y), \quad \eta = u - v = g(\nu) = g(x - y), \quad \text{i.e.}$$

$$u(z) = \frac{f(x + y) + g(x - y)}{2}, \quad v(z) = \frac{f(x + y) - g(x - y)}{2}, \tag{1.11}$$

in which $f(t), g(t)$ are two arbitrary real continuously differentiable functions on $[0, 2]$. Noting the boundary condition (1.9), we have

$$au + bv = 0 \text{ on } L_1 \text{ or } L_2, \quad [av - bu]|_{z=z_1} = 0, \text{ i.e.}$$

$$[a((1-i)x) + b((1-i)x)]f(0)$$

$$+ [a((1-i)x) - b((1-i)x)]g(2x) = 0 \text{ on } [0, 1], \quad \text{or}$$

$$[a((1+i)x - 2i) + b((1+i)x - 2i)]f(2x - 2)$$ \tag{1.12}

$$+ [a((1+i)x - 2i) - b((1+i)x - 2i)]g(2) = 0 \text{ on } [1, 2],$$

$$w(z_1) = 0, \quad [u+v]|_{z_1} = f(0) = 0, \quad \text{or } [u-v]|_{z_1} = g(2) = 0.$$

The second formula in (1.12) can be rewritten as

$$[a((1-i)t/2) + b((1-i)t/2)]f(0) + [a((1-i)t/2) - b((1-i)t/2)]g(t) = 0,$$

$$f(0) = g(t) = 0, \quad \text{or } [a((1+i)t/2 + 1 - i) + b((1-i)t/2 + 1 - i)]f(t)$$ \tag{1.13}

$$+ [a((1-i)t/2 + 1 - i) - b((1-i)t/2 + 1 - i)]g(2) = 0,$$

$$g(t) = f(t) = 0, \quad t \in [0, 2].$$

Thus the solution (1.11) becomes

$$u(z) - v(z) = \frac{1}{2}f(x + y), \quad g(x - y) = 0, \text{ or}$$

$$u(z) = -v(z) = \frac{1}{2}g(x - y), \quad f(x + y) = 0, \tag{1.14}$$

if $a(z) - b(z) \neq 0$ on L_1 or $a(z) + b(z) \neq 0$ on L_2 respectively. In particular, we have

$$u(x) = v(x) = \frac{1}{2}f(x), \quad x \in [0, 2], \quad \text{or}$$

$$u(x) = -v(x) = \frac{1}{2}g(x), \quad x \in [0, 2].$$

(1.15)

Next, due to $f(0) = 0$, or $g(2) = 0$, from (1.15), we can derive that

$$u(x) - v(x) = 0, \quad \text{i.e. } \text{Re}\,[(1+i)w(x)] = 0, \quad x \in [0, 2], \quad \text{or}$$

$$u(x) + v(x) = 0, \quad \text{i.e. } \text{Re}\,[(1-i)w(x)] = 0, \quad x \in [0, 2].$$

(1.16)

Noting the index $K = -1/2$ of Problem A for (1.2) in D^+ and according to the result in Section 1, Chapter III, and [85]11),[86]1), we know that $w(z) = 0$ in $\overline{D^+}$. Thus

$$u(z) + v(z) = \text{Re}\,[(1-i)w(z)] = f(x+y) = 0, \quad g(x-y) = 0, \quad \text{or}$$

$$u(z) - v(z) = \text{Re}\,[(1+i)w(z)] = g(x-y) = 0, \quad f(x+y) = 0,$$

(1.17)

obviously

$$w(z) = u(z) + iv(z) = w_1(z) - w_2(z) = 0 \text{ on } D^-.$$

(1.18)

This proves the uniqueness of solutions of Problem A for (1.2).

1.3   Existence of solutions of the Riemann–Hilbert problem for the simplest complex equation of mixed type

Now, we prove the existence of solutions of the Riemann–Hilbert problem (Problem A) for (1.2).

Theorem 1.2   *Problem A for (1.2) has a solution.*

Proof   As stated before, the general solution of (1.2) in D^- can be expressed as

$$u(z) = \frac{f(x+y) + g(x-y)}{2}, \quad v(z) = \frac{f(x+y) - g(x-y)}{2},$$

$$\text{i.e. } w(z) = \frac{(1+i)f(x+y) + (1-i)g(x-y)}{2},$$

(1.19)

in which $f(t), g(t)$ are two arbitrary real continuously differentiable functions on $[0, 2]$. Taking into account the boundary condition (1.4), we have

$au + bv = r(x)$ on L_1 or L_2, i.e.

$$[a((1-i)x)+b((1-i)x)]f(0)$$

$$+ [a((1-i)x)-b((1-i)x)]g(2x)$$

$$= 2r((1-i)x) \text{ on } [0,1], f(0)$$

$$= [a(z_1) + b(z_1)]r(z_1) + [a(z_1) - b(z_1)]b_1, \text{ or} \tag{1.20}$$

$$[a((1+i)x-2i)+b((1+i)x-2i)]f(2x-2)$$

$$+[a((1+i)x-2i)-b((1+i)x-2i)]g(2)$$

$$= 2r((1+i)x-2i) \text{ on } [1,2], g(2)$$

$$= [a(z_1) - b(z_1)]r(z_1) - [a(z_1) + b(z_1)]b_1.$$

The second and third formulas in (1.20) can be rewritten as

$$[a((1-i)t/2) - b((1-i)t/2)]g(t)$$

$$= 2r((1-i)t/2) - [a((1-i)t/2) + b((1-i)t/2)]f(0), \quad t \in [0,2], \text{ or}$$

$$[a((1+i)t/2+1-i)+b((1+i)t/2+1-i)]f(t) \tag{1.21}$$

$$= 2r((1+i)t/2+1-i)$$

$$-[a((1+i)t/2+1-i)-b((1+i)t/2+1-i)]g(2), \quad t\in [0,2],$$

thus the solution (1.19) possesses the form

$$u(z) = \frac{1}{2}\{f(x+y) + g(x-y)\}, \quad v(z) = \frac{1}{2}\{f(x+y) - g(x-y)\},$$

$$g(x-y) = \frac{2r((1-i)(x-y)/2) - [a((1-i)(x-y)/2)+b((1-i)(x-y)/2)]f(0)}{a((1-i)(x-y)/2)-b((1-i)(x-y)/2)},$$

$$u(z) = \frac{1}{2}\{g(x-y)+f(x+y)\}, \quad v(z) = \frac{1}{2}\{-g(x-y)+f(x+y)\}, \tag{1.22}$$

$$f(x+y) = \frac{2r((1+i)(x+y)/2+1-i)}{a((1+i)(x+y)/2+1-i) + b((1+i)(x+y)/2+1-i)}$$

$$- \frac{[a((1+i)(x+y)/2+1-i) - b((1+i)(x+y)/2+1-i)]g(2)}{a((1+i)(x+y)/2+1-i) + b((1+i)(x+y)/2+1-i)},$$

if $a(z) - b(z) \neq 0$ on L_1, or $a(z) + b(z) \neq 0$ on L_2 respectively. In particular, we get

$$u(x) = \frac{1}{2}\{f(x) + g(x)\}, \quad v(x) = \frac{1}{2}\{f(x) - g(x)\},$$

$$g(x) = \left\{\frac{2r((1-i)x/2) - [a((1-i)x/2) + b((1-i)x/2)]f(0)}{a((1-i)x/2) - b((1-i)x/2)}\right\},$$

$$u(x) = \frac{1}{2}\{g(x) + f(x)\}, \quad v(x) = \frac{1}{2}\{-g(x) + f(x)\},$$

$$f(x) = \frac{2r((1+i)x/2+1-i)}{a((1+i)x/2+1-i)+b((1+i)x/2+1-i)}$$

$$-\frac{[a((1+i)x/2+1-i)-b((1+i)x/2+1-i)]g(2)}{a((1+i)x/2+1-i)+b((1+i)x/2+1-i)}, \quad x \in [0,2]. \tag{1.23}$$

From the above formulas, it follows that

$$u(x)-v(x) = \frac{2r((1-i)x/2)-[a((1-i)x/2)+b((1-i)x/2)]f(0)}{a((1-i)x/2)-b((1-i)x/2)}, \text{ or}$$

$$u(x)+v(x) = \frac{2r((1+i)x/2+1-i)}{a((1+i)x/2+1-i)+b((1+i)x/2+1-i)}$$

$$-\frac{[a((1+i)x/2+1-i)-b((1+i)x/2+1-i)]g(2)}{a((1+i)x/2+1-i)+b((1+i)x/2+1-i)}, \quad x \in [0,2] \tag{1.24}$$

$$f(0) = [a(1-i)+b(1-i)]r(1-i)+[a(1-i)-b(1-i)]b_1, \text{ or}$$

$$g(2) = [a(1-i)-b(1-i)]r(1-i)-[a(1-i)+b(1-i)]b_1,$$

i.e.

$$\text{Re}\,[(1+i)w(x)] = s(x),$$

$$s(x) \doteq \frac{2r((1-i)x/2)-[a((1-i)x/2)+b((1-i)x/2)]f(0)}{a((1-i)x/2)-b((1-i)x/2)},$$

$$x \in [0,2], \text{ or}$$

$$\text{Re}\,[(1-i)w(x)] = s(x) = \frac{2r((1+i)x/2+1-i)}{a((1+i)x/2+1-i)-b((1+i)x/2+\ -i)} \tag{1.25}$$

$$-\frac{[a((1+i)x/2+1-i)-b((1+i)x/2+1-i)]g(2)}{a((1+i)x/2+1-i)-b((1+i)x/2+1-i)},$$

$$x \in [0,2],$$

if $a((1-i)x) - b((1-i)x) \neq 0$ on $[0,1]$ or $a((1+i)x-2i)+b((1+i)x-2i) \neq 0$ on $[1,2]$ respectively. We introduce a conformal mapping $\zeta = \zeta(z)$ from the domain D^+ onto the upper half-plane $G = \{\text{Im}\,\zeta > 0\}$, such that the three points $z = 0,1,2$ map to $\zeta = -1,0,1$ respectively, it is not difficult to derive that the conformal mapping and its inverse mapping can be expressed by the elementary functions, namely

$$\zeta(z) = \frac{5(z-1)}{(z-1)^2+4}, \quad z(\zeta) = 1 + \frac{5}{2\zeta}(1 - \sqrt{1 - 16\zeta^2/25}).$$

Denoting $W(\zeta) = w[z(\zeta)]$ and

$$\Lambda(\zeta) = \begin{cases} \lambda[z(\zeta)], & \zeta \in \Gamma_1 = \zeta(\Gamma), \\[2mm] \dfrac{1-i}{\sqrt{2}}, & \zeta \in \Gamma_2 = \{-1 \leq \text{Re}\,\zeta \leq 1, \text{Im}\,\zeta = 0\}, \text{ or} \\[2mm] \dfrac{1+i}{\sqrt{2}}, & \zeta \in \Gamma_2 = \{-1 \leq \text{Re}\,\zeta \leq 1, \text{Im}\,\zeta = 0\}, \end{cases} \tag{1.26}$$

in which the points $\zeta_1 = 1$, $\zeta_2 = -1$ are the discontinuous points of $\Lambda(\zeta)$ on $\partial G = \{\text{Im}\,\zeta = 0\}$, from (1.6),(1.7), it can be seen that the index of $\Lambda(\zeta)$ on $\partial G = \{\text{Im}\,\zeta = 0\}$ is $K = -1/2$. Hence, according to the result of Theorem 1.1, Chapter III, we know that the discontinuous Riemann–Hilbert boundary value problem for analytic functions $W(\zeta)$ in G with the boundary condition

$$\text{Re}\,[\overline{\Lambda(z)}W(\zeta)] = R(\zeta) = \begin{cases} r[z(\zeta)], & \zeta \in \Gamma_1, \\ s[z(\zeta)]/\sqrt{2}, & \zeta \in \Gamma_2, \end{cases} \tag{1.27}$$

has a unique solution $W(\zeta)$ in G as follows:

$$W(\zeta) = \frac{X(\zeta)}{\pi i}\left[\int_{-\infty}^{\infty} \frac{\Lambda(t)R(t)}{(t-\zeta)X(t)}dt + ic_*\frac{2+\zeta}{2-\zeta}\right] \text{ in } G, \tag{1.28}$$

and

$$X(\zeta) = i\frac{\zeta-2}{\zeta-i}\Pi(\zeta)e^{iS(\zeta)}, \ \Pi(\zeta) = \left(\frac{\zeta-1}{\zeta+i}\right)^{\gamma_1}\left(\frac{\zeta+1}{\zeta+i}\right)^{\gamma_2},$$

$$c_* = \frac{2i+1}{2+i}\int_{-\infty}^{\infty}\frac{\Lambda(t)R(t)}{X(t)(t-i)}dt,$$

and $S(\zeta)$ is an analytic function in $\text{Im}\,\zeta > 0$ with the boundary condition

$$\text{Re}\,[S(t)] = \arg[\Lambda_1(t)\left(\frac{t-2}{t+i}\right)] \text{ on } \text{Im}\,t = 0, \ \text{Im}\,[S(i)] = 0, \tag{1.29}$$

where $\gamma_j(j=1,2)$ are as stated in (1.7) and $\Lambda_1(t) = \lambda(t)\overline{\Pi(t)}(t-2)|x+i|/[|\Pi(t)| \times |t-2|(x+i)]$, and the boundedness of $w(z)$ or boundedness of integral of the solution $w(z)$ in the neighborhood $\subset \overline{D}\backslash\{0,2\}$ of $t_1 = 2$ and $t_2 = 0$ is determined by $J_j = 0, \gamma_j > 0$, or $J_j = 0, \gamma_j = 0$ and $J_j = 1(j=1,2)$ respectively. Hence Problem A for (1.2) has a solution $w(z)$ in the form

$$w(z) = \begin{cases} W[\zeta(z)], \ z \in \overline{D^+}\backslash\{0,2\}, \\[2mm] \frac{1}{2}\Big\{(1+i)f(x+y)+(1-i) \\[2mm] \qquad \times \frac{2r((1-i)(x-y)/2)-[a((1-i)(x-y)/2)+b((1-i)(x-y)/2)]f(0)}{a((1-i)(x-y)/2)-b((1-i)(x-y)/2)}\Big\}, \ \text{or} \\[4mm] \frac{1}{2}\Big\{(1-i)g(x-y)+\frac{2(1+i)r((1+i)(x+y)/2+1-i)}{a((1+i)(x+y)/2+1-i)+b((1+i)(x+y)/2+1-i)} \\[3mm] \qquad -\frac{[a((1+i)(x+y)/2+1-i)+b((1+i)(x+y)/2+1-i)]g(2)}{a((1+i)(x+y)/2+1-i)+b((1+i)(x+y)/2+1-i)}\Big\}, \\[3mm] \qquad z = x+iy \in \overline{D^-}\backslash\{0,2\}, \end{cases}$$

$$\tag{1.30}$$

in which $f(0), g(2)$ are as stated in (1.24), $W(\zeta)$ in $\overline{D^+}\backslash\{0,2\}$ is as stated in (1.28), and from (1.23), we derive that

$$f(x+y) = u(x+y) + v(x+y) = \text{Re}\left[(1-i)W(\zeta(x+y))\right],$$

$$g(x-y) = u(x-y) - v(x-y) = \text{Re}\left[(1+i)W(\zeta(x-y))\right],$$ (1.31)

where $W[\zeta(x+y)]$ and $W[\zeta(x-y)]$ are the values of $W[\zeta(z)]$ on $0 \le z = x+y \le 2$ and $0 \le z = x-y \le 2$ respectively.

From the foregoing representation of the solution $w(z)$ of Problem A for (1.2) and the mapping $\zeta(z)$, we can derive that $w(z)$ satisfies the estimate

$$C_\beta[w(z)X(z), \overline{D^+}] + C_\beta[w^\pm(z)Y^\pm(z), \overline{D^-}] \le M_1,$$ (1.32)

in which $X(z) = \Pi_{j=1}^2 |z-t_j|^{2|\gamma_j|+\delta}$, $Y^\pm(z) = |x \pm y - t_j|^{2|\gamma_j|+\delta}$, $w^\pm(z) = \text{Re}\, w \pm \text{Im}\, w$, $\beta(0 < \beta < \delta)$, δ are sufficiently small positive constants, and $M_1 = M_1(\beta, k_0, k_2, D)$ is a non-negative constant [85]15).

Finally, we mention that if the index K is an arbitrary even integer or $2K$ is an arbitrary odd integer, the above Riemann–Hilbert problem for (1.2) can be considered, but in general the boundary value problem for $K \le -1$ have some solvability conditions or its solution for $K \ge 0$ is not unique.

2 The Riemann–Hilbert Problem for First Order Linear Complex Equations of Mixed Type

In this section we discuss the Riemann–Hilbert boundary value problem for first order linear complex equations of mixed (elliptic-hyperbolic) type in a simply connected domain. Firstly, we give the representation theorem and prove the uniqueness of solutions for the above boundary value problem, secondly by using the method of successive iteration, the existence of solutions for the above problem is proved.

2.1 Formulation of Riemann–Hilbert problem of first order complex equations of mixed type

Let D be a simply connected bounded domain in the complex plane $\mathbb{C}$ with the boundary $\partial D = \Gamma \cup L$, where $\Gamma, L = L_1 \cup L_2$, $D^+ = D \cap \{y > 0\}$, $D^- = D \cap \{y < 0\}$ and $z_1 = 1 - i$ are as stated in Section 1.

We discuss the first order linear system of mixed (elliptic-hyperbolic) type equations

$$\begin{cases} u_x - v_y = au + bv + f, \\ v_x + \text{sgny}\, u_y = cu + dv + g, \end{cases} \quad z = x + iy \in D,$$ (2.1)

in which a, b, c, d, f, g are functions of $(x, y)(\in D)$, its complex form is the following complex equation of first order

$$\left\{ \begin{matrix} w_{\bar{z}} \\ w_{\overline{z^*}} \end{matrix} \right\} = F(z, w), \quad F = A_1(z)w + A_2(z)\overline{w} + A_3(z) \text{ in } \left\{ \begin{matrix} D^+ \\ D^- \end{matrix} \right\}, \qquad (2.2)$$

where

$$w = u + iv, \quad z = x + iy, \quad w_{\bar{z}} = \frac{1}{2}[w_x + iw_y], \quad w_{\overline{z^*}} = \frac{1}{2}[w_x - i\bar{w}_y],$$

$$A_1 = \frac{a - ib + ic + d}{4}, \quad A_2 = \frac{a + ib + ic - d}{4}, \quad A_3 = \frac{f + ig}{2}.$$

Suppose that the complex equation (2.2) satisfies the following conditions.

Condition C

$A_j(z) \, (j = 1, 2, 3)$ are measurable in $z \in D^+$ and continuous in $\overline{D^-}$ in $D^* = \bar{D}\backslash(\{0, 2\} \cup \{x \pm y = 2, \operatorname{Im} z \le 0\})$ or $D^* = \bar{D}\backslash(\{0, 2\} \cup \{x \pm y = 0, \operatorname{Im} z \le 0\})$, and satisfy

$$L_p[A_j, \overline{D^+}] \le k_0, \quad j = 1, 2, \quad L_p[A_3, \overline{D^+}] \le k_1, \qquad (2.3)$$

$$C[A_j, \overline{D^-}] \le k_0, \quad j = 1, 2, \quad C[A_3, \overline{D^-}] \le k_1. \qquad (2.4)$$

where $p \, (> 2)$, k_0, k_1 are non-negative constants.

2.2 The representation and uniqueness of solutions of the Riemann–Hilbert problem for mixed complex equations

We first introduce a lemma, which is a special case of Theorem 2.1, Chapter III.

Lemma 2.1 *Suppose that the complex equation (2.2) satisfies Condition C. Then any solution of Problem A for (2.2) in D^+ with the boundary conditions (1.3) and*

$$\operatorname{Re}\left[\overline{\lambda(x)}w(x)\right] = s(x), \quad \lambda(x) = 1 - i \text{ or } 1 + i, \quad x \in L_0, \quad C_\alpha[s(x), L_0] \le k_3, \quad (2.5)$$

can be expressed as

$$w(z) = \Phi(z)e^{\phi(z)} + \psi(z), \quad z \in \overline{D^+}, \qquad (2.6)$$

where $\operatorname{Im}[\phi(z)] = 0$, $z \in L_0 = (0, 2)$, and $\phi(z)$, $\psi(z)$ satisfies the estimates

$$C_\beta[\phi, \overline{D^+}] + L_{p_0}[\phi_{\bar{z}}, \overline{D^+}] \le M_2, \quad C_\beta[\psi, \overline{D^+}] + L_{p_0}[\psi_{\bar{z}}, \overline{D^+}] \le M_2, \qquad (2.7)$$

in which k_3, $\beta \, (0 < \beta \le \alpha)$, $p_0 \, (2 < p_0 \le 2)$, $M_2 = M_2(p_0, \beta, k, D^+)$ are non-negative constants, $k = (k_0, k_1, k_2, k_0)$, $\Phi(z)$ is analytic in D^+ and $w(z)$ satisfies the estimate

$$C_\beta[w(z)X(z), \overline{D^+}] \le M_3(k_1 + k_2 + k_3), \qquad (2.8)$$

in which

$$X(z) = |z - t_1|^{\eta_1}|z - t_2|^{\eta_2}, \quad \eta_j = \begin{cases} 2|\gamma_j| + \delta, & \text{if } \gamma_j < 0, \\ \delta, & \gamma_j \geq 0, \end{cases} \quad j = 1, 2, \qquad (2.9)$$

here $\gamma_j(j = 1, 2)$ are real constants as stated in (1.7) and δ is a sufficiently small positive constant, and $M_3 = M_3(p_0, \beta, k_0, D^+)$ is a non-negative constant.

Theorem 2.2 *If the complex equation (2.2) satisfies Condition C in D, then any solution of Problem A with the boundary conditions (1.3), (1.4) for (2.2) can be expressed as*

$$w(z) = w_0(z) + W(z), \qquad (2.10)$$

where $w_0(z)$ is a solution of Problem A for the complex equation (1.2) and $W(z)$ possesses the form

$$W(z) = w(z) - w_0(z), \quad w(z) = \tilde{\Phi}(z)e^{\tilde{\phi}(z)} + \tilde{\psi}(z) \text{ in } D^+,$$

$$\tilde{\phi}(z) = \tilde{\phi}_0(z) + Tg = \tilde{\phi}_0(z) - \frac{1}{\pi}\iint_{D^+}\frac{g(\zeta)}{\zeta - z}d\sigma_\zeta, \quad \tilde{\psi}(z) = Tf \text{ in } D^+, \qquad (2.11)$$

$$W(z) = \Phi(z) + \Psi(z), \quad \Psi(z) = \int_2^\nu g_1(z)d\nu e_1 + \int_0^\mu g_2(z)d\mu e_2 \text{ in } z \in D^-,$$

in which $\tilde{\phi}(z) = 0$ on L_0, $e_1 = (1 + i)/2$, $e_2 = (1 - i)/2$, $\mu = x + y$, $\nu = x - y$, $\tilde{\phi}_0(z)$ is an analytic function in D^+, and

$$g(z) = \begin{cases} A_1 + A_2\overline{w}/w, & w(z) \neq 0, \\ 0, & w(z) = 0, \end{cases} \quad f = A_1\tilde{\psi} + A_2\overline{\tilde{\psi}} + A_3 \text{ in } D^+, \qquad (2.12)$$

$$g_1(z) = A\xi + B\eta + E, \quad g_2(z) = C\xi + D\eta + F \text{ in } D^-,$$

where $\xi = \text{Re}\,w + \text{Im}\,w$, $\eta = \text{Re}\,w - \text{Im}\,w$, $A = \text{Re}\,A_1 + \text{Im}\,A_1$, $B = \text{Re}\,A_2 + \text{Im}\,A_2$, $C = \text{Re}\,A_2 - \text{Im}\,A_2$, $D = \text{Re}\,A_1 - \text{Im}\,A_1$, $E = \text{Re}\,A_3 + \text{Im}\,A_3$, $F = \text{Re}\,A_3 - \text{Im}\,A_3$, and $\tilde{\phi}(z), \tilde{\psi}(z)$ satisfy the estimates

$$C_\beta[\tilde{\phi}(z), \overline{D^+}] + L_{p_0}[\tilde{\phi}_{\bar{z}}, \overline{D^+}] \leq M_4, \quad C_\beta[\tilde{\psi}(z), \overline{D^+}] + L_{p_0}[\tilde{\psi}_{\bar{z}}, \overline{D^+}] \leq M_4, \qquad (2.13)$$

where $M_4 = M_4(p_0, \beta, k, D^+)$ is a non-negative constant, $\tilde{\Phi}(z)$ is analytic in D^+ and $\Phi(z)$ is a solution of equation (1.2) in D^- satisfying the boundary conditions

$$\text{Re}\,[\overline{\lambda(z)}(e^{\tilde{\phi}(z)}\tilde{\Phi}(z) + \tilde{\psi}(z))] = r(z), \quad z \in \Gamma,$$

$$\text{Re}\,[\overline{\lambda(x)}(\tilde{\Phi}(x)e^{\tilde{\phi}(x)} + \tilde{\psi}(x))] = s(x), \quad x \in L_0,$$

$$\text{Re}\,[\lambda(x)\Phi(x)] = \text{Re}\,[\lambda(x)(W(x) - \Psi(x))], \quad z \in L_0, \qquad (2.14)$$

$$\text{Re}\,[\overline{\lambda(z)}\Phi(z)] = -\text{Re}\,[\overline{\lambda(z)}\Psi(z)], \quad z \in L_1 \text{ or } L_2,$$

$$\text{Im}\,[\overline{\lambda(z_1)}\Phi(z_1)] = -\text{Im}\,[\overline{\lambda(z_1)}\Psi(z_1)].$$

Moreover the solution $w_0(z)$ of Problem A for (1.2) satisfies the estimate (1.32), namely

$$C_\beta[w_0(z)X(z), \overline{D^+}] + C_\beta[w_0^\pm(z)Y^\pm(z), \overline{D^-}] \le M_5(k_1 + k_2) \qquad (2.15)$$

where $w_0^\pm(z) = \operatorname{Re} w_0(z) \pm \operatorname{Im} w_0(z)$, $Y^\pm(z) = \prod_{j=1}^2 |x \pm y - t_j|^{\eta_j}, j = 1, 2, X(z), \eta_j = 2|\gamma_j| + \delta\,(j = 1, 2)$, β are as stated in (1.32), and $M_5 = M_5(p_0, \beta, k_0, D)$ is a non-negative constant.

Proof Let the solution $w(z)$ be substituted in the position of w in the complex equation (2.2) and (2.12), thus the functions $g_1(z)$, $g_2(z)$ and $\Psi(z)$ in $\overline{D^-}$ in (2.11),(2.12) can be determined. Moreover we can find the solution $\Phi(z)$ of (1.2) with the boundary condition (2.14), where

$$s(x) = \begin{cases} \dfrac{2r((1-i)x/2) - 2R(1-i)x/2)}{a((1-i)x/2) - b((1-i)x/2)} + \operatorname{Re}\left[\overline{\lambda(x)}\Psi(x)\right], & x \in L_0, \text{ or} \\[3mm] \dfrac{2r((1+i)x/2+1-i) - 2R((1+i)x/2+1-i)}{a((1+i)x/2 + 1 - i) + b((1+i)x/2 + 1 - i)} + \operatorname{Re}\left[\overline{\lambda(x)}\Psi(x)\right], & x \in L_0, \end{cases} \qquad (2.16)$$

here and later $R(z) = \operatorname{Re}\left[\overline{\lambda(z)}\Psi(z)\right]$ on L_1 or L_2, thus

$$w(z) = w_0(z) + W(z) = \begin{cases} \tilde{\Phi}(z)e^{\tilde{\phi}(z)} + \tilde{\psi}(z) & \text{in } D^+, \\ w_0(z) + \Phi(z) + \Psi(z) & \text{in } D^-, \end{cases} \qquad (2.17)$$

is the solution of Problem A for the complex equation

$$\begin{Bmatrix} w_{\bar{z}} \\ w_{\bar{z}^*} \end{Bmatrix} = A_1 w + A_2 \bar{w} + A_3 \text{ in } \begin{Bmatrix} D^+ \\ D^- \end{Bmatrix}, \qquad (2.18)$$

which can be expressed as in (2.10) and (2.11).

2.3 The unique solvability of the Riemann–Hilbert problem for first order complex equations of mixed type

Theorem 2.3 *Let the mixed complex equation (2.2) satisfy Condition C. Then Problem A for (2.2) has a solution in D.*

Proof In order to find a solution $w(z)$ of Problem A in D, we express $w(z)$ in the form (2.10)- (2.12). In the following, we shall find a solution of Problem A by using the successive iteration. First of all, denoting the solution $w_0(z) = (\zeta_0 e_1 + \eta_0 e_2)$ of Problem A for (1.2), and substituting them into the positions of $w = (\xi e_1 + \eta e_2)$ in the right-hand side of (2.2), similarly to (2.10)–(2.12), we have the corresponding functions $g_0(z)$, $f_0(z)$ in D^+ and the functions

$$W_1(z) = w_1(z) - w_0(z), \quad w_1(z) = \tilde{\Phi}_1(z)e^{\tilde{\phi}_1(z)} + \tilde{\psi}_1(z),$$

$$\tilde{\phi}_1(z) = \tilde{\phi}_0(z) - \frac{1}{\pi}\iint_{D^+}\frac{g_0(\zeta)}{\zeta - z}d\sigma_\zeta, \quad \tilde{\psi}_1(z) = Tf_0 \text{ in } D^+,$$

$$w_1(z) = w_0(z) + W_1(z), \quad W_1(z) = \Phi_1(z) + \Psi_1(z), \tag{2.19}$$

$$\Psi_1(z) = \int_2^\nu [A\xi_0 + B\eta_0 + E]e_1 d\nu + \int_0^\mu [C\xi_0 + D\eta_0 + F]e_2 d\mu \text{ in } D^-$$

can be determined, where $\mu = x + y, \nu = x - y$, and the solution $w_0(z)$ satisfies the estimate (2.15), i.e.

$$C_\beta[w_0(z)X(z),\overline{D^+}] + C_\beta[w_0^\pm(z)Y^\pm(z),\overline{D^-}] \le M_6 = M_5(k_1 + k_2), \tag{2.20}$$

where $\beta, X(z), Y^\pm(z)$ are as stated in (2.15). Moreover, we find an analytic function $\tilde{\Phi}_1(z)$ in D^+ and a solution $\Phi_1(z)$ of (1.2) in D^- satisfying the boundary conditions

$$\text{Re}\,[\overline{\lambda(z)}(e^{\tilde{\phi}_1(z)}\tilde{\Phi}_1(z) + \tilde{\psi}_1(z))] = r(z), \quad z \in \Gamma,$$

$$\text{Re}\,[\overline{\lambda(x)}(\tilde{\Phi}_1(x)e^{\tilde{\phi}_1(x)} + \tilde{\psi}_1(x))] = s_1(x), \quad x \in L_0,$$

$$\text{Re}\,[\lambda(x)\Phi_1(x)] = -\text{Re}\,[\lambda(x)\Psi_1(x)], \quad z \in L_0, \tag{2.21}$$

$$\text{Re}\,[\overline{\lambda(z)}\Phi_1(z)] = -\text{Re}\,[\overline{\lambda(z)}\Psi_1(z)], \quad z \in L_1 \text{ or } L_2,$$

$$\text{Im}\,[\overline{\lambda(z_1)}\Phi_1(z_1)] = -\text{Im}\,[\overline{\lambda(z_1)}\Psi_1(z_1)],$$

in which

$$s_1(x) = \begin{cases} \dfrac{2r((1-i)x/2) - 2R_1((1-i)x/2)}{a((1-i)x/2) - b((1-i)x/2)} + \text{Re}\,[\overline{\lambda(x)}\Psi_1(x)], & x \in L_0, \text{ or} \\[3mm] \dfrac{2r((1+i)x/2+1-i) - 2R_1((1+i)x/2+1-i)}{a((1+i)x/2 + 1 - i) + b((1 + i)x/2 + 1 - i)} + \text{Re}\,[\overline{\lambda(x)}\Psi_1(x)], & x \in L_0, \end{cases}$$

here and later $R_1(z) = \text{Re}\,[\overline{\lambda(z)}\Psi_1(z)]$ on L_1 or L_2, and

$$w_1(z) = w_0(z) + W_1(z) = \begin{cases} \tilde{\Phi}_1(z)e^{\tilde{\phi}_1(z)} + \tilde{\psi}_1(z) \text{ in } D^+, \\ w_0(z) + \Phi_1(z) + \Psi_1(z) \text{ in } \overline{D^-} \end{cases} \tag{2.22}$$

satisfies the estimate

$$C_\beta[w_1(z)X(z),\overline{D^+}] + C[w_1^\pm(z)Y^\pm(z),\overline{D^-}] \le M_7 = M_7(p_0, \beta, k, D), \tag{2.23}$$

where $\tilde{\phi}_1(z), \tilde{\psi}_1(z), \tilde{\Phi}_1(z)$ are similar to the functions in Theorem 2.2. Furthermore we substitute $w_1(z) = w_0(z) + W_1(z)$ and the corresponding functions $w_1^+(z) = \xi_1(z) = \text{Re}\,w_1(z) + \text{Im}\,w(z), w_1^-(z) = \eta_1(z) = \text{Re}\,w_1(z) - \text{Im}\,w(z)$ into the positions of w, ξ, η in (2.11),(2.12), and similarly to (2.19)–(2.22), we can find the corresponding functions

$\tilde{\phi}_2(z), \tilde{\psi}_2(z), \tilde{\Phi}_2(z)$ in D^+ and $\Psi_2(z), \Phi_2(z)$ and $W_2(z) = \Phi_2(z) + \Psi_2(z)$ in $\overline{D^-}$, and the function

$$w_2(z) = w_0(z) + W_2(z) = \begin{cases} \tilde{\Phi}_2(z)e^{\tilde{\phi}_2(z)} + \tilde{\psi}_2(z) \text{ in } D^+, \\ w_0(z) + \Phi_2(z) + \Psi_2(z) \text{ in } \overline{D^-} \end{cases} \tag{2.24}$$

satisfies a similar estimate of the form (2.23). Thus there exists a sequence of functions $\{w_n(z)\}$ as follows

$$w_n(z) = w_0(z) + W_n(z) = \begin{cases} \tilde{\Phi}_n(z)e^{\tilde{\phi}_n(z)} + \tilde{\psi}_n(z) \text{ in } D^+, \\ w_0(z) + \Phi_n(z) + \Psi_n(z) \text{ in } D^-, \\ \Psi_n(z) = \int_2^{\nu} [A\xi_{n-1} + B\eta_{n-1} + E]e_1 d\nu \\ \quad + \int_0^{\mu} [C\xi_{n-1} + D\eta_{n-1} + F]e_2 d\mu \text{ in } \overline{D^-}, \end{cases} \tag{2.25}$$

and then

$$|[w_1^{\pm}(z) - w_0^{\pm}(z)]Y^{\pm}(z)| \le |\Phi_1^{\pm}(z)Y^{\pm}(z)| + \sqrt{2}\left[|Y^+(z) \int_2^{\nu} [A\xi_0 + B\eta_0 + E]e_1 d\nu| \right.$$

$$\left. + |Y^-(z) \int_0^{\mu} [C\xi_0 + D\eta_0 + F]e_2 d\mu| \right] \le 2M_8M(4m+1)R' \text{ in } \overline{D^-}, \tag{2.26}$$

where $m = \max\{C[w_0^+(z)Y^+(z), \overline{D^-}] + C[w_0^-(z)Y^-(z), \overline{D^-}]\}$, $M_8 = \max_{z\in\overline{D^-}}(|A|, |B|, |C|, |D|, |E|, |F|)$, $R' = 2$, $M = 1 + 4k_0^2(1 + k_0^2)$, M_5 is a constant as stated in (2.20). It is clear that $w_n(z) - w_{n-1}(z)$ satisfies

$$w_n(z) - w_{n-1}(z) = \Phi_n(z) - \Phi_{n-1}(z) + \int_2^{\nu} [A(\xi_n - \xi_{n-1}) + B(\eta_n - \eta_{n-1})]e_1 d\nu$$

$$+ \int_0^{\mu} [C(\xi_n - \xi_{n-1}) + D(\eta_n - \eta_{n-1})]e_2 d\mu \text{ in } \overline{D^-}, \tag{2.27}$$

where $n = 1, 2, \ldots$. From the above equality, we can obtain

$$|[w_n^{\pm} - w_{n-1}^{\pm}]Y^{\pm}(z)| \le [2M_8M(4m+1)]^n$$

$$\times \int_0^{R'} \frac{R'^{n-1}}{(n-1)!} dR' \le \frac{[2M_8M(4m+1)R']^n}{n!} \text{ in } \overline{D^-}, \tag{2.28}$$

and then we can see that the sequence of functions $\{w_n^{\pm}(z)Y^{\pm}(z)\}$, i.e.

$$w_n^{\pm}(z)Y^{\pm}(z) = \{w_0^{\pm}(z) + [w_1^{\pm}(z) - w_0^{\pm}(z)] + \cdots + [w_n^{\pm}(z) - w_{n-1}^{\pm}(z)]\}Y^{\pm}(z) \tag{2.29}$$

$(n = 1, 2, \ldots)$ in $\overline{D^-}$ uniformly converge to functions $w_*^{\pm}(z)Y^{\pm}(z)$, and $w_*(z)$ satisfies the equality

$$w_*(z) = w_0(z) + \Phi_*(z) + \Psi_*(z),$$

$$\Psi_*(z) = \int_2^{\nu} [A\xi_* + B\eta_* + E]e_1 d\nu$$

$$+ \int_0^{\mu} [C\xi_* + D\eta_* + F]e_2 d\mu \text{ in } \overline{D^-}, \tag{2.30}$$

where $\xi^* = \operatorname{Re} w^* + \operatorname{Im} w^*$, $\eta = \operatorname{Re} w^* - \operatorname{Im} w^*$, and $w_*(z)$ satisfies the estimate

$$C[w_*^{\pm}(z)Y^{\pm}(z), \overline{D^-}] \leq e^{2M_8M(4m+1)R'}. \tag{2.31}$$

Moreover, we can find a sequence of functions $\{w_n(z)\}(w_n(z) = \tilde{\Phi}_n(z)e^{\tilde{\phi}_n(z)} + \tilde{\psi}_n(z))$ in D^+ and $\tilde{\Phi}_n(z)$ is an analytic function in $\bar{D}^+$ satisfying the boundary conditions

$$\begin{aligned}
\operatorname{Re}\left[\overline{\lambda(z)}(\tilde{\Phi}_n(z)e^{\tilde{\phi}_n(z)} + \tilde{\psi}_n(z))\right] &= r(z), \quad z \in \Gamma, \\
\operatorname{Re}\left[\overline{\lambda(x)}(\tilde{\Phi}_n(x)e^{\tilde{\phi}_n(x)} + \tilde{\psi}_n(x))\right] &= s(x), \quad x \in L_0,
\end{aligned} \tag{2.32}$$

in which

$$s_n(x) = \begin{cases} \dfrac{2r((1-i)x/2) - 2R_n((1-i)x/2)}{a((1-i)x/2) - b((1-i)x/2)} + \operatorname{Re}\left[\overline{\lambda(x)}\Psi_n(x))\right], & x \in L_0, \text{ or} \\[3mm] \dfrac{2r((1+i)x/2+1-i) - 2R_n((1-i)x/2+1-i)}{a((1+i)x/2+1-i) + b((1+i)x/2+1-i)} + \operatorname{Re}\left[\overline{\lambda(x)}\Psi_n(x)\right], & x \in L_0, \end{cases} \tag{2.33}$$

here and later $R_n(z) = \operatorname{Re}\left[\overline{\lambda(z)}\Psi_n(z)\right]$ on L_1 or L_2. From (2.31), it follows that

$$C_\beta[s_n(x)X(x), L_0] \leq 2k_2k_0 + \frac{[2M_8M(4m+1)R']^n}{n!} = M_9, \tag{2.34}$$

and then the estimate

$$C_\beta[w_n(z)X(z), \overline{D^+}] \leq M_3(k_1 + k_2 + M_9), \tag{2.35}$$

thus from $\{w_n(z)X(x)\}$, we can choose a subsequence which uniformly converge a function $w_*(z)X(z)$ in $\overline{D^+}$. Combining (2.31) and (2.35), it is obvious that the solution $w_*(z)$ of Problem A for (2.2) in $\bar{D}$ satisfies the estimate

$$C_\beta[w_*(z)X(z), \overline{D^+}] + C[w_*^{\pm}(z)Y^{\pm}(z), \overline{D^-}] \leq M_{10} = M_{10}(p_0, \beta, k, D), \tag{2.36}$$

where M_{10} is a non-negative constant.

Theorem 2.4 *Suppose that the complex equation (2.2) satisfies Condition C. Then Problem A for (2.2) has at most one solution in D.*

Proof Let $w_1(z), w_2(z)$ be any two solutions of Problem A for (2.2). By Condition C, we see that $w(z) = w_1(z) - w_2(z)$ satisfies the homogeneous complex equation and boundary conditions

$$Lw = \tilde{A}_1 w + \tilde{A}_2 \bar{w} \quad \text{in} \quad D, \tag{2.37}$$

$$\begin{aligned}
\operatorname{Re}\left[\overline{\lambda(z)}w(z)\right] &= 0, \quad z \in \Gamma, \quad \operatorname{Re}\left[\overline{\lambda(x)}w(x)\right] = s(x), \quad x \in L_0, \\
\operatorname{Re}\left[\overline{\lambda(z)}w(z)\right] &= 0, \quad z \in L_1 \text{ or } L_2, \quad \operatorname{Re}\left[\overline{\lambda(z_1)}w(z_1)\right] = 0.
\end{aligned} \tag{2.38}$$

From Theorem 2.2, the solution $w(z)$ can be expressed in the form

$$w(z) = \begin{cases} \tilde{\Phi}(z)e^{\tilde{\phi}(z)}, \ \tilde{\phi}(z) = \tilde{T}g \text{ in } D^+, \\ g(z) = \begin{cases} A_1 + A_2\bar{w}/w, & w(z) \neq 0, \ z \in D^+, \\ 0, \ w(z) = 0, & z \in D^+, \end{cases} \\ \Phi(z) + \Psi(z), \\ \Psi(z) = \int_2^\nu [A\xi + B\eta]e_1 d\nu + \int_0^\mu [C\xi + D\eta]e_2 d\mu \text{ in } \overline{D^-}, \end{cases}$$

(2.39)

where $\tilde{\Phi}(z)$ is analytic in D^+ and $\Phi(z)$ is a solution of (1.2) in D^- satisfying the boundary condition (2.14), but $\tilde{\psi}(z) = 0, z \in D^+, r(z) = 0, z \in \Gamma$, and

$$s(x) = \begin{cases} \dfrac{-2R((1-i)x/2)}{a((1-i)x/2) - b((1-i)x/2)} + \text{Re}\,[\overline{\lambda(x)}\Psi(x)], & x \in L_0, \text{ or} \\ \dfrac{-2R((1-i)x/2 + 1 - i)}{a((1+i)x/2 + 1 - i) + b((1+i)x/2 + 1 - i)} + \text{Re}\,[\overline{\lambda(x)}\Psi(x)], & x \in L_0. \end{cases}$$

By using the method in the proof of Theorem 2.3, we can derive that

$$|w^\pm(z)Y^\pm(z)| \leq \frac{[2M_8 M(4m+1)R']^n}{n!} \text{ in } \overline{D^-}.$$

(2.40)

Let $n \to \infty$, we get $w^\pm(z) = 0$, i.e. $w(z) = w_1(z) - w_2(z) = 0$, $\Psi(z) = \Phi(z) = 0$ in D^-. Noting that $w(z) = \tilde{\Phi}(z)e^{\tilde{\phi}(z)}$ satisfies the boundary conditions in (2.38), we see that the analytic function $\tilde{\Phi}(z)$ in D^+ satisfies the boundary conditions

$$\text{Re}\,[\overline{\lambda(z)}e^{\tilde{\phi}(z)}\tilde{\Phi}(z)] = 0, \quad z \in \Gamma, \quad \text{Re}\,[\overline{\lambda(x)}\tilde{\Phi}(x)] = 0, \quad x \in L_0, \quad (2.41)$$

and the index of the boundary value problem (2.41) is $K = -1/2$, hence $\tilde{\Phi}(z) = 0$ in D^+, and then $w(z) = \tilde{\Phi}(z)e^{\tilde{\phi}(z)} = 0$ in D^+, namely $w(z) = w_1(z) - w_2(z) = 0$ in D^+. This proves the uniqueness of solutions of Problem A for (2.2).

From Theorems 2.3 and 2.4, we see that under Condition C, Problem A for equation (2.2) has a unique solution $w(z)$, which can be found by using successive iteration, and $w(z)$ of Problem A satisfies the estimates

$$C_\beta[w(z)X(z), \overline{D^+}] \leq M_{11}, \ C[w^\pm(z)Y^\pm(z), \overline{D^-}] \leq M_{12}, \quad (2.42)$$

where $w^\pm(z) = \text{Re}\,w(z) \pm \text{Im}\,w(z)$, $X(z), Y^\pm(z)$ are as stated in (1.32), and $\beta(0 < \beta < \delta)$, $M_j = M_j(p_0, \beta, k, D)$ $(j = 11, 12)$ are non-negative constants, $k = (k_0, k_1, k_2)$. Moreover, we can derive the following theorem.

Theorem 2.5 *Suppose that equation (2.2) satisfies Condition C. Then any solution $w(z)$ of Problem A for (2.2) satisfies the estimates*

$$C_\beta[w(z)X(z), \overline{D^+}] \leq M_{13}(k_1 + k_2), \ C[w^\pm(z)Y^\pm(z), \overline{D^-}] \leq M_{14}(k_1 + k_2), \quad (2.43)$$

in which $M_j = M_j(p_0, \beta, k_0, D)(j = 13, 14)$ *are non-negative constants.*

Proof When $k_1 + k_2 = 0$, from Theorem 2.3, it is easy to see that (2.43) holds. If $k_1 + k_2 > 0$, then it is seen that the function $W(z) = w(z)/(k_1 + k_2)$ is a solution of the homogeneous boundary value problem

$$Lw = F(z, w)/k, \quad F/k = A_1 W + A_2 \overline{W} + A_3/k \text{ in } D,$$

$$\text{Re}\,[\overline{\lambda(z)}W(z)] = r(z)/k, \quad z \in \Gamma, \quad \text{Im}\,[\overline{\lambda(z_1)}W(z_1)] = b_1/k,$$

$$\text{Re}\,[\overline{\lambda(z)}W(z)] = r(z)/k, \quad z \in L_j, \quad j = 1 \text{ or } 2,$$

where $L_p[A_3/k, \overline{D^+}] \leq 1, C[A_3/k, \overline{D^-}] \leq 1, C_\alpha[r(z)/k, \Gamma] \leq 1, C_\alpha[r(z)/k, L_j] \leq 1, j = 1$ or 2, $|b_1/k| \leq 1$. On the basic of the estimate (2.42), we can obtain the estimates

$$C_\beta[w(z)X(z), \overline{D^+}] \leq M_{13}, \quad C[w^\pm(z)Y^\pm(z), \overline{D^-}] \leq M_{14}, \tag{2.44}$$

where $M_j = M_j(p_0, \beta, k_0, D)$ $(j = 13, 14)$ are non-negative constants. From (2.44), it follows the estimate (2.43).

From the estimates (2.43),(2.44), we can see the regularity of solutions of Problem A for (2.2). In the next section, we shall give the Hölder estimate of solutions of Problem A for first order quasilinear complex equation of mixed type with the more restrictive conditions than Condition C, which includes the linear complex equation (2.2) as a special case.

3   The Riemann–Hilbert Problem for First Order Quasilinear Complex Equations of Mixed Type

In this section we discuss the Riemann–Hilbert boundary value problem for first order quasilinear complex equations of mixed (elliptic-hyperbolic) type in a simply connected domain. We first give the representation theorem and prove the uniqueness of solutions for the above boundary value problem, and then by using the successive iteration, the existence of solutions for the above problem is proved.

3.1   Representation and uniqueness of solutions of Riemann–Hilbert problem for first order quasilinear complex equations of mixed type

Let D be a simply connected bounded domain as stated in Subsection 2.1. We discuss the quasilinear mixed (elliptic-hyperbolic) system of first order equations

$$\begin{cases} u_x - v_y = au + bv + f, \\ v_x + \text{sgny}\, u_y = cu + dv + g, \end{cases} \quad z = x + iy \in D, \tag{3.1}$$

in which a, b, c, d, f, g are functions of $(x, y)\,(\in D)$, $u, v\,(\in \mathbb{R})$, its complex form is the following complex equation of first order

$$\begin{Bmatrix} w_{\bar{z}} \\ w_{\bar{z}^*} \end{Bmatrix} = F(z, w), \quad F = A_1 w + A_2 \bar{w} + A_3 \text{ in } \begin{Bmatrix} D^+ \\ D^- \end{Bmatrix}, \tag{3.2}$$

where $A_j = A_j(z, w)$, $j = 1, 2, 3$, and the relations between $A_j\ (j = 1, 2, 3)$ and a, b, c, d, f, g are the same as those in (2.2).

Suppose that the complex equation (3.2) satisfies the following conditions.

Condition C

1) $A_j(z, w)\,(j = 1, 2, 3)$ are continuous in $w \in \mathbb{C}$ for almost every point $z \in D^+$, and are measurable in $z \in D^+$ and continuous on $\overline{D^-}$ for all continuous functions $w(z)$ in $D^* = \bar{D}\backslash(\{0, 2\} \cup \{x \pm y = 2, y \le 0\})$ or $D^* = \bar{D}\backslash(\{0, 2\} \cup \{x \pm y = 0, y \le 0\})$, and satisfy

$$L_p[A_j, \overline{D^+}] \le k_0, \quad j = 1, 2, \quad L_p[A_3, \overline{D^+}] \le k_1,$$
$$C[A_j, \overline{D^-}] \le k_0, \quad j = 1, 2, \quad C[A_3, \overline{D^-}] \le k_1. \tag{3.3}$$

2) For any continuous functions $w_1(z), w_2(z)$ on D^*, the following equality holds:

$$F(z, w_1) - F(z, w_2) = \tilde{A}_1(z, w_1, w_2)(w_1 - w_2) + \tilde{A}_2(z, w_1, w_2)(\overline{w_1} - \overline{w_2}) \text{ in } D, \tag{3.4}$$

where

$$L_p[\tilde{A}_j, \overline{D^+}] \le k_0, \quad C[\tilde{A}_j, \overline{D^-}] \le k_0, \quad j = 1, 2 \tag{3.5}$$

in (3.3),(3.5), $p\,(> 2), k_0, k_1$ are non-negative constants. In particular, when (3.2) is a linear equation (2.2), the condition (3.4) is obviously valid.

The boundary conditions of Riemann–Hilbert problem for the complex equation (3.2) are as stated in (1.3),(1.4). Let the solution $w(z)$ of Problem A be substituted in the coefficients of (3.2). Then the equation can be viewed as a linear equation (2.2). Hence we have the same representation theorems as Lemma 2.1 and Theorem 2.2.

Theorem 3.1 *Suppose that the quasilinear complex equation (3.2) satisfies Condition C. Then Problem A for (3.2) has a unique solution in D.*

Proof We first prove the uniqueness of the solution of Problem A for (3.2). Let $w_1(z), w_2(z)$ be any two solutions of Problem A for (3.2). By Condition C, we see that $w(z) = w_1(z) - w_2(z)$ satisfies the homogeneous complex equation and boundary conditions

$$w_{\bar{z}} = \tilde{A}_1 w + \tilde{A}_2 \bar{w} \text{ in } D, \tag{3.6}$$

$$\text{Re}\,[\overline{\lambda(z)}w(z)] = 0, \quad z \in L_1 \text{ or } L_2, \quad \text{Re}\,[\overline{\lambda(\tau_1)}w(\tau_1)] = 0, \tag{3.7}$$

where the conditions on the coefficients $\tilde{A}_j(j = 1, 2)$ are the same as in the proof of Theorem 2.4 for the linear equation (2.2). Besides the remaining proof is the same in the proof of Theorems 2.3 and 2.4.

Next noting the conditions (3.3),(3.4), by using the same method, the existence of solutions of Problem A for (3.2) can be proved, and any solution $w(z)$ of Problem A for (3.2) satisfies the estimate (2.43).

In order to give the Hölder estimate of solutions for (3.2), we need to add the following condition.

3) For any complex numbers z_1, $z_2(\in \bar{D})$, w_1, w_2, the above functions satisfy

$$|A_j(z_1, w_1) - A_j(z_2, w_2)| \le k_0[|z_1 - z_2|^\alpha + |w_1 - w_2|], \quad j = 1, 2,$$
$$|A_3(z_1, w_1) - A_3(z_2, w_2)| \le k_1[|z_1 - z_2|^\alpha + |w_1 - w_2|], \quad z \in \overline{D^-}, \tag{3.8}$$

in which $\alpha(1/2 < \alpha < 1)$, k_0, k_1 are non-negative constants.

On the basis of the results of Theorem 4.4 in Chapter I and Theorem 2.3 in Chapter III, we can derive the following theorem.

Theorem 3.2 *Let the quasilinear complex equation (3.2) satisfy Condition C and (3.8). Then any solution $w(z)$ of Problem A for (3.2) satisfies the following estimates*

$$C_\delta[X(z)w(z), \overline{D^+}] \le M_{15}, \quad C_\delta[Y^\pm(z)w^\pm(z), \overline{D^-}] \le M_{16}, \tag{3.9}$$

in which $w^\pm(z) = \operatorname{Re} w(z) \pm \operatorname{Im} w(z)$ and

$$X(z) = \prod_{j=1}^{2} |z - t_j|^{\eta_j}, \quad Y^\pm(x) = \prod_{j=1}^{2} |x \pm y - t_j|^{\eta_j}, \quad \eta_j = \begin{cases} 2|\gamma_j| + 2\delta, & \text{if } \gamma_j < 0, \\ 2\delta, & \gamma_j \ge 0, \end{cases} \tag{3.10}$$

here $\gamma_j(j = 1, 2)$ are real constants as stated in (1.7) and δ is a sufficiently small positive constant, and $M_j = M_j(p_0, \beta, k, D)\,(j = 15, 16)$ are non-negative constants, $k = (k_0, k_1, k_2)$.

3.2 Existence of solutions of Problem A for general first order complex equations of mixed type

Now, we consider the general quasilinear mixed complex equation of first order

$$Lw = \begin{Bmatrix} w_{\bar{z}} \\ \overline{w}_{\bar{z}^*} \end{Bmatrix} = F(z, w_z) + G(z, w), \quad z \in \begin{Bmatrix} D^+ \\ \overline{D^-} \end{Bmatrix},$$
$$F = A_1 w + A_2 \bar{w} + A_3, \quad G = A_4 |w|^\sigma, \quad z \in D, \tag{3.11}$$

in which $F(z, w)$ satisfies Condition C, σ is a positive constant, and $A_4(z, w)$ satisfies the same conditions as $A_j(j = 1, 2)$, where the main condition is

$$C[A_4(z, w), \bar{D}] \le k_0, \tag{3.12}$$

and denote the above conditions by Condition C'.

Theorem 3.3   *Let the mixed complex equation* (3.11) *satisfy Condition C'.*

(1) *When $0 < \sigma < 1$, Problem A for* (3.11) *has a solution $w(z)$.*

(2) *When $\sigma > 1$, Problem A for* (3.11) *has a solution $w(z)$, provided that*

$$M_{17} = k_1 + k_2 + |b_1| \tag{3.13}$$

is sufficiently small.

Proof   (1) Consider the algebraic equation for t :

$$(M_{13} + M_{14})\{k_1 + k_2 + 2k_0 t^\sigma + |b_1|\} = t, \tag{3.14}$$

in which M_{13}, M_{14} are constants stated in (2.43). It is not difficult to see that the equation (3.14) has a unique solution $t = M_{18} \geq 0$. Now, we introduce a closed and convex subset B^* of the Banach space $C(\bar{D})$, whose elements are the function $w(z)$ satisfying the condition

$$C[w(z)X(z), \overline{D^+}] + C[w^\pm(z)Y^\pm(z), \overline{D^-}] \leq M_{18}. \tag{3.15}$$

We arbitrarily choose a function $w_0(z) \in B^*$ for instance $w_0(z) = 0$ and substitute it into the position of w in the coefficients of (3.11) and $G(z, w)$. From Theorem 3.1, it is clear that problem A for

$$Lw - A_1(z, w_0)w - A_2(z, w_0)\bar{w} - A_3(z, w_0) = G(z, w_0), \tag{3.16}$$

has a unique solution $w_1(z)$. By (2.43), we see that the solution $w_1(z)$ satisfies the estimate in (3.15). By using successive iteration, we obtain a sequence of solutions $w_m(z)(m = 1, 2, ...)$ of Problem A, which satisfy the equations

$$Lw_{m+1} - A_1(z, w_m)w_{m+1z} - A_2(z, w_m)\bar{w}_{m+1}$$
$$+A_3(z, w_m) = G(z, w_m) \text{ in } \bar{D}, \quad m = 1, 2, \ldots \tag{3.17}$$

and $w_{m+1}(z)X(z) \in B^*$, $m = 1, 2, \ldots$. From (3.17), we see that $\tilde{w}_{m+1}(z) = w_{m+1}(z) - w_m(z)$ satisfies the complex equation and boundary conditions

$$L\tilde{w}_{m+1} - \tilde{A}_1\tilde{w}_{m+1} - \tilde{A}_2\bar{\tilde{w}}_{m+1} = \tilde{G}, \tilde{G}(z) = G(z, w_m) - G(z, w_{m-1}) \text{ in } \bar{D},$$
$$\text{Re}\,[\overline{\lambda(z)}\tilde{w}_{m+1}(z)] = 0 \text{ on } \Gamma \cup L_j, \quad j = 1 \text{ or } 2, \quad \text{Im}\,[\overline{\lambda(z_1)}\tilde{w}_{m+1}(z_1)] = 0, \tag{3.18}$$

where $m = 1, 2, \ldots$. Noting that $C[X(z)\tilde{G}(z), \bar{D}] \leq 2k_0 M_{18}$, M_{18} is a solution of the algebraic equation (3.14) and according to the proof of Theorem 2.3,

$$C[\tilde{w}_{m+1}X(z), \overline{D^+}] + C[\tilde{w}_{m+1}^\pm(z)Y^\pm(z), \overline{D^-}] \leq M_{18} \tag{3.19}$$

can be obtained. The function $\tilde{w}_{m+1}$ can be expressed as

$$w_{m+1}(z) = w_0(z) + \Phi_{m+1}(z) + \Psi_{m+1}(z),$$
$$\Psi_{m+1}(z) = \int_2^{x-y} [\tilde{A}\xi_{m+1} + \tilde{B}\eta_{m+1} + \tilde{E}]e_1 d(x-y)$$
$$+ \int_0^{x+y} [\tilde{C}\tilde{\xi}_{m+1} + \tilde{D}\eta_{m+1} + \tilde{F}]e_2 d(x+y) \text{ in } D^-, \tag{3.20}$$

in which the relation between $\tilde{A}, \tilde{B}, \tilde{C}, \tilde{D}, \tilde{E}, \tilde{F}$ and $\tilde{A}_1, \tilde{A}_2, \tilde{G}$ is the same as that of A, B, C, D, E, F and A_1, A_2, A_3 in (2.12). By using the method from the proof of Theorem 2.5, we can obtain

$$C[\tilde{w}_{m+1}X(z), \overline{D^+}] + C[\tilde{w}_{m+1}^{\pm}(z)Y^{\pm}(z), \overline{D^-}] \le \frac{(M_{20}R')^m}{m!},$$

where $M_{20} = 2M_{19}M(M_5+1)(4m+1)$, $R' = 2$, $m = C[w_0^{\pm}(z)Y^{\pm}(z), \tilde{D}]$, herein $M_{19} = \max\{C[\tilde{A},Q], C[\tilde{B},Q], C[\tilde{C},Q], C[\tilde{D},Q], C[\tilde{E},Q], C[\tilde{F},Q]\}$, $M = 1 + 4k_0^2(1 + k_0^2)$. From the above inequality, it is seen that the sequence of functions $\{w_m(z)X(z)\}$, i.e.

$$w_m^{\pm}(z)Y^{\pm}(z) = \{w_0^{\pm}(z) + [w_1^{\pm}(z) - w_0^{\pm}(z)] + \cdots + [w_m^{\pm}(z) - w_{m-1}^{\pm}(z)]\}Y^{\pm}(z) \quad (3.21)$$

$(m = 1, 2, \ldots)$ uniformly converge to $w_*^{\pm}(z)Y^{\pm}(z)$, and similarly to (2.30), the corresponding function $w_*(z)$ satisfies the equality

$$w_*(z) = w_0(z) + \Phi_*(z) + \Psi_*(z),$$

$$\Psi_*(z) = \int_2^{x-y}[A\xi_* + B\eta_* + E]e_1 d(x-y) \qquad (3.22)$$

$$+ \int_0^{x+y}[C\xi_* + D\eta_* + F]e_2 d(x+y) \text{ in } D^-,$$

and the function $w_*(z)$ is just a solution of Problem A for the quasilinear equation (3.11) in the closure of the domain D.

(2) Consider the algebraic equation

$$(M_{13} + M_{14})\{k_1 + k_2 + 2k_0 t^{\sigma} + |b_1|\} = t \qquad (3.23)$$

for t. It is not difficult to see that equation (3.23) has a solution $t = M_{18} \ge 0$, provided that M_{17} in (3.13) is small enough. Now, we introduce a closed and convex subset B_* of the Banach space $C(\bar{D})$, whose elements are the functions $w(z)$ satisfying the conditions

$$C[w(z)X(z), \overline{D^+}] + C[w^{\pm}(z)Y^{\pm}(z), \overline{D^-}] \le M_{18}. \qquad (3.24)$$

By using the same method as in (1), we can find a solution $u(z) \in B_*$ of Problem A for equation (3.11) with $\sigma > 1$.

4   The Riemann–Hilbert Problem for First Order Quasilinear Equations of Mixed type in General Domains

This section deals with the Riemann–Hilbert boundary value problem for quasilinear first order equations of mixed (elliptic-hyperbolic) type in general domains.

4.1 Formulation of the oblique derivative problem for second order equations of mixed type in general domains

Let D be a simply connected bounded domain D in the complex plane $\mathbb{C}$ with the boundary $\partial D = \Gamma \cup L$, where Γ, L are as stated in Section 1. Now, we consider the domain D' with the boundary $\Gamma \cup L_1' \cup L_2'$, where the parameter equations of the curves L_1', L_2' are as follows:

$$L_1' = \{\gamma_1(x) + y = 0, \, 0 \le x \le l\}, \quad L_2' = \{x - y = 2, \, l \le x \le 2\}, \quad (4.1)$$

in which $\gamma_1(x)$ on $0 \le x \le l = \gamma_1(l) + 2$ is continuous and $\gamma_1(0) = 0$, $\gamma_1(x) > 0$ on $0 \le x \le l$, and $\gamma_1(x)$ is differentiable on $0 \le x \le l$ except finitely many points and $1 + \gamma_1'(x) > 0$. Denote $D'^+ = D' \cap \{y > 0\} = D^+$, $D'^- = D' \cap \{y < 0\}$ and $z_1' = l - i\gamma_1(l)$. Here we mention that in [12]1),3), the author assumes that the derivative of $\gamma(x)$ satisfies $\gamma_1'(x) > 0$ on $0 \le x \le l$ and other conditions.

We consider the first order quasilinear complex equation of mixed type as stated in (3.2) in D', and assume that (3.2) satisfies Condition C in $\overline{D'}$.

The oblique derivative boundary value problem for equation (3.2) may be formulated as follows:

Problem A' Find a continuous solution $w(z)$ of (3.2) in $D_* = \bar{D} \backslash \{0, L_2'\}$, which satisfies the boundary conditions

$$\operatorname{Re}[\overline{\lambda(z)}w(z)] = r(z), \quad z \in \Gamma, \qquad (4.2)$$

$$\operatorname{Re}[\overline{\lambda(z)}w(z)] = r(z), \quad z \in L_1',$$
$$\operatorname{Im}[\overline{\lambda(z)}u_{\bar{z}}]|_{z=z_1'} = b_1, \qquad (4.3)$$

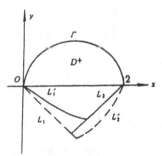

Figure 4.1

where $\lambda(z) = a(x) + ib(x)$ and $|\lambda(z)| = 1$ on $\Gamma \cup L_1'$, and b_0, b_1 are real constants, and $\lambda(z), r(z), b_0, b_1$ satisfy the conditions

$$C_\alpha[\lambda(z), \Gamma] \le k_0, \quad C_\alpha[r(z), \Gamma] \le k_2, \quad |b_1| \le k_2,$$

$$C_\alpha[\lambda(z), L_1'] \le k_0, \quad C_\alpha[r(z), L_1'] \le k_2, \quad \max_{z \in L_1'} \frac{1}{|a(x) - b(x)|} \le k_0, \qquad (4.4)$$

in which $\alpha \, (1/2 < \alpha < 1), k_0, k_2$ are non-negative constants. The boundary value problem for equation (3.2) with $A_3(z, u, u_z) = 0$, $z \in D, u \in \mathbb{R}, u_z \in \mathbb{C}$, $r(z) = 0$, $z \in \Gamma \cup L_1'$ and $b_0 = b_1 = 0$ will be called Problem A_0'. The number

$$K = \frac{1}{2}(K_1 + K_2) \qquad (4.5)$$

is called the index of Problem A' and Problem A_0' as stated in Section 1. Similarly we only discuss the case of $K = -1/2$ on ∂D^+, because in this case the solution of Problem A' is unique. Besides we choose $\gamma_1 > 0$. In the following, we first discuss the domain D' and then discuss another general domain D''.

4.2   The existence of solutions of Problem A for first order equations of mixed type in general domains

1.   By the conditions in (4.1), the inverse function $x = \sigma(\nu)$ of $x + \gamma_1(x) = \nu = x - y$ can be found and $\sigma'(\nu) = 1/[1 + \gamma_1'(x)]$. Hence the curve L_1' can be expressed by $x = \sigma(\nu) = (\mu + \nu)/2$, i.e. $\mu = 2\sigma(\nu) - \nu$, $0 \le \nu \le l + \gamma_1(l)$. We make a transformation

$$\tilde{\mu} = 2\left[\frac{\mu - 2\sigma(\nu) + \nu}{2 - 2\sigma(\nu) + \nu}\right], \quad \tilde{\nu} = \nu, \quad 2\sigma(\nu) - \nu \le \mu \le 2, \quad 0 \le \nu \le 2, \tag{4.6}$$

where μ, ν are real variables, its inverse transformation is

$$\mu = [2 - 2\sigma(\nu) + \nu]\tilde{\mu}/2 + 2\sigma(\nu) - \nu, \quad \nu = \tilde{\nu}, \quad 0 \le \tilde{\mu} \le 2, \quad 0 \le \tilde{\nu} \le 2. \tag{4.7}$$

It is not difficult to see that the transformation in (4.6) maps the domain D'^- onto D^-. The transformation (4.6) and its inverse transformation (4.7) can be rewritten as

$$\begin{cases} \tilde{x} = \dfrac{1}{2}(\tilde{\mu} + \tilde{\nu}) = \dfrac{4x - (2 + x - y)[2\sigma(x + \gamma_1(x)) - x - \gamma_1(x)]}{4 - 4\sigma(x + \gamma_1(x)) + 2x + 2\gamma_1(x)}, \\[4mm] \tilde{y} = \dfrac{1}{2}(\tilde{\mu} - \tilde{\nu}) = \dfrac{4y - (2 - x + y)[2\sigma(x + \gamma_1(x)) - x - \gamma_1(x)]}{4 - 4\sigma(x + \gamma_1(x)) + 2x + 2\gamma_1(x)}, \end{cases} \tag{4.8}$$

and

$$\begin{cases} x = \dfrac{1}{2}(\mu + \nu) = \dfrac{[2 - 2\sigma(x + \gamma_1(x)) + x + \gamma_1(x)](\tilde{x} + \tilde{y})}{4} \\[4mm] \qquad\qquad\quad + \sigma(x + \gamma_1(x)) - \dfrac{x + \gamma_1(x) - \tilde{x} + \tilde{y}}{2}, \\[4mm] y = \dfrac{1}{2}(\mu - \nu) = \dfrac{[2 - 2\sigma(x + \gamma_1(x)) + x + \gamma_1(x)](\tilde{x} + \tilde{y})}{4} \\[4mm] \qquad\qquad\quad + \sigma(x + \gamma_1(x)) - \dfrac{x + \gamma_1(x) + \tilde{x} - \tilde{y}}{2}. \end{cases} \tag{4.9}$$

Denote by $\tilde{z} = \tilde{x} + j\tilde{y} = f(z)$, $z = x + jy = f^{-1}(\tilde{z})$ the transformation (4.8) and the inverse transformation (4.9) respectively. In this case, the system of equations is

$$\xi_\nu = A\xi + B\eta + E, \quad \eta_\mu = C\xi + D\eta + F, \quad z \in D'^-, \tag{4.10}$$

which is another form of (3.2) in D'^-. Suppose that (3.2) in D' satisfies Condition C, through the transformation (4.7), we obtain $\xi_{\tilde{\nu}} = \xi_\nu$, $\eta_{\tilde{\mu}} = [2 - 2\sigma(\nu) + \nu]\eta_\mu/2$ in D'^-, and then

$$\xi_{\tilde{\nu}} = A\xi + B\eta + E$$

$$\eta_{\tilde{\mu}} = \frac{[2 - 2\sigma(\nu) + \nu][C\xi + D\eta + F]}{2} \quad \text{in } D^-, \tag{4.11}$$

and through the transformation (4.8), the boundary condition (4.3) is reduced to

$$\text{Re}\,[\lambda(f^{-1}(\tilde{z}))w(f^{-1}(\tilde{z}))] = r(f^{-1}(\tilde{z})), \quad \tilde{z} \in L_1,$$

$$\text{Im}\,[\lambda(f^{-1}(\tilde{z}_1))w(f^{-1}(\tilde{z}_1)] = b_1, \tag{4.12}$$

in which $\tilde{z}_1 = f(z_1')$. Therefore the boundary value problem (4.10),(4.3) is transformed into the boundary value problem (4.11),(4.12), i.e. the corresponding Problem A in D. On the basis of Theorem 3.1, we see that the boundary value problem (3.2)(in D^+),(4.11),(4.2),(4.12) has a unique solution $w(\tilde{z})$, and $w(z)$ is just a solution of Problem A for (3.2) in D' with the boundary conditions (4.2),(4.3).

Theorem 4.1 *If the mixed equation* (3.2) *in D' satisfies Condition C in the domain D' with the boundary $\Gamma \cup L_1' \cup L_2'$, where L_1', L_2' are as stated in* (4.1), *then Problem A' for* (3.2) *with the boundary conditions* (4.2), (4.3) *has a unique solution $w(z)$.*

2. Next let the domain D'' be a simply connected domain with the boundary $\Gamma \cup L_1'' \cup L_2''$, where Γ is as stated before and

$$L_1'' = \{\gamma_1(x) + y = 0, 0 \le x \le l\}, \quad L_2'' = \{\gamma_2(x) + y = 0, l \le x \le 2\}, \tag{4.13}$$

in which $\gamma_1(0) = 0, \gamma_2(2) = 0, \gamma_1(x) > 0, 0 \le x \le l, \gamma_2(x) > 0, l \le x \le 2, \gamma_1(x)$ on $0 \le x \le l, \gamma_2(x)$ on $l \le x \le 2$ are continuous, and differentiable except at isolated points, and $1 + \gamma_1'(x) > 0, 1 - \gamma_2'(x) > 0$. Denote $D''^+ = D'' \cap \{y > 0\} = D^+$ and $D''^- = D'' \cap \{y < 0\}$ and $z_1'' = l - i\gamma_1(l) = l - i\gamma_2(l)$. We consider the Riemann–Hilbert problem (Problem A') for equation (3.2) in D'' with the boundary conditions (4.2) and

$$\text{Re}\,[\overline{\lambda(z)}w(z)] = r(z), \quad z \in L_2'', \quad \text{Im}\,[\overline{\lambda(z_1'')}w(z_1'')] = b_1, \tag{4.14}$$

where $z_1'' = l - i\gamma_1(l) = l - i\gamma_2(l)$, and $\lambda(z)$, $r(z)$ satisfy the corresponding condition

$$C_\alpha[\lambda(z), \Gamma] \le k_0, \quad C_\alpha[r(z), \Gamma] \le k_2, \quad |b_1| \le k_2, \quad C_\alpha[\lambda(z), L_2''] \le k_0,$$

$$C_\alpha[r(z), L_2''] \le k_2, \quad \max_{z \in L_1''} \frac{1}{|a(x) - b(x)|}, \quad \max_{z \in L_2''} \frac{1}{|a(x) + b(x)|} \le k_0, \tag{4.15}$$

in which $\alpha\,(1/2 < \alpha < 1), k_0, k_2$ are non-negative constants. By the conditions in (4.13), the inverse function $x = \tau(\mu)$ of $x - \gamma_2(x) = \mu$ can be found, namely

$$\nu = 2\tau(\mu) - \mu, \quad 0 \le \mu \le 2. \tag{4.16}$$

We make a transformation

$$\tilde{\mu} = \mu, \quad \tilde{\nu} = \frac{2\nu}{2\tau(\mu) - \mu}, \quad 0 \le \mu \le 2, \quad 0 \le \nu \le 2\tau(\mu) - \mu, \tag{4.17}$$

where μ, ν are real variables, its inverse transformation is

$$\mu = \tilde{\mu} = \tilde{x} + \tilde{y},$$

$$\nu = \frac{[2\tau(\mu) - \mu]\tilde{\nu}}{2}$$

$$= \frac{[2\tau(x - \gamma_2(x)) - x + \gamma_2(x)](\tilde{x} - \tilde{y})}{2}, \quad 0 \le \tilde{\mu} \le 2, 0 \le \tilde{\nu} \le 2. \tag{4.18}$$

Hence we have

$$
\tilde{x} = \frac{1}{2}(\tilde{\mu} + \tilde{\nu}) = \frac{2(x-y)+(x+y)[2\tau(x-\gamma_2(x))-x+\gamma_2(x)]}{2[2\tau(x-\gamma_2(x))-x+\gamma_2(x)]},
$$

$$
\tilde{y} = \frac{1}{2}(\tilde{\mu} - \tilde{\nu}) = \frac{-2(x-y)+(x+y)[2\tau(x-\gamma_2(x))-x+\gamma_2(x)]}{2[2\tau(x-\gamma_2(x))-x+\gamma_2(x)]},
$$

$$
x = \frac{1}{2}(\mu+\nu) = \frac{1}{4}[(2\tau(x-\gamma_2(x))-x+\gamma_2(x))(\tilde{x}-\tilde{y})+2(\tilde{x}+\tilde{y})],
$$

$$
y = \frac{1}{2}(\mu-\nu) = \frac{1}{4}[(-2\tau(x-\gamma_2(x))+x-\gamma_2(x))(\tilde{x}-\tilde{y})+2(\tilde{x}+\tilde{y})].
$$

(4.19)

Denote by $\tilde{z} = \tilde{x} + j\tilde{y} = g(z)$, $z = x + jy = g^{-1}(\tilde{z})$ the transformation (4.18) and its inverse transformation in (4.19) respectively. Through the transformation (4.18), we obtain

$$
(u+v)_{\tilde{\nu}} = [\tau(\mu)-\mu/2](u+v)_\nu, \quad (u-v)_{\tilde{\mu}} = (u-v)_\mu \text{ in } D'^-. \tag{4.20}
$$

System (4.10) in D''^- is reduced to

$$
\begin{aligned}
\xi_{\tilde{\nu}} &= [\tau(\mu)-\mu/2][A\xi+B\eta+E] \\
\eta_{\tilde{\mu}} &= C\xi+D\eta+F
\end{aligned}
\quad \text{in } D'^-. \tag{4.21}
$$

Moreover, through the transformation (4.19), the boundary condition (4.14) on L''_2 is reduced to

$$
\mathrm{Re}\,[\overline{\lambda(g^{-1}(\tilde{z}))}w(g^{-1}(\tilde{z}))] = r[g^{-1}(\tilde{z})], z \in L'_2,
$$

$$
\mathrm{Im}\,[\overline{\lambda(g^{-1}(z'_1))}w(g^{-1}(z'_1))] = b_1,
$$

(4.22)

in which $z'_1 = \tilde{z}''_1 = g(z''_1)$. Therefore the boundary value problem (4.10),(4.14) is transformed into the boundary value problem (4.21),(4.22). According to the method in the proof of Theorem 4.1, we can see that the boundary value problem (3.2) (in D^+), (4.21), (4.2), (4.22) has a unique solution $w(\tilde{z})$, and then $w(z)$ is a solution of the boundary value problem (3.2),(4.2),(4.14). But we mention that through the transformation (4.17) or (4.19), the boundaries L''_1, L''_2 are reduced to L'_1, L'_2 respectively, such that L'_1, L'_2 satisfy the condition as stated in (4.1). In fact, if the intersection z''_1 of L''_1 and L''_2 belongs to L_2 and $\gamma_1(x) \geq 2(1-l)+x, 2-2l \leq x \leq l$, then the above requirement can be satisfied. If $z''_1 \in L_1 = \{x+y=0\}$, $\gamma_2(x) \geq 2l-x, l \leq x \leq 2l$, then we can proceed similarly.

Theorem 4.2 *If the mixed equation (3.2) satisfies Condition C in the domain D'' with the boundary $\Gamma \cup L''_1 \cup L''_2$, where L''_1, L''_2 are as stated in (4.13), then Problem A' for (3.2), (4.2), (4.14) in D'' has a unique solution $w(z)$.*

5 The Discontinuous Riemann–Hilbert Problem for Quasilinear Mixed Equations of First Order

This section deals with the discontinuous Riemann–Hilbert problem for quasilinear mixed (elliptic-hyperbolic) complex equations of first order in a simply connected domain. Firstly, we give the representation theorem and prove the uniqueness of solutions for the above boundary value problem. Afterwards by using the method of successive iteration, the existence of solutions for the above problem is proved.

5.1 Formulation of the discontinuous Riemann–Hilbert problem for complex equations of mixed type

Let D be a simply connected domain with the boundary $\Gamma \cup L_1 \cup L_2$ as stated as before, where $D^+ = \{|z-1| < 1, \operatorname{Im} z > 0\}$. We discuss the first order quasilinear complex equations of mixed type as stated in (3.2) with Condition C.

In order to introduce the discontinuous Riemann-Hilbert boundary value problem for the complex equation (3.2), let the functions $a(z), b(z)$ possess discontinuities of first kind at $m-1$ distinct points $z_1, z_2, \ldots, z_{m-1} \in \Gamma$, which are arranged according to the positive direction of Γ, and $Z = \{z_0 = 2, z_1, \ldots, z_m = 0\} \cup \{x \pm y = 0, x \pm y = 2, y \leq 0\}$, where m is a positive integer, and $r(z) = O(|z-z_j|^{-\beta_j})$ in the neighborhood of $z_j (z = 0, 1, \ldots, m)$ on $\overline{\Gamma}$, in which $\beta_j (j = 0, 1, \ldots, m)$ are sufficiently small positive numbers. Denote $\lambda(z) = a(z) + ib(z)$ and $|a(z)| + |b(z)| \neq 0$, there is no harm in assuming that $|\lambda(z)| = 1, z \in \Gamma^* = \Gamma \backslash Z$. Suppose that $\lambda(z), r(z)$ satisfy the conditions

$$\lambda(z) \in C_\alpha(\Gamma_j), \quad |z-z_j|^{\beta_j} r(z) \in C_\alpha(\Gamma_j), \quad j = 0, 1, \ldots, m, \tag{5.1}$$

herein Γ_j is an arc from the point z_{j-1} to z_j on Γ and $\Gamma_j (j = 1, \ldots, m)$ does not include the end points, $\alpha(0 < \alpha < 1)$ is a constant.

Problem A_* Find a continuous solution $w(z)$ of (3.2) in $D^* = \bar{D} \backslash Z$ and $w(z)$ on Z maybe become infinite of an order lower than unity, which satisfies the boundary conditions

$$\operatorname{Re}[\overline{\lambda(z)}w(z)] = r(z), \quad z \in \Gamma, \tag{5.2}$$

$$\operatorname{Re}[\overline{\lambda(z)}w(z)] = r(z), \quad z \in L_j \ (j = 1 \text{ or } 2), \quad \operatorname{Im}[\overline{\lambda(z)}w(z)]|_{z=z_1} = b_1, \tag{5.3}$$

where b_1 are real constants, $\lambda(z) = a(x) + ib(x)(|\lambda(z)| = 1)$, $z \in \Gamma \cup L_j (j = 1 \text{ or } 2)$, and $\lambda(z), r(z), b_1$ satisfy the conditions

$$C_\alpha[\lambda(z), \Gamma_j] \leq k_0, \quad C_\alpha[r(z), \Gamma_j] \leq k_2, \quad |b_1| \leq k_2,$$

$$C_\alpha[\lambda(z), L_j] \leq k_0, \quad C_\alpha[r(z), L_j] \leq k_2, \quad j = 1 \text{ or } 2,$$

$$\max_{z \in L_1} \frac{1}{|a(z) - b(z)|} \leq k_0, \text{ or } \max_{z \in L_2} \frac{1}{|a(z) + b(z)|} \leq k_0, \tag{5.4}$$

in which $\alpha \, (1/2 < \alpha < 1), k_0, k_2$ are non-negative constants. The above discontinuous Riemann–Hilbert boundary value problem for (3.2) is called Problem A_*.

Denote by $\lambda(z_j - 0)$ and $\lambda(z_j + 0)$ the left limit and right limit of $\lambda(z)$ as $z \to z_j$ ($j = 0, 1, \ldots, m$) on Γ, and

$$e^{i\phi_j} = \frac{\lambda(z_j - 0)}{\lambda(z_j + 0)}, \quad \gamma_j = \frac{1}{\pi i} \ln \left[\frac{\lambda(z_j - 0)}{\lambda(z_j + 0)}\right] = \frac{\phi_j}{\pi} - K_j,$$

$$K_j = \left[\frac{\phi_j}{\pi}\right] + J_j, \; J_j = 0 \text{ or } 1, \quad j = 0, 1, \ldots, m, \tag{5.5}$$

in which $z_m = 0, z_0 = 2, \lambda(z) = \exp(-i\pi/4)$ on $L_0 = (0, 2)$ and $\lambda(z_0 - 0) = \lambda(z_n + 0) = \exp(-i\pi/4)$, or $\lambda(z) = \exp(i\pi/4)$ on L_0 and $\lambda(z_0 - 0) = \lambda(z_n + 0) = \exp(i\pi/4)$, and $0 \le \gamma_j < 1$ when $J_j = 0$, and $-1 < J_j < 0$ when $J_j = 1, j = 0, 1, \ldots, m$, and

$$K = \frac{1}{2}(K_0 + K_2 + \cdots + K_m) = \sum_{j=0}^{m}\left[\frac{\phi_j}{2\pi} - \frac{\gamma_j}{2}\right] \tag{5.6}$$

is called the index of Problem A_*. Now the function $\lambda(z)$ on $\Gamma \cup L_0$ is not continuous, we can choose $J_j = 0$ or $1 \, (0 \le j \le m)$, hence the index K is not unique. Here we choose the index $K = -1/2$. Let $\beta_j + \gamma_j < 1, j = 0, 1, \ldots, m$. We can require that the solution $u(z)$ satisfy the conditions

$$u_z = O(|z - z_j|^{-\delta}), \delta = \begin{cases} \beta_j + \tau, \text{ for } \gamma_j \ge 0, \text{ and } \gamma_j < 0, \quad \beta_j > |\gamma_j|, \\ |\gamma_j| + \tau, \text{ for } \gamma_j < 0, \quad \beta_j \le |\gamma_j|, \quad j = 1, \ldots, m, \end{cases} \tag{5.7}$$

in the neighborhood of z_j in D^*, where $\tau(<\alpha)$ is a small positive number.

For Problem A_* of the quasilinear complex equation (3.2), we can prove that there exists a unique solution by using a similar method as stated in the last section.

Next we discuss the more general discontinuous Riemann–Hilbert problem. As stated before, denote $L = L_1 \cup L_2, L_1 = \{x = -y, 0 \le x \le 1\}, L_2 = \{x = y + 2, 1 \le x \le 2\}$, and $D^+ = D \cap \{y > 0\}, D^- = D \cap \{y < 0\}$. Here, there are n points $E_1 = a_1, E_2 = a_2, \ldots, E_n = a_n$ on the segment $AB = (0, 2) = L_0$, where $a_0 = 0 < a_1 < a_2 < \cdots < a_n < a_{n+1} = 2$, and denote by $A = A_0 = 0, A_1 = (1 - i)a_1/2, A_2 = (1-i)a_2/2, \ldots, A_n = (1-i)a_n/2, A_{n+1} = C = 1-i$ and $B_1 = 1-i+(1+i)a_1/2, B_2 = 1 - i + (1 + i)a_2/2, \ldots, B_n = 1 - i + (1 + i)a_n/2, B = B_{n+1} = 2$, on the segments AC, CB respectively. Moreover, we denote $D_1^- = \overline{D^-} \cap \{\cup_{j=0}^{[n/2]}(a_{2j} \le x - y \le a_{2j+1})\}$, $D_2^- = \overline{D^-} \cap \{\cup_{j=1}^{[(n+1)/2]}(a_{2j-1} \le x + y \le a_{2j})\}$ and $\tilde{D}_{2j+1}^- = \overline{D^-} \cap \{a_{2j} \le x - y \le a_{2j+1}\}$, $j = 0, 1, \ldots, [n/2]$, $\tilde{D}_{2j}^- = \overline{D^-} \cap \{a_{2j-1} \le x + y \le a_{2j}\}, j = 1, \ldots, [(n + 1)/2]$, and $D_*^- = \overline{D^-}\backslash Z, Z = \{\cup_{j=0}^{n+1}(x \pm y = a_j, y \le 0)\}, D_* = D^+ \cup D_*^-$.

Problem A^*   Find a continuous solution $w(z)$ of (3.2) in $D^* = \bar{D}\backslash Z$, where $Z = \{z_0, z_1, \ldots, z_m, a_1, \ldots, a_n\} \cup \{x \pm y = a_j, y \le 0, j = 1, \ldots, n\}$, and the above solution $w(z)$ satisfies the boundary conditions (5.2) and

$$\text{Re}\left[\overline{\lambda(z)}w(z)\right]=r(z),$$

$$z\in L_3=\sum_{j=0}^{[n/2]}A_{2j}A_{2j+1},$$

$$\text{Re}\left[\overline{\lambda(z)}w(z)\right]=r(z),$$

$$z\in L_4=\sum_{j=1}^{[(n+1)/2]}B_{2j-1}B_{2j},$$

$$\text{Im}\left[\overline{\lambda(z)}w(z)\right]\big|_{z=A_{2j+1}}=c_{2j+1},$$

$$j=0,1,\ldots,[n/2],$$

$$\text{Im}\left[\overline{\lambda(z)}w(z)\right]\big|_{z=B_{2j-1}}=c_{2j},$$

$$j=1,\ldots,[(n+1)/2],$$

(5.8)

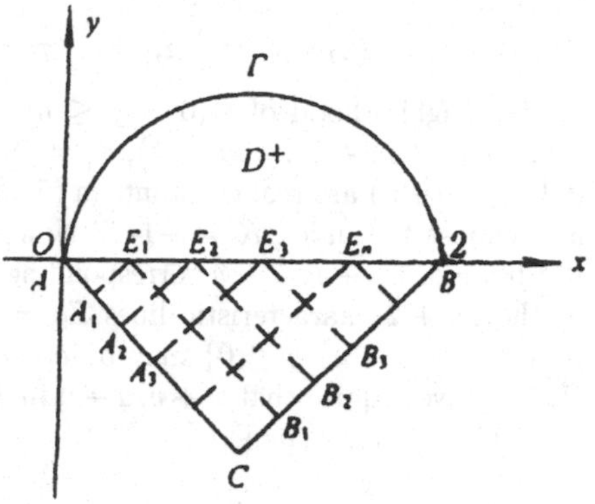

Figure 5.1

where $c_j(j=1,\ldots,n+1)$ are real constants, $\lambda(z)=a(x)+ib(x)$, $|\lambda(z)|=1$, $z\in\Gamma$, and $\lambda(z)$, $r(z)$, $c_j(j=1,\ldots,n+1)$ satisfy the conditions

$$C_\alpha[\lambda(z),\Gamma]\le k_0,\quad C_\alpha[r(z),\Gamma]\le k_2,\quad |c_j|\le k_2,\quad j=0,1,\ldots,n+1,$$

$$C_\alpha[\lambda(z),L_j]\le k_0,\quad C_\alpha[r(z),L_j]\le k_2,\quad j=3,4,$$

$$\max_{z\in L_3}\frac{1}{|a(x)-b(x)|},\quad\text{and}\quad\max_{z\in L_4}\frac{1}{|a(x)+b(x)|}\le k_0,$$

(5.9)

where $\alpha\,(1/2<\alpha<1)$, k_0,k_2 are non-negative constants. The above discontinuous Riemann–Hilbert boundary value problem for (3.2) is called Problem A^*.

Denote by $\lambda(t_j-0)$ and $\lambda(t_j+0)$ the left limit and right limit of $\lambda(z)$ as $z\to t_j=z_j(j=0,1,\ldots,m,\,z_{m+k}=a_k,k=1,\ldots,n,\,z_{n+m+1}=2)$ on $\Gamma\cup L_0\,(L_0=(0,2))$, and

$$e^{i\phi_j}=\frac{\lambda(t_j-0)}{\lambda(t_j+0)},\quad \gamma_j=\frac{1}{\pi i}\ln\left(\frac{\lambda(t_j-0)}{\lambda(t_j+0)}\right)=\frac{\phi_j}{\pi}-K_j,$$

$$K_j=\left[\frac{\phi_j}{\pi}\right]+J_j,\quad J_j=0\text{ or }1,\quad j=0,1,\ldots,m+n,$$

(5.10)

in which $[a]$ is the largest integer not exceeding the real number a, $\lambda(z)=\exp(-i\pi/4)$ on $L_1'=AB\cap\overline{D_1^-}$ and $\lambda(a_{2j}+0)=\lambda(a_{2j+1}-0)=\exp(-i\pi/4)$, $j=0,1,\ldots,[n/2]$, and $\lambda(z)=\exp(i\pi/4)$ on $L_2'=AB\cap\overline{D_2^-}$ and $\lambda(a_{2j-1}+0)=\lambda(a_{2j}-0)=\exp(i\pi/4)$, $j=1,\ldots,[(n+1)/2]$, and $0\le\gamma_j<1$ when $J_j=0$, and $-1<\gamma_j<0$ when $J_j=1$, $j=0,1,\ldots,m+n$, and

$$K=\frac{1}{2}(K_0+K_1+\cdots+K_{m+n})=\sum_{j=0}^{m+n}\left(\frac{\phi_j}{2\pi}-\frac{\gamma_j}{2}\right)$$

(5.11)

is called the index of Problem A^* and Problem A_0^*. We can require that the solution $w(z)$ in D^+ satisfy the conditions

$$w(z) = O(|z - z_j|^{-\tau}), \quad \tau = \gamma'_j + \delta, \quad j = 0, 1, \ldots, m + n, \qquad (5.12)$$

in the neighborhood of $z_j (0 \leq j \leq m + n)$ in D^+, where $\gamma'_j = \max(0, -\gamma_j) (j = 1, \ldots, m - 1, m + 1, \ldots, m + n), \gamma'_m = \max(0, -2\gamma_m), \gamma'_0 = \max(0, -2\gamma_0)$ and $\gamma_j (j = 0, 1, \ldots, m + n)$ are real constants in (5.10), δ is a sufficiently small positive number, and choose the index $K = -1/2$. Now we explain that in the closed domain $\overline{D^-}$, the functions $u + v, u - v$ corresponding to the solution $w(z)$ in the neighborhoods of the $2n + 2$ characteristic lines $Z_0 = \{x + y = 0, x - y = 2, x \pm y = a_j (j = m + 1, \ldots, m + n), y \leq 0\}$ may be not bounded if $\gamma_j \leq 0 (j = m, \ldots, m + n + 1)$. Hence if we require that $u + v, u - v$ in $\overline{D^-}\backslash Z_0$ is bounded, then it needs to choose $\gamma_j > 0 (j = 0, 1, \ldots, m + n + 1)$.

5.2 Representation of solutions for the discontinuous Riemann–Hilbert problem

We first introduce a lemma.

Lemma 5.1 *Suppose that the complex equation (3.2) satisfies Condition C. Then there exists a solution of Problem A^* for (3.2) in D^+ with the boundary conditions (5.2) and*

$$\text{Re}\,[\overline{\lambda(z)}w(z)]|_{z=x} = s(x), \quad C_\beta[s(x), L'_j] \leq k_3, \quad j = 1, 2,$$

$$\lambda(x) = \begin{cases} 1 - i \text{ on } L'_1 = \overline{D^-_1} \cap AB, \\ 1 + i \text{ on } L'_2 = \overline{D^-_2} \cap AB, \end{cases} \qquad (5.13)$$

and $w(z)$ satisfies the estimate

$$C_\beta[w(z)X(z), \overline{D^+}] \leq M_{21}(k_1 + k_2 + k_3), \qquad (5.14)$$

in which k_3 is a non-negative constant, $s(x)$ is as stated in the form (5.25) below, $X(z) = \Pi_{j=0}^{m+n}|z - z_j|^{\gamma'_j + \delta}$, herein $\gamma'_j = \max(0, -\gamma_j)(j = 1, \ldots, m - 1, m + 1, m + n), \gamma'_0 = \max(0, -2\gamma_0), \gamma'_m = \max(0, -2\gamma_m)$ and $\gamma_j (j = 0, 1, \ldots, m + n)$ are real constants in (5.10), $\beta (0 < \beta < \delta), \delta$ are sufficiently small positive numbers, and $M_{21} = M_{21}(p_0, \beta, k_0, D^+)$ is a non-negative constant.

By using the method as in the proofs of Theorems 2.1–2.3, Chapter III, Theorem 1.2 and Lemma 2.1, we can prove the lemma.

Theorem 5.2 *Problem A^* for equation (1.2) in D has a unique solution $w(z)$.*

Proof First of all, similarly to Theorem 1.2, the solution $w(z) = u(z) + iv(z)$ of equation (1.2) in $\overline{D^-}$ can be expressed as (1.19). According to the proof of Theorem 1.2, we can obtain $f(x + y)$ on $L'_1 = \overline{D^-_1} \cap AB$ and $g(x - y)$ on $L'_2 = \overline{D^-_2} \cap AB$ in the form

$$g(x) = k(x) = \frac{2r((1-i)x/2) - [a((1-i)x/2) + b((1-i)x/2)]h_{2j}}{a((1-i)x/2) - b((1-i)x/2)}$$

$$\text{on } \tilde{L}_{2j+1} = \overline{\tilde{D}_{2j+1}^-} \cap AB,$$

$$f(x) = h(x) = \frac{2r((1+i)x/2 + 1 - i)}{a((1+i)x/2 + 1 - i) + b((1+i)x/2 + 1 - i)}$$

$$- \frac{[a((1+i)x/2 + 1 - i) - b((1+i)x/2 + 1 - i)]k_{2j-1}}{a((1+i)x/2 + 1 - i) + b((1+i)x/2 + 1 - i)}$$

$$\text{on } \tilde{L}_{2j} = \overline{\tilde{D}_{2j}^-} \cap AB,$$

where $\tilde{D}_j^- (j = 1, 2, \ldots, 2n + 1)$ are as stated in Subsection 5.1, and

$$h_{2j} = \text{Re}\left[\lambda(A_{2j+1})(r(A_{2j+1}) + ic_{2j+1})\right] + \text{Im}\left[\lambda(A_{2j+1})(r(A_{2j+1}) + ic_{2j+1})\right]$$

$$\text{on } \tilde{L}_{2j+1}, \quad j = 0, 1, \ldots, [n/2],$$

$$k_{2j-1} = \text{Re}\left[\lambda(B_{2j-1})(r(B_{2j-1}) + ic_{2j})\right] - \text{Im}\left[\lambda(B_{2j-1})(r(B_{2j-1}) + ic_{2j})\right]$$

$$\text{on } \tilde{L}_{2j}, \quad j = 1, \ldots, [(n+1)/2],$$

where $\tilde{L}_{2j+1} = \tilde{D}_{2j+1}^- \cap AB, j = 0, 1, \ldots, [n/2], \tilde{L}_{2j} = \tilde{D}_{2j}^- \cap AB, j = 1, \ldots, [(n+1)/2]$.
By using Theorem 2.2, Chapter III, choosing an appropriate index $K = -1/2$, there
exists a unique solution $w(z)$ of Problem A^* in $\overline{D^+}$ with the boundary conditions
(5.2) and

$$\text{Re}\left[\overline{\lambda(x)}w(x)\right] = \begin{cases} k(x), \\ h(x), \end{cases} \quad \lambda(x) = \begin{cases} 1 - i \text{ on } L_1' = \overline{D_1^-} \cap AB, \\ 1 + i \text{ on } L_2' = \overline{D_2^-} \cap AB, \end{cases}$$

and denote

$$\text{Re}\left[\lambda(x)w(x)\right] = \begin{cases} h(x) \text{ on } L_1', \\ k(x) \text{ on } L_2', \end{cases} \tag{5.16}$$

$$C_\beta[X(x)k(x), L_1'] \le k_2, \quad C_\beta[X(x)h(x), L_2'] \le k_2,$$

herein $\beta(0 < \beta \le \alpha < 1)$, k_2 are non-negative constants.

Next we find a solution $w(z)$ of Problem A^* for (1.2) in $\overline{D^-}$ with the boundary
conditions

$$\text{Re}\left[\overline{\lambda(z)}w(z)\right] = r(z) \text{ on } L_3 \cup L_4,$$

$$\text{Im}\left[\overline{\lambda(z)}w(z)\right]|_{z=A_{2j+1}} = c_{2j+1}, \quad j = 0, 1, \ldots, [n/2], \tag{5.17}$$

$$\text{Im}\left[\overline{\lambda(z)}w(z)\right]|_{z=B_{2j-1}} = c_{2j}, \quad j = 1, \ldots, [(n+1)/2],$$

and (5.16), where $c_j(j = 1, \ldots, n + 1)$ are as stated in (5.8). By the result and the
method in Chapters I and II, we can find the solution of Problem A^* for (1.2) in $\tilde{D}_1^-$
in the form

$$w(z) = \tilde{w}(z) + \lambda(A_{2j+1})[r(A_{2j+1}) + ic_{2j+1}],$$

$$\tilde{w}(z) = \frac{1}{2}[(1+i)f_{2j+1}(x+y) + (1-i)g_{2j+1}(x-y)],$$

$$f_{2j+1}(x+y) = \operatorname{Re}[\lambda(x+y)w(x+y)], \quad g_{2j+1}(x-y)$$

$$= \frac{2r((1-i)(x-y)/2)}{a((1-i)(x-y)/2) - b((1-i)(x-y)/2)} \quad \text{in } \tilde{D}_{2j+1}^-,$$

$$j = 0, 1, \ldots, [n/2],$$

$$w(z) = \tilde{w}(z) + \lambda(B_{2j-1})[r(B_{2j-1}) + ic_{2j}],$$

$$\tilde{w}(z) = \frac{1}{2}[(1+i)f_{2j}(x+y) + (1-i)g_{2j}(x-y)], \quad f_{2j}(x+y)$$

$$= h(x+y) = \frac{2r((1+i)(x+y)/2 + 1 - i)}{a((1+i)(x+y)/2 + 1 - i) + b((1+i)(x+y)/2 + 1 - i)},$$

$$g_{2j}(x-y) = \operatorname{Re}[\lambda(x-y)w(x-y)] \text{ in } \tilde{D}_{2j}^-, \quad j = 1, \ldots, \left[\frac{n+1}{2}\right].$$

Furthermore, from the above solution we can find the solution of Problem A^* for (1.2) in $D^- \backslash \{D_1^- \cup D_2^-\}$, and the solution $w(z)$ of Problem A^* for (1.2) in $\overline{D^-}$ possesses the form

$$w(z) = \frac{1}{2}[(1+i)f(x+y) + (1-i)g(x-y)],$$

$$f(x+y) = \operatorname{Re}[(1-i)w(x+y)] \text{ in } D^- \backslash \overline{D_2^-},$$

$$g(x-y) = k(x-y) \text{ in } D_1^-, \tag{5.18}$$

$$f(x+y) = h(x+y) \text{ in } D_2^-,$$

$$g(x-y) = \operatorname{Re}[(1+i)w(x-y)] \text{ in } D^- \backslash \overline{D_1^-},$$

where $k(x), h(x)$ are as stated in (5.15), and $w(z)$ is the solution of Problem A^* for (1.2) with the boundary conditions (5.2),(5.16).

Theorem 5.3 *Let the complex equation (3.2) satisfy Condition C. Then any solution of Problem A^* for (3.2) can be expressed as*

$$w(z) = w_0(z) + W(z) \text{ in } D, \tag{5.19}$$

where $w_0(z)$ is a solution of Problem A^* for equation (1.2), and $W(z)$ possesses the form

$$W(z) = w(z) - w_0(z) \text{ in } D, \quad w(z) = \tilde{\Phi}(z)e^{\tilde{\phi}(z)} + \tilde{\psi}(z),$$

$$\tilde{\phi}(z) = \tilde{\phi}_0(z) + Tg = \tilde{\phi}_0(z) - \frac{1}{\pi}\int\int_{D^+}\frac{g(\zeta)}{\zeta - z}d\sigma_\zeta, \quad \tilde{\psi}(z) = Tf \text{ in } D^+,$$

$$W(z) = \Phi(z) + \Psi(z) \text{ in } D^-,$$

$$\Psi(z) = \begin{cases} \int_{a_{2j+1}}^{\nu} g^1(z)d\nu e_1 + \int_0^{\mu} g^2(z)d\mu e_2 \text{ in } D^-_{2j+1}, & j=0,1,\ldots,[n/2], \\ \int_2^{\nu} g^1(z)d\nu e_1 + \int_{a_{2j-1}}^{\mu} g^2(z)d\mu e_2 \text{ in } D^-_{2j}, & j=1,\ldots,[(n+1)/2], \end{cases} \quad (5.20)$$

in which $e_1 = (1+i)/2$, $e_2 = (1-i)/2$, $\mu = x+y$, $\nu = x-y$, $\tilde{\phi}_0(z)$ *is an analytic function in* D^+, $\text{Im}\,[\tilde{\phi}(z)] = 0$ *on* L_0, *and*

$$g(z) = \begin{cases} A_1 + A_2\overline{w}/(w), & w(z) \neq 0, \\ 0, \, w(z) = 0, & z \in D^+, \end{cases}$$

$$f = A_1 Tf + A_2 \overline{Tf} + A_3 \text{ in } D^+, \quad (5.21)$$

$$g^1(z) = A\xi + B\eta + E, \, g^2(z) = C\xi + D\eta + F \text{ in } D^-,$$

where $\xi = \text{Re}\,w + \text{Im}\,w$, $\eta = \text{Re}\,w - \text{Im}\,w$, *and* $\tilde{\phi}(z)$, $\tilde{\psi}(z)$ *satisfy the estimates*

$$C_\beta[\tilde{\phi}(z), \overline{D^+}] + L_{p_0}[\tilde{\phi}_{\tilde{z}}, \overline{D^+}] \le M_{22},$$

$$C_\beta[\tilde{\psi}(z), \overline{D^+}] + L_{p_0}[\tilde{\psi}_{\tilde{z}}, \overline{D^+}] \le M_{22}, \quad (5.22)$$

where $p_0 \, (0 < p_0 \le p)$, $M_{22} = M_{22}\,(p_0, \alpha, k, D^+)$ *are non-negative constants,* $\tilde{\Phi}(z)$ *is analytic in* D^+ *and* $\Phi(z)$ *is a solution of equation* (1.2) *in* D^- *satisfying the boundary conditions*

$$\text{Re}\,[\overline{\lambda(z)}e^{\tilde{\phi}(z)}\tilde{\Phi}(z)] = r(z) - \text{Re}\,[\overline{\lambda(z)}\tilde{\psi}(z)], \quad z \in \Gamma,$$

$$\text{Re}\,[\overline{\lambda(x)}(\tilde{\Phi}(x)e^{\tilde{\phi}(x)} + \tilde{\psi}(x))] = s(x), \quad x \in L_0,$$

$$\text{Re}\,[\lambda(x)\Phi(x)] = \text{Re}\,[\lambda(x)(W(x) - \Psi(x))], \quad z \in L_0,$$

$$\text{Re}\,[\overline{\lambda(z)}\Phi(z)] = -\text{Re}\,[\overline{\lambda(z)}\Psi(z)], \quad z \in L_3 \cup L_4, \quad (5.23)$$

$$\text{Im}\,[\overline{\lambda(z)}(\Phi(z) + \Psi(z))]|_{z=A_{2j+1}} = 0, \quad j = 0,1,\ldots,[n/2],$$

$$\text{Im}\,[\overline{\lambda(z)}(\Phi(z) + \Psi(z))]|_{z=B_{2j-1}} = 0, \quad j = 1,\ldots,[(n+1)/2].$$

Moreover the solution $w_0(z)$ *of Problem* A^* *for* (1.2) *satisfies the estimate in the form*

$$C_\beta[X(z)w_0(z), \overline{D^+}] + C[w_0^{\pm}(\mu,\nu)Y^{\pm}(\mu,\nu), \overline{D^-}] \le M_{23}(k_1 + k_2) \quad (5.24)$$

in which $X(z) = \Pi_{j=0}^{m+n}|z - z_j|^{\gamma_j + \delta}$, $Y^{\pm}(z) = Y^{\pm}(\mu,\nu) = \Pi_{j=0}^{m+n}|x \pm y - z_j|^{\gamma'_j + \delta}$, $w_0^{\pm}(\mu,\nu) = \text{Re}\,w_0(z)\pm \text{Im}\,w_0(z)$, $w_0(z) = w_0(\mu,\nu)$, $\mu = x+y$, $\nu = x-y$, $\gamma'_j(j = 0,1,\ldots,m+n)$ *are as stated in* (5.14), $M_{23} = M_{23}\,(p_0, \beta, k_0, D)$ *is a non-negative constant.*

Proof   Let the solution $w(z)$ of Problem A^* be substituted into the complex equation (3.2) and the solution $w_0(z) = \xi_0 e_1 + \eta_0 e_2$ of Problem A^* for equation (1.2) be substituted in the position of w in (5.21). Thus the functions $f(z), g(z)$ in D^+ and $g_1(z)$, $g_2(z)$ and $\Psi(z)$ in $\overline{D^-}$ in (5.20),(5.21) can be determined. Moreover, by Theorem 5.2, we can find an analytic function $\tilde{\Phi}(z)$ in D^+ and a solution $\Phi(z)$ of (1.2) in $\overline{D^-}$ with the boundary conditions (5.23), where

$$
s(x) = \begin{cases}
\dfrac{2r((1-i)x/2) - 2R((1-i)x/2)}{a((1-i)x/2) - b((1-i)x/2)} \\[2mm]
- \dfrac{[a((1-i)x/2) + b((1-i)x/2)]h_{2j}}{a((1-i)x/2) - b((1-i)x/2)} + \mathrm{Re}\,[\overline{\lambda(x)}\Psi(x)], \\[2mm]
x \in (a_{2j}, a_{2j+1}), \quad j = 0, 1, \ldots, [n/2], \\[2mm]
\dfrac{2r((1+i)x/2+1-i) - 2R((1+i)x/2+1-i)}{a((1+i)x/2+1-i) + b((1+i)x/2+1-i)} \\[2mm]
- \dfrac{[a((1+i)x/2+1-i) - b((1+i)x/2+1-i)]k_{2j-1}}{a((1+i)x/2+1-i) + b((1+i)x/2+1-i)} \\[2mm]
+ \mathrm{Re}\,[\overline{\lambda(x)}\Psi(x)], x \in (a_{2j-1}, a_{2j}), j = 1, \ldots, [(n+1)/2],
\end{cases} \tag{5.25}
$$

in which the real constants $h_{2j}(j = 0, 1, ..., [n/2])$, $k_{2j-1}(j = 1, ..., [(n+1)/2])$ are as stated in (5.15), thus

$$
w(z) = w_0(z) + W(z) = \begin{cases}
\tilde{\Phi}(z)^{\tilde{\phi}(z)} + \tilde{\psi}(z) \text{ in } D^+, \\
w_0(z) + \Phi(z) + \Psi(z) \text{ in } D^-,
\end{cases}
$$

is the solution of Problem A^* for the complex equation

$$
\begin{Bmatrix} w_{\bar{z}} \\ \bar{w}_{\overline{z^*}} \end{Bmatrix} = A_1 w + A_2 \bar{w} + A_3 \text{ in } \begin{Bmatrix} D^+ \\ D^- \end{Bmatrix}, \tag{5.26}
$$

and the solution $w_0(z)$ of Problem A^* for (1.2) satisfies the estimate (5.24).

5.3   Existence and uniqueness of solutions of the Riemann–Hilbert problem for (3.2)

Theorem 5.4   *Suppose that the complex equation (3.2) satisfies Condition C. Then Problem A^* for (3.2) is solvable.*

Proof   In order to find a solution $w(z)$ of Problem A^* in D, we express $w(z)$ in the form (5.19)–(5.21). On the basis of Theorem 5.2, we see that Problem A^* for (1.2) has a unique solution $w_0(z)(= \xi_0 e_1 + \eta_0 e_2)$, and substitute it into the position of $w = \xi e_1 + \eta e_2$ in the right-hand side of (3.2). Similarly to (2.19), from (5.19)–(5.21), we obtain the corresponding functions $g_0(z), f_0(z)$ in $\overline{D^+}$, $g_0^1(z), g_0^2(z)$ in $\overline{D^-}$ and the functions

$$\tilde{\phi}_1(z) = \tilde{\phi}_0(z) - \frac{1}{\pi} \iint_{D^+} \frac{g_0(\zeta)}{\zeta - z} d\sigma_\zeta, \quad \tilde{\psi}(z) = Tf_0 \text{ in } D^+,$$

$$\Psi_1(z) = \begin{cases} \int_{a_{2j+1}}^{\nu} g_0^1(z) d\nu e_1 + \int_0^{\mu} g_0^2(z) d\mu e_2 \text{ in } D_{2j+1}^-, \quad j = 0, 1, \dots, [n/2], \\ \int_2^{\nu} g_0^1(z) d\nu e_1 + \int_{a_{2j-1}}^{\mu} g_0^2(z) d\mu e_2 \text{ in } D_{2j}^-, \quad j = 1, \dots, [(n+1)/2], \end{cases} \tag{5.27}$$

can be determined, where $\mu = x + y, \nu = x - y$, and the solution $w_0(z)$ satisfies the estimate (5.24), i.e.

$$C_\beta[w_0(z)X(z), \overline{D^+}] + C_\beta[w_0^\pm(z)Y^\pm(z), \overline{D^-}] \le M_{23}(k_1 + k_2). \tag{5.28}$$

Moreover, by Theorem 5.2, we can find an analytic function $\tilde{\Phi}(z)$ in D^+ and a solution $\Phi_1(z)$ of (1.2) in D^- satisfying the boundary conditions

$$\begin{aligned} &\text{Re}\,[\overline{\lambda(z)}(\tilde{\Phi}_1(z)e^{\tilde{\phi}_1(z)} + \tilde{\psi}_1(z))] = r(z), \quad z \in \Gamma, \\ &\text{Re}\,[\overline{\lambda(x)}(\tilde{\Phi}_1(x) + \tilde{\psi}_1(x))] = s(z), \quad x \in L_0, \\ &\text{Re}\,[\lambda(x)\Phi_1(x)] = \text{Re}\,[\lambda(x)(W_1(x) - \Psi_1(x))], \quad z \in L_0, \\ &\text{Re}\,[\overline{\lambda(z)}\Phi_1(z)] = -\text{Re}\,[\overline{\lambda(z)}\Psi_1(z)], \quad z \in L_3 \cup L_4, \\ &\text{Im}\,[\overline{\lambda(z)}(\Phi_1(z) + \Psi_1(z))]|_{z = A_{2j+1}} = 0, \quad j = 0, 1, \dots, [n/2], \\ &\text{Im}\,[\overline{\lambda(z)}(\Phi_1(z) + \Psi_1(z))]|_{z = B_{2j-1}} = 0, \quad j = 1, \dots, [(n+1)/2], \end{aligned} \tag{5.29}$$

where

$$s_1(x) = \begin{cases} \dfrac{2r((1-i)x/2) - 2R_1((1-i)x/2)}{a((1-i)x/2) - b((1-i)x/2)} \\ \quad -\dfrac{[a((1-i)x/2) + b((1-i)x/2)]h_{2j}}{a((1-i)x/2) - b((1-i)x/2)} + \text{Re}\,[\overline{\lambda(x)}\Psi_1(x)], \\ \quad x \in (a_{2j}, a_{2j+1}), \quad j = 0, 1, \dots, [n/2], \\ \dfrac{2r((1+i)x/2+1-i) - 2R_1((1+i)x/2+1-i)}{a((1+i)x/2+1-i) + b((1+i)x/2+1-i)} \\ \quad -\dfrac{[a((1+i)x/2+1-i) - b((1+i)x/2+1-i)]k_{2j-1}}{a((1+i)x/2+1-i) + b((1+i)x/2+1-i)} \\ \quad +\text{Re}\,[\overline{\lambda(x)}\Psi_1(x)], \quad x \in (a_{2j-1}, a_{2j}), \quad j = 1, \dots, [(n+1)/2], \end{cases}$$

in which the real constants h_{2j}, k_{2j-1} are as stated in (5.15), and

$$w_1(z) = w_0(z) + W_1(z) = w_0(z) + \Phi_1(z) + \Psi_1(z) \text{ in } D^- \tag{5.30}$$

satisfies the estimate

$$C_\beta[w_1(z)X(z), \overline{D^+}] + C[w_1^\pm(\mu, \nu)Y^\pm(\mu, \nu), \overline{D^-}] \le M_{24} = M_{24}(p_0, \beta, k, D^-), \quad (5.31)$$

here $w_1^\pm(\mu, \nu) = \operatorname{Re} w_1(\mu, \nu) \pm \operatorname{Im} w_1(\mu, \nu)$, $Y^\pm(\mu, \nu)$, $X(z), Y^\pm(z)$ are as stated in (5.24). Furthermore we substitute $w_1(z) = w_0(z) + W_1(z)$ and corresponding functions $w_1(z)$, $\xi_1(z) = \operatorname{Re} w_1(z) + \operatorname{Im} w_1(z)$, $\eta_1(z) = \operatorname{Re} w_1(z) - \operatorname{Im} w_1(z)$ into the positions $w(z)$, $\xi(z)$, $\eta(z)$ in (5.20), (5.21), and similarly to (5.27)–(5.30), we can find the corresponding functions $\tilde{\phi}_2(z), \tilde{\psi}_2(z), \tilde{\Phi}_2(z)$ in D^+ and $\Psi_2(z), \Phi_2(z)$ and $W_2(z) = \Phi_2(z) + \Psi_2(z)$ in $\overline{D^-}$. The function

$$w_2(z) = w_0(z) + W_2(z) = \begin{cases} \tilde{\Phi}_2(z)e^{\tilde{\phi}_2(z)} + \tilde{\psi}_2(z) & \text{in } D^+, \\ w_0(z) + \Phi_2(z) + \Psi_2(z) & \text{in } \overline{D^-} \end{cases} \quad (5.32)$$

satisfies the similar estimate in the form (5.31). Thus there exists a sequence of functions $\{w_n(z)\}$ as follows

$$w_n(z) = w_0(z) + W_n(z) = \begin{cases} \tilde{\Phi}_n(z)e^{\tilde{\phi}_n(z)} + \tilde{\psi}_n(z) & \text{in } D^+, \\ w_0(z) + \Phi_n(z) + \Psi_n(z) & \text{in } D^-, \end{cases}$$

$$\Psi_n(z) = \begin{cases} \int_{a_{2j+1}}^\nu g_{n-1}^1(z)e_1 d\nu + \int_0^\mu g_{n-1}^2(z)e_2 d\mu & \text{in } D_{2j+1}^-, \quad j = 0, 1, \ldots, [n/2], \\ \int_2^\nu g_{n-1}^1(z)e_1 d\nu + \int_{a_{2j-1}}^\mu g_{n-1}^2(z)e_2 d\mu & \text{in } D_{2j}^-, \quad j = 1, \ldots, [(n+1)/2], \end{cases} \quad (5.33)$$

$$g_{n-1}^1(z) = A\xi_{n-1} + B\eta_{n-1} + E, \quad g_{n-1}^2(z) = C\xi_{n-1} + D\eta_{n-1} + F \text{ in } D^-,$$

and then

$$|[w_1^\pm(\mu, \nu) - w_0^\pm(\mu, \nu)]Y^\pm(\mu, \nu)| \le |\Phi_1^\pm(\mu, \nu)Y^\pm(\mu, \nu)|$$

$$+ \sqrt{2}\left\{|Y^-(\mu, \nu)|\left[\max_{1 \le j \le n+1} |\int_{a_{2j+1}}^\nu g_0^1(z)e_1 d\nu| + |\int_2^\nu g_0^2(z)e_1 d\nu|\right]\right.$$

$$+ |Y^+(\mu, \nu)|\left[|\int_0^\mu g_0^1(z)e_2 d\mu| + \max_{1 \le j \le n+1} |\int_{a_{2j-1}}^\mu g_0^2(z)e_2 d\mu|\right]\right\} \quad (5.34)$$

$$\le 2M_{25}M(4m+1)R' \text{ in } D^-,$$

where $m = C[w_0^+(\mu, \nu)Y^+(\mu, \nu), \overline{D^-}] + C[w_0^-(\mu, \nu)Y^-(\mu, \nu), \overline{D^-}]$, $M = 1 + 4k_0^2(1 + k_0^2)$,

$R' = 2$, $M_{25} = \max_{z \in \overline{D^-}}(|\tilde{A}|, |\tilde{B}|, |\tilde{C}|, |\tilde{D}|)$. It is clear that $w_n(z) - w_{n-1}(z)$ satisfies

$w_n(z) - w_{n-1}(z)$

$$= \Phi_n(z) - \Phi_{n-1}(z) + \begin{cases} \int_{a_{2j+1}}^\nu [g_{n-1}^1 - g_{n-2}^1]e_1 d\nu + \int_0^\mu [g_{n-1}^2 - g_{n-2}^2]e_2 d\mu \text{ in } D_{2j+1}^-, \\ j = 0, 1, \ldots, [n/2], \\ \int_2^\nu [g_{n-1}^1 - g_{n-2}^1]e_1 d\nu + \int_{a_{2j-1}}^\mu [g_{n-1}^2 - g_{n-2}^2]e_2 d\mu \text{ in } D_{2j}^-, \\ j = 1, \ldots, [(n+1)/2]. \end{cases} \tag{5.35}$$

Moreover, we can find the solution $w(z)$ of Problem A^* for (3.2) in the set $D^- \backslash \{(\cup_{j=0}^{[n/2]} \tilde{D}_{2j+1}^-) \cup (\cup_{j=1}^{[(n+1)/2]} \tilde{D}_{2j}^-)\} = D^- \backslash \{D_1^- \cup D_2^-\}$. From the above result,

$$|[w_n^\pm - w_{n-1}^\pm]Y^\pm| \leq [2M_{25}M(4m+1)]^n \int_0^{R'} \frac{R'^{m-1}}{(n-1)!} dR'$$
$$\leq \frac{[2M_{25}M(4m+1)R']^n}{n!} \text{ in } D^-, \tag{5.36}$$

can be obtained, and then we see that the sequence of functions $\{w_n^\pm(\mu, \nu)Y^+(\mu, \nu)\}$, i.e.

$$w_n^\pm(\mu, \nu)Y^\pm(\mu, \nu) = \{w_0^\pm + [w_1^\pm - w_0^\pm] + \ldots + [w_n^\pm - w_{n-1}^\pm]\}Y^\pm(\mu, \nu)(n = 1, 2, \ldots) \tag{5.37}$$

in D^- uniformly converge to $w_*^\pm(\mu, \nu)Y^\pm(\mu, \nu)$, and $w_*(z) = [w^+(\mu, \nu) + w^-(\mu, \nu) - i(w^+(\mu, \nu) - w^-(\mu, \nu))]/2$ satisfies the equality

$$w_*(z) = w_0(z) + \Phi_*(z) + \Psi_*(z),$$

$$\Psi_*(z) = \begin{cases} \int_{a_{2j+1}}^\nu g_*^1(z)e_1 d\nu + \int_0^\mu g_*^2(z)e_2 d\mu \text{ in } D_{2j+1}^-, \quad j = 0, 1, \ldots, [n/2], \\ \int_2^\nu g_*^1(z)e_1 d\nu + \int_{a_{2j-1}}^\mu g_*^2(z)e_2 d\mu \text{ in } D_{2j}^-, \quad j = 1, \ldots, [(n+1)/2], \end{cases} \tag{5.38}$$

$$g_*^1(z) = A\xi_* + B\eta_* + E, g_*^2(z) = C\xi_* + D\eta_* + F \text{ in } D^-,$$

and the corresponding function $u_*(z)$ is just a solution of Problem A^* for equation (3.2) in the domain D^- and $w_*(z)$ satisfies the estimate

$$C[w_*^\pm(\mu, \nu)Y^\pm(\mu, \nu), \overline{D^-}] \leq M_{26} = e^{4M_{25}M(2m+1)R'}. \tag{5.39}$$

In addition, we can find a sequence of functions $\{w_n(z)\}(w_n(z) = \tilde{\Phi}_n(z)e^{\tilde{\phi}_n(z)} + \tilde{\psi}_n(z))$ in D^+ and $\tilde{\Phi}_n(z)$ is an analytic function in D^+ satisfying the boundary conditions

$$\text{Re}\,[\overline{\lambda(z)}(\tilde{\Phi}_n(z)e^{\tilde{\phi}_n(z)} + \tilde{\psi}_n(z))] = r(z), \; z \in \Gamma,$$
$$\text{Re}\,[\overline{\lambda(x)}(\tilde{\Phi}_n(x)e^{\tilde{\phi}_n(x)} + \tilde{\psi}_n(x))] = s(x), \; x \in L_0, \tag{5.40}$$

in which

$$
s_n(x) = \begin{cases}
\dfrac{2r((1-i)x/2) - 2R_n((1-i)x/2)}{a((1-i)x/2) - b((1-i)x/2)} \\[2mm]
\quad -\dfrac{[a((1-i)x/2) + b((1-i)x/2)]h_{2j}}{\sqrt{2}[a((1-i)x/2) - b((1-i)x/2)]} + \mathrm{Re}\,[\overline{\lambda(x)}\Psi_n(x)], \\[2mm]
\quad x \in (a_{2j}, a_{2j+1}), \quad j = 0,1,\ldots,[n/2], \\[2mm]
\dfrac{2r((1+i)x/2+1-i) - 2R_n((1+i)x/2+1-i)}{a((1+i)x/2+1-i) + b((1+i)x/2+1-i)} \\[2mm]
\quad -\dfrac{[a((1+i)x/2+1-i) - b((1+i)x/2+1-i)]k_{2j-1}}{a((1+i)x/2+1-i) + b((1+i)x/2+1-i)} \\[2mm]
\quad +\mathrm{Re}\,[\overline{\lambda(x)}\Psi_n(x)], \quad x \in (a_{2j-1}, a_{2j}), \quad j = 1,\ldots,[(n+1)/2],
\end{cases}
\tag{5.41}
$$

in which the real constants h_{2j}, k_{2j-1} are as stated in (5.15). From (5.31),(5.39), it follows that

$$
C_\beta[X(x)s_n(x), L_0] \le M_{27} = M_{27}(p_0, \beta, k, D),
\tag{5.42}
$$

and then the estimate

$$
C_\beta[w_n(z)X(z), \overline{D^+}] \le M_{21}(k_1 + k_2 + M_{27}).
\tag{5.43}
$$

Thus from $\{w_n(z)X(x)\}$, we can choose a subsequence which uniformly converges to a function $w_*(z)X(z)$ in $\overline{D^+}$. Combining (5.43) and (5.39), it is obvious that the solution $w_*(z)$ of Problem A^* for (3.2) in $\bar{D}$ satisfies the estimate

$$
C_\beta[w_*(z)X(z), \overline{D^+}] + C[w_*^\pm(z)Y^\pm(z), \overline{D^-}] \le M_{28} = M_{28}(p_0, \beta, k, D),
\tag{5.44}
$$

where M_{28} is a non-negative constant.

Theorem 5.5 *If the complex equation (3.2) satisfies Condition C, then Problem A^* for (3.2) has at most one solution in D.*

Proof Let $w_1(z), w_2(z)$ be any two solutions of Problem A^* for (3.2). By Condition C, we see that $w(z) = w_1(z) - w_2(z)$ satisfies the homogeneous complex equation

$$
\begin{Bmatrix} w_{\bar z} \\ \bar{w}_{z^*} \end{Bmatrix} = A_1 w + A_2 \bar{w} \text{ in } \begin{Bmatrix} D^+ \\ D^- \end{Bmatrix}
\tag{5.45}
$$

and boundary conditions

$$
\begin{aligned}
&\mathrm{Re}\,[\overline{\lambda(z)}w(z)] = 0, \quad z \in \Gamma, \\
&\mathrm{Re}\,[\overline{\lambda(x)}w(x)] = s(x), \quad x \in L_0, \\
&\mathrm{Re}\,[\overline{\lambda(z)}w(z)] = 0, \quad z \in L_3 \cup L_4, \\
&\mathrm{Im}\,[\overline{\lambda(z)}w(z)]|_{z=A_{2j+1}} = 0, \quad j = 0,1,\ldots,[n/2], \\
&\mathrm{Im}\,[\overline{\lambda(z)}w(z)]|_{z=B_{2j-1}} = 0, \quad j = 1,\ldots,[(n+1)/2],
\end{aligned}
\tag{5.46}
$$

in which

$$
s(x) = \begin{cases} \dfrac{2r((1-i)x/2) - 2R((1-i)x/2)}{a((1-i)x/2) - b((1-i)x/2)} \\ \quad +\text{Re}\,[\overline{\lambda(x)}\Psi(x)], \quad x \in (a_{2j}, a_{2j+1}), \quad j = 0, 1, \ldots, [n/2], \\[2mm] \dfrac{2r((1+i)x/2+1-i) - 2R((1+i)x/2+1-i)}{a((1+i)x/2+1-i) + b((1+i)x/2+1-i)} \\ \quad +\text{Re}\,[\overline{\lambda(x)}\Psi(x)], \quad x \in (a_{2j-1}, a_{2j}), \quad j = 1, \ldots, [(n+1)/2], \end{cases}
\tag{5.47}
$$

From Theorem 5.3, the solution $w(z)$ can be expressed in the form

$$
w(z) = \begin{cases} \tilde{\Phi}(z)e^{\tilde{\phi}(z)}, \quad \tilde{\phi}(z) = \tilde{T}g \text{ in } D^+, \\[2mm] g(z) = \begin{cases} A_1 + A_2\bar{w}/w, \quad w(z) \neq 0, \quad z \in D^+, \\ 0, \quad w(z) = 0, \quad z \in D^+, \end{cases} \\[4mm] \Phi(z) + \Psi(z), \\[2mm] \Psi(z) = \begin{cases} \displaystyle\int_{a_{2j+1}}^{\nu} [\tilde{A}\xi + \tilde{B}\eta]e_1 d\nu + \int_0^{\mu} [\tilde{C}\xi + \tilde{D}\eta]e_2 d\mu \\ \quad \text{in } D^-_{2j+1}, j = 0, 1, \ldots, [n/2], \\[2mm] \displaystyle\int_2^{\nu} [\tilde{A}\xi + \tilde{B}\eta]e_1 d\nu + \int_{a_{2j-1}}^{\mu} [\tilde{C}\xi + \tilde{D}\eta]e_2 d\mu \\ \quad \text{in } D^-_{2j}, j = 1, \ldots, [(n+1)/2]. \end{cases} \end{cases}
\tag{5.48}
$$

By using the method in the proof of Theorem 5.4, we can get that

$$
|w^{\pm}(z)Y^{\pm}(z)| \leq \frac{[2M_{25}M(4m+1)R']^n}{n!} \text{ in } D^-.
\tag{5.49}
$$

Let $n \to \infty$, we get $w^{\pm}(z) = 0$, i.e. $w(z) = w_1(z) - w_2(z) = 0$, and then $\Phi(z) = \Psi(z) = 0$ in $\overline{D^-}$, thus $s(x) = 0$ on L_0. Noting that $w(z) = \tilde{\Phi}(z)e^{\tilde{\phi}(z)}$ satisfies the boundary conditions in (5.46), we see that the analytic function $\tilde{\Phi}(z)$ in D^+ satisfies the boundary conditions

$$
\text{Re}\,[\overline{\lambda(z)}e^{\tilde{\phi}(z)}\tilde{\Phi}(z)] = 0, \quad z \in \Gamma, \quad \text{Re}\,[\overline{\lambda(x)}\tilde{\Phi}(x)] = 0, \quad x \in L_0,
\tag{5.50}
$$

and the index of the boundary value problem (5.50) is $K = -1/2$, hence $\tilde{\Phi}(z) = 0$ in D^+, and then $w(z) = \tilde{\Phi}(z)e^{\tilde{\phi}(z)} = 0$ in D^+, namely $w(z) = w_1(z) - w_2(z) = 0$ in D^+. This proves the uniqueness of solutions of Problem A^* for (3.2).

From Theorems 5.4 and 5.5, we see that under Condition C, Problem A^* for equation (3.2) has a unique solution $w(z)$, which can be found by using successive iteration, and the solution $w(z)$ satisfies the estimate (5.44), i.e.

$$
C_{\beta}[w(z)X(z), \overline{D^+}] + C[w^{\pm}(\mu,\nu)Y^{\pm}(\mu,\nu), \overline{D^-}] \leq M_{29},
\tag{5.51}
$$

where $k = (k_0, k_1, k_2)$, $M_{29} = M_{29}(p_0, \beta, k, D)$ is a non-negative constant. Moreover we have

Theorem 5.6   *Suppose that equation (3.2) satisfies Condition C. Then any solution $w(z)$ of Problem A^* for (3.2) satisfies the estimates (5.44) and*

$$C_\beta[w(z)X(z), \overline{D^+}] + C[w(z), \overline{D^-}] \leq M_{30}(k_1 + k_2), \qquad (5.52)$$

where $X(z)$, $Y(z)$ are as stated in (5.24) respectively, and $M_{30} = M_{30}(p_0, \beta, \delta, k_0, D)$ is a non-negative constant.

From the estimates (5.51) and (5.52), we can see the regularity of solutions of Problem A^* for (3.2).

Finally, we mention that if the index K is an arbitrary even integer or $2K$ is an arbitrary odd integer, the above Problem A^* for (3.2) can be considered. But in general Problem A^* for (3.2) with $K \leq -1$ has $-2K - 1$ solvability conditions or when $K \geq 0$ its general solution includes $2K + 1$ arbitrary real conditions.

For more general first order complex equations of mixed type, the discontinuous Riemann–Hilbert boundary value problem remains to be discussed.

The references for this chapter are [3],[8],[12],[16],[20],[25],[35],[36],[42],[44],[52], [55],[60],[63],[73],[75],[83],[85],[95],[98].

CHAPTER V

SECOND ORDER LINEAR EQUATIONS OF MIXED TYPE

In this chapter, we discuss the oblique derivative boundary value problem for second order linear equation of mixed (elliptic-hyperbolic) type in a simply connected domain. We first prove uniqueness and existence of solutions for the above boundary value problem, and then give a priori estimates of solutions for the problem, finally discuss the existence of solutions for the above problem in general domains. In books [12]1),3), the author investigated the Dirichlet problem (Tricomi problem) for the mixed equation of second order, i.e. $u_{xx} + \text{sgn}y\, u_{yy} = 0$. In [69], the author discussed the Tricomi problem for the generalized Lavrent'ev-Bitsadze equation $u_{xx} + \text{sgn}y\, u_{yy} + Au_x + Bu_y + Cu = 0$, i.e. $u_{\xi\eta} + au_{\xi} + bu_{\eta} + cu = 0$ with the conditions: $a \geq 0, a_{\xi} + ab - c \geq 0, c \geq 0$ in the hyperbolic domain. In this section, we cancel the above assumption in [69] and obtain the solvability result on the discontinuous Poincaré problem, which includes the corresponding results in [12]1),3),[69] as special cases.

1 Oblique Derivative Problems for Simplest Second Order Equation of Mixed Type

In this section, we introduce the oblique derivative boundary value problem for simplest mixed equation of second order in a simply connected domain, and verify the uniqueness and existence of solutions for the above boundary value problem.

Let D be a simply connected bounded domain in the complex plane $\mathcal{C}$ with the boundary $\partial D = \Gamma \cup L$, where $\Gamma (\subset \{y > 0\}) \in C_{\alpha}^2 (0 < \alpha < 1)$ with the end points $z = 0, 2$ and $L = L_1 \cup L_2, L_1 = \{x = -y, 0 \leq x \leq 1\}, L_2 = \{x = y + 2, 1 \leq x \leq 2\}$, and denote $D^+ = D \cap \{y > 0\}$, $D^- = D \cap \{y < 0\}$ and $z_1 = 1 - i$. We may assume that $\Gamma = \{|z - 1| = 1, y \geq 0\}$, otherwise through a conformal mapping, the requirement can be realized.

1.1 The oblique derivative problem for simplest second order equation of mixed type

In A. V. Bitsadze's books [12]1),3), the author discussed the solvability of several boundary value problems including the Dirichlet problem or Tricomi problem

(Problem D or Problem T) for the second order equation of mixed type

$$u_{xx} + \text{sgn} y\, u_{yy} = 0 \text{ in } D, \tag{1.1}$$

the equation is so–called Lavrent'ev-Bitsadze equation, its complex form is as follows:

$$\left\{ \begin{array}{c} u_{\bar{z}z} \\ u_{z^{\bullet}\bar{z}^{\bullet}} \end{array} \right\} = 0 \text{ in } \left\{ \begin{array}{c} D^{+} \\ D^{-} \end{array} \right\}, \tag{1.2}$$

where

$$u_{z^{\bullet}} = u_{\bar{z}},\ w_{\overline{z^{\bullet}}} = \frac{1}{2}[w_x - i\bar{w}_y].$$

Now we formulate the oblique derivative boundary value problem as follows:

Problem P Find a continuously differentiable solution $u(z)$ of (1.2) in $D^{\bullet} = \bar{D} \backslash \{0, x - y = 2\}$ or $D^{\bullet} = \bar{D} \backslash \{x + y = 0, 2\}$, which is continuous in $\bar{D}$ and satisfies the boundary conditions

$$\frac{1}{2}\frac{\partial u}{\partial l} = \text{Re}\,[\overline{\lambda(z)}u_z] = r(z), \quad z \in \Gamma, \quad u(0) = b_0, \quad u(2) = b_2, \tag{1.3}$$

$$\frac{1}{2}\frac{\partial u}{\partial l} = \text{Re}\,[\overline{\lambda(z)}u_{\bar{z}}] = r(z), \quad z \in L_j\ (j = 1 \text{ or } 2), \quad \text{Im}\,[\overline{\lambda(z)}u_{\bar{z}}]|_{z=z_1} = b_1, \tag{1.4}$$

where l is a given vector at every point on $\Gamma \cup L_j$ $(j = 1 \text{ or } 2)$, $\lambda(z) = a(x) + ib(x) = \cos(l, x) - i\cos(l, y)$, if $z \in \Gamma$, and $\lambda(z) = a(z) + ib(z) = \cos(l, x) + i\cos(l, y)$, if $z \in L_j$ $(j = 1 \text{ or } 2)$, b_0, b_1 are real constants, and $\lambda(z)$, $r(z)$, b_0, b_1, b_2 satisfy the conditions

$$C_\alpha[\lambda(z), \Gamma] \le k_0, \quad C_\alpha[r(z), \Gamma] \le k_2, \quad C_\alpha[\lambda(z), L_j] \le k_0, \quad C_\alpha[r(z), L_j] \le k_2, \quad j = 1 \text{ or } 2,$$

$$\cos(l, n) \ge 0 \text{ on } \Gamma, \quad |b_j| \le k_2, \quad j = 0, 1, 2, \quad \max_{z \in L_1}\frac{1}{|a(z) - b(z)|} \text{ or } \max_{z \in L_2}\frac{1}{|a(z) + b(z)|} \le k_0, \tag{1.5}$$

in which n is the outward normal vector at every point on Γ, $\alpha\,(1/2 < \alpha < 1)$, k_0, k_2 are non-negative constants. For convenience, we may assume that $u_z(z_1) = 0$, otherwise through a transformation of function $U_{\bar{z}}(z) = u_{\bar{z}}(z) - \lambda(z_1)[r(z_1) + ib_1]$, the requirement can be realized.

The boundary value problem for (1.2) with $r(z) = 0$, $z \in \Gamma \cup L_j$ $(j = 1 \text{ or } 2)$ and $b_0 = b_1 = b_2 = 0$ will be called Problem P_0. The number

$$K = \frac{1}{2}(K_1 + K_2), \tag{1.6}$$

is called the index of Problem P and Problem P_0, where

$$K_j = \left[\frac{\phi_j}{\pi}\right] + J_j, \quad J_j = 0 \text{ or } 1, \quad e^{i\phi_j} = \frac{\lambda(t_j - 0)}{\lambda(t_j + 0)}, \quad \gamma_j = \frac{\phi_j}{\pi} - K_j, \quad j = 1, 2, \tag{1.7}$$

in which $t_1 = 2$, $t_2 = 0$, $\lambda(t) = e^{i\pi/4}$ on $L_0 = (0,2)$ and $\lambda(t_1 - 0) = \lambda(t_2 + 0) = \exp(i\pi/4)$, or $\lambda(t) = e^{i7\pi/4}$ on L_0 and $\lambda(t_1 - 0) = \lambda(t_2 + 0) = \exp(i7\pi/4)$. Here we choose $K = 0$, or $K = -1/2$ on the boundary ∂D^+ of D^+ if $\cos(\nu,n) \equiv 0$ on Γ and the condition $u(2) = b_2$ can be canceled. In this case the solution of Problem P for (1.2) is unique. In order to ensure that the solution $u(z)$ of Problem P is continuously differentiable in D^*, we need to choose $\gamma_1 > 0$. If we require that the solution $u(z)$ in $\bar{D}$ is only continuous, it is suffices to choose $-2\gamma_1 < 1$, $-2\gamma_2 < 1$. Problem P in this case still includes the Dirichlet problem as a special case. If the boundary condition $\mathrm{Re}\,[\overline{\lambda(z)}u_{\bar{z}}] = r(z)$ on $L_j (j = 1$ or $2)$ in (1.4) is replaced by $\mathrm{Re}\,[\overline{\lambda(z)}u_z] = r(z)$ on $L_j (j = 1$ or $2)$, then Problem P does not include the above Dirichlet problem (Problem D) as a special case.

Setting that $u_z = w(z)$, it is clear that Problem P for (1.2) is equivalent to the Riemann–Hilbert boundary value problem (Problem A) for the first order complex equation of mixed type

$$\left\{ \begin{matrix} w_{\bar{z}} \\ \bar{w}_{\overline{z^*}} \end{matrix} \right\} = 0 \text{ in } \left\{ \begin{matrix} D^+ \\ D^- \end{matrix} \right\}, \tag{1.8}$$

with the boundary conditions

$$\mathrm{Re}\,[\overline{\lambda(z)}w(z)] = r(z), \quad z \in \Gamma, \quad u(2) = b_2,$$
$$\mathrm{Re}\,[\overline{\lambda(z)}w(z)] = r(z), \quad z \in L_j \,(j = 1 \text{ or } 2), \quad \mathrm{Im}[\overline{\lambda(z_1)}w(z_1)] = b_1, \tag{1.9}$$

and the relation

$$u(z) = 2\mathrm{Re}\int_0^z w(z)dz + b_0 \text{ in } D, \tag{1.10}$$

in which the integral path in $\overline{D^-}$ is as stated in Chapter II. On the basis of the result in Section 1, Chapter IV, we can find a solution $w(z)$ of Problem A for the mixed complex equation (1.8) as stated in (1.30), Chapter IV, but the function $\lambda(x)$ in the integral formula in D^+ should be replaced by $\overline{\lambda(x)}$ on $L_0 = (0,2)$, the function $w(z)$ in D^- should be replaced by $\overline{w(z)}$ in the second formula in (1.30), Chapter IV. Hence we have the following theorem.

Theorem 1.1 *Problem P for the mixed equation (1.2) has a unique solution in the form (1.10), where*

$$w(z) = \tilde{w}(z) + \overline{\lambda(z_1)}[r(z_1) - ib_1],$$

$$\tilde{w}(z) = \begin{cases} W[\zeta(z)], \ z \in \overline{D^+}\backslash\{0,2\}, \\ \dfrac{1}{2}(1-i)f(x+y) + \dfrac{1}{2}(1+i)g(x-y), \end{cases}$$

$$g(x-y) = \frac{2r((1-i)(x-y)/2)}{a((1-i)(x-y)/2) - b((1-i)(x-y)/2)}$$
$$- \frac{[a((1-i)(x-y)/2) + b((1-i)(x-y)/2)]f(0)}{a((1-i)(x-y)/2) - b((1-i)(x-y)/2)}, \text{ or}$$

$$\tilde{w}(z) = \frac{1}{2}(1+i)g(x-y) + \frac{1}{2}(1-i)f(x+y),$$

$$f(x+y) = \frac{2r((1+i)(x+y)/2+1-i)}{a((1+i)(x+y)/2+1-i)+b((1+i)(x+y)/2+1-i)}$$

$$-\frac{[a((1+i)(x+y)/2+1-i)-b((1+i)(x+y)/2+1-i)]g(2)}{a((1+i)(x+y)/2+1-i)+b((1+i)(x+y)/2+1-i)}, \qquad (1.11)$$

$$z = x + iy \in \overline{D^-}\backslash\{0,2\},$$

in which $f(0) = [a(z_1)+b(z_1)]r(z_1) + [a(z_1)-b(z_1)]b_1$, $g(2) = [a(z_1)-b(z_1)]r(z_1) - [a(z_1)+b(z_1)]b_1$, $W(\zeta)$ *on* $\overline{D^+}\backslash\{0,2\}$ *is as stated in* (1.28), *Chapter IV, but where the function* $\lambda(z)$ *on* L_0 *is as stated in* (1.7), *and* $\lambda(z), r(z), b_1$ *are as stated in* (1.3), (1.4). *Moreover the functions*

$$f(x+y) = U(x+y,0) - V(x+y,0) = \mathrm{Re}\,[(1+i)W(\zeta(x+y))],$$

$$g(x-y) = U(x-y,0) + V(x-y,0) = \mathrm{Re}\,[(1-i)W(\zeta(x-y))], \qquad (1.12)$$

where $U = u_x/2$, $V = -u_y/2$, $W[\zeta(x+y)]$ *and* $W[\zeta(x-y)]$ *are the values of* $W[\zeta(z)]$ *on* $0 \le z = x+y \le 2$ *and* $0 \le z = x-y \le 2$ *respectively.*

From the above representation of the solution $u(z)$ of Problem P for (1.2), we can derive that $u(z)$ satisfies the estimate

$$C_\beta[u(z),\overline{D}] + C_\beta[w(z)X(z),\overline{D^+}] + C[w^\pm(z)Y^\pm(z),\overline{D^-}] \le M_1 k_2, \qquad (1.13)$$

in which k_2 are as stated in (1.5), $w^\pm(z) = \mathrm{Re}\,w \mp \mathrm{Im}\,w$ and

$$X(z) = \prod_{j=1}^{2}|z-t_j|^{\eta_j}, \quad Y^\pm(z) = \prod_{j=1}^{2}|x\pm y-t_j|^{\eta_j},$$

$$\eta_j = \begin{cases} 2|\gamma_j|+\delta, & \text{if } \gamma_j < 0, \\ \delta, & \gamma_j \ge 0, \end{cases} \quad j = 1,2, \qquad (1.14)$$

here γ_j $(j=1,2)$ are real constants as stated in (1.7) and β, δ $(\beta < \delta)$ are sufficiently small positive constant, and $M_1 = M_1(p_0, \beta, k_0, D^+)$ is a non-negative constant. From the estimate (1.13), we can see the regularity of solutions of Problem P for (1.2).

Finally, we mention that if the index K is an arbitrary even integer or $2K$ is an arbitrary odd integer, the above oblique derivative problem for (1.1) or (1.2) can be considered, but in general these boundary value problems for $K \le -1$ have some solvability conditions or their solutions for $K \ge 0$ are not unique.

1.2   The Dirichlet boundary value problem for simplest second order equation of mixed type

The Dirichlet problem (Problem D or Problem T) for (1.2) is to find a solution of (1.2) with the boundary conditions

$$u(z) = \phi(z) \text{ on } \Gamma \cup L_j (j = 1 \text{ or } 2) \qquad (1.15)$$

where $\phi(z)$ satisfies the condition

$$C^1[\phi(z), \Gamma \cup L_j] \le k_2, \quad j = 1 \text{ or } 2, \tag{1.16}$$

In the following we shall explain that Problem D is a special case of Problem P. In fact, denote $w = u_z$ in D, Problem D for the mixed equation (1.2) is equivalent to Problem A for the mixed equation (1.8) with the boundary condition (1.9) and the relation (1.10), in which

$$\lambda(z) = a + ib = \begin{cases} \overline{i(z-1)}, & \theta = \arg(z-1) \text{ on } \Gamma, \\ \dfrac{1-i}{\sqrt{2}} \text{ on } L_1, \text{ or } \dfrac{1+i}{\sqrt{2}} \text{ on } L_2, \end{cases}$$

$$r(z) = \begin{cases} \phi_\theta \text{ on } \Gamma, \\ \dfrac{\phi_x}{\sqrt{2}} \text{ on } L_1, \text{ or } \dfrac{\phi_x}{\sqrt{2}} \text{ on } L_2, \end{cases} \tag{1.17}$$

$$b_1 = \operatorname{Im}\left[\frac{1+i}{\sqrt{2}} u_{\bar{z}}(z_1 - 0)\right] = \frac{\phi_x - \phi_x}{\sqrt{2}}\Big|_{z=z_1-0} = 0, \text{ or }$$

$$b_1 = \operatorname{Im}\left[\frac{1-i}{\sqrt{2}} u_{\bar{z}}(z_1 + 0)\right] = 0; \ b_0 = \phi(0),$$

in which $a = 1/\sqrt{2} \ne b = -1/\sqrt{2}$ on L_1 or $a = 1/\sqrt{2} \ne -b = -1/\sqrt{2}$ on L_2.

As for the index $K = -1/2$ of Problem D on ∂D^+, we can derive as follows: According to (1.17), the boundary conditions of Problem D in D^+ possess the form

$$\operatorname{Re}\left[i(z-1)w(z)\right] = r(z) = \phi_\theta, \quad \text{on } \Gamma,$$

$$\operatorname{Re}\left[\frac{1-i}{\sqrt{2}} w(x)\right] = s(x) = \frac{\phi'((1-i)x/2)}{\sqrt{2}}, \quad x \in [0, 2], \text{ or}$$

$$\operatorname{Re}\left[\frac{1+i}{\sqrt{2}} w(x)\right] = s(x) = \frac{\phi'((1+i)x/2 + 1 - i)}{\sqrt{2}}, \quad x \in [0, 2],$$

it is clear that the possible discontinuous points of $\lambda(z)$ on ∂D^+ are $t_1 = 2$, $t_2 = 0$, and

$$\lambda(t_1 + 0) = e^{3\pi i/2}, \quad \lambda(t_2 - 0) = e^{\pi i/2},$$

$$\lambda(t_1 - 0) = \lambda(t_2 + 0) = e^{\pi i/4}, \quad \text{or } \lambda(t_1 - 0) = \lambda(t_2 + 0) = e^{7\pi i/4},$$

$$\frac{\lambda(t_1 - 0)}{\lambda(t_1 + 0)} = e^{-5\pi i/4} = e^{i\phi_1}, \quad \frac{\lambda(t_2 - 0)}{\lambda(t_2 + 0)} = e^{\pi i/4} = e^{i\phi_2} \text{ or}$$

$$\frac{\lambda(t_1 - 0)}{\lambda(t_1 + 0)} = e^{\pi i/4} = e^{i\phi_1}, \quad \frac{\lambda(t_2 - 0)}{\lambda(t_2 + 0)} = e^{-5\pi i/4} = e^{i\phi_2}.$$

In order to insure the uniqueness of solutions of Problem D, we choose that

$$-1 < \gamma_1 = \frac{\phi_1}{\pi} - K_1 = -\frac{5}{4} - (-1) = -\frac{1}{4} < 0,$$

$$0 \le \gamma_2 = \frac{\phi_2}{\pi} - K_2 = \frac{1}{4} < 1, \text{ or}$$

$$0 \le \gamma_1 = \frac{\phi_1}{\pi} - K_1 = \frac{1}{4} < 1,$$

$$-1 < \gamma_2 = \frac{\phi_2}{\pi} - K_2 = -\frac{5}{4} - (-1) = -\frac{1}{4} < 0,$$

thus

$$K_1 = -1, \quad K_2 = 0, \quad K = \frac{K_1 + K_2}{2} = -\frac{1}{2}, \text{ or}$$

$$K_1 = 0, \quad K_2 = -1, \quad K = \frac{K_1 + K_2}{2} = -\frac{1}{2}.$$

In this case, the unique solution $w(z)$ is continuous in $D^* = \bar{D}\backslash\{0, x - y = 2\}$ or $D^* = \bar{D}\backslash\{x + y = 0, 2\}$; for the first case, $w(z)$ in the neighborhood of $t_2 = 0$ is bounded, and $w(z)$ in the neighborhood of $t_1 = 2$ possesses the singularity in the form $|z - 2|^{-1/2}$ and its integral (1.10) is bounded; for the second case, $w(z)$ in the neighborhood of $t_1 = 2$ is bounded, and $w(z)$ in the neighborhood of $t_2 = 0$ possesses the singularity of $|z|^{-1/2}$ and its integral is bounded. If we require that the solution $w(z) = u_z$ is bounded in $\overline{D^+}\backslash\{0, 2\}$, then it suffices to choose the index $K = -1$, in this case the problem has one solvability condition.

From Theorem 1.1, it follows that the following theorem holds.

Theorem 1.2 *Problem D for the mixed equation (1.2) has a unique continuous solution in $\bar{D}$ as stated in (1.10), where $\lambda(z), r(z), b_1$ are as stated in (1.17) and $W(\zeta)$ in $\overline{D^+}\backslash\{0, 2\}$ is as stated in (1.28), Chapter IV, but in which $\lambda(x) = (1 + i)/\sqrt{2}$ or $(1 - i)/\sqrt{2}$ on L_0 and $f(x + y), g(x - y)$ are as stated in (1.12).* [85]15)

2 Oblique Derivative Problems for Second Order Linear Equations of Mixed Type

In this section, we mainly discuss the uniqueness and existence of solutions for second order linear equations of mixed type.

2.1 Formulation of the oblique derivative problem for mixed equations of second order

Let D be a simply connected bounded domain D in the complex plane $\mathbb{C}$ with the boundary $\partial D = \Gamma \cup L$ as stated in Section 1. We consider the linear mixed equation of second order

$$u_{xx} + \text{sgn}y\, u_{yy} = au_x + bu_y + cu + d \text{ in } D, \tag{2.1}$$

where a, b, c, d are functions of $z(\in D)$, its complex form is the following equation of second order

$$\begin{cases} u_{z\bar{z}} = \text{Re}\,[A_1(z)u_z] + A_2(z)u + A_3(z) \text{ in } D^+, \\ u_{z^{\bullet}\overline{z^{\bullet}}} = \text{Re}\,[A_1(z)u_z] + A_2(z)u + A_3(z) \text{ in } D^-, \end{cases} \tag{2.2}$$

where

$$z = x + iy, \quad u_z = \frac{1}{2}[u_x - iu_y], \quad u_{\bar{z}} = \frac{1}{2}[u_x + iu_y], \quad u_{z\bar{z}} = \frac{1}{4}[u_{xx} + u_{yy}],$$

$$u_{z^{\bullet}} = \frac{1}{2}[u_x + iu_y] = u_{\bar{z}}, \quad u_{\bar{z}\overline{z^{\bullet}}} = \frac{1}{2}[(u_{\bar{z}})_x - i(\overline{u_{\bar{z}}})_y] = \frac{1}{4}[u_{xx} - u_{yy}],$$

$$A_1 = \frac{a+ib}{2}, \quad A_2 = \frac{c}{4}, \quad A_3 = \frac{d}{4} \text{ in } D.$$

Suppose that equation (2.2) satisfies the following conditions: **Condition C**

The coefficients $A_j(z)\,(j = 1, 2, 3)$ in (2.2) are measurable in $z \in D^+$ and continuous in $\overline{D^-}$ and satisfy

$$L_p[A_j, \overline{D^+}] \le k_0, \quad j = 1, 2, \quad L_p[A_3, \overline{D^+}] \le k_1, \quad A_2 \ge 0 \text{ in } D^+, \tag{2.3}$$

$$C[A_j, \overline{D^-}] \le k_0, \quad j = 1, 2, \quad C[A_3, \overline{D^-}] \le k_1. \tag{2.4}$$

where $p\,(> 2), k_0, k_1$ are non-negative constants. If the condition (2.4) is replaced by

$$C_\alpha[A_j, \overline{D^-}] \le k_0, \quad j = 1, 2, \quad C_\alpha[A_3, \overline{D^-}] \le k_1,$$

in which $\alpha(0 < \alpha < 1)$ is a real constant, then the conditions will be called Condition C'.

The oblique derivative boundary value problem (Problem P) for equation (2.2) is to find a continuously differentiable solution $u(z)$ of (2.2) in $D^* = \bar{D}\backslash\{0, x - y = 2\}$ or $D^* = \bar{D}\backslash\{x + y = 0, 2\}$, which is continuous in $\bar{D}$ and satisfies the boundary conditions (1.3) and (1.4), in which b_0, b_2 is a real constant satisfying the condition $|b_0|, |b_2| \le k_2$. The index K is defined as stated in Section 1, now we discuss the case

$$K = \frac{1}{2}(K_1 + K_2) = 0, \tag{2.5}$$

where

$$K_j = \left[\frac{\phi_j}{\pi}\right] + J_j, \; J_j = 0 \text{ or } 1, \quad e^{i\phi_j} = \frac{\lambda(t_j - 0)}{\lambda(t_j + 0)}, \quad \gamma_j = \frac{\phi_j}{\pi} - K_j, \quad j = 1, 2, \tag{2.6}$$

in which $t_1 = 2$, $t_2 = 0$, $\lambda(t) = e^{i\pi/4}$ on $L_0 = (0, 2)$ and $\lambda(t_1 - 0) = \lambda(t_2 + 0) = \exp(i\pi/4)$, or $\lambda(t) = e^{i7\pi/4}$ on L_0 and $\lambda(t_1 - 0) = \lambda(t_2 + 0) = \exp(i7\pi/4)$. We mention that if $A_2 = 0$ in D, or $\cos(l,n) \equiv 0$ on Γ, then we do not need the point condition $u(2) = b_2$ in (1.3) and only choose the index $K = -1/2$. Because, if $\cos(l,n) \equiv 0$ on Γ, from the boundary condition (1.3), we can determine the value $u(2)$ by the value $u(0)$, namely

$$u(2) = 2\mathrm{Re}\int_0^2 u_z dz + u(0) = 2\int_0^2 \mathrm{Re}\,[i(z-1)u_z]d\theta + b_0 = 2\int_\pi^0 r(z)d\theta + b_0,$$

in which $\overline{\lambda(z)} = i(z-1)$, $\theta = \arg(z-1)$ on Γ. In brief, we choose that

$$K = \begin{cases} 0, \\ -\dfrac{1}{2}, \end{cases} \text{ the point conditions are } \begin{cases} u(0) = b_0, u(2) = b_2 \\ u(0) = b_0 \end{cases}, \text{ if } \begin{cases} \cos(l,n) \not\equiv 0 \\ \cos(l,n) \equiv 0 \end{cases} \text{ on } \Gamma.$$

In order to ensure that the solution $u(z)$ of Problem P is continuously differentiable in D^*, we need to choose $\gamma_1 > 0$ or $\gamma_2 > 0$. If we only require that the solution is continuous, it suffices to choose $-2\gamma_2 < 1$, $-2\gamma_1 < 1$ respectively. In the following, we shall only discuss the case: $K = 0$, and the case: $K = -1/2$ can be similarly discussed.

2.2   The representation and uniqueness of solutions for the oblique derivative problem for (2.2)

Now we give the representation theorems of solutions for equation (2.2).

Theorem 2.1   *Let equation (2.2) satisfy Condition C in D^+, $u(z)$ be a continuous solution of (2.2) in $\overline{D^+}$ and continuously differentiable in $D_*^+ = \overline{D^+}\backslash\{0,2\}$. Then $u(z)$ can be expressed as*

$$u(z) = U(z)\Psi(z) + \psi(z) \text{ in } D^+,$$

$$U(z) = 2\mathrm{Re}\int_0^z w(z)dz + b_0, \; w(z) = \Phi(z)e^{\phi(z)} \text{ in } D^+, \tag{2.7}$$

where $\psi(z)$, $\Psi(z)$ are the solutions of equation (2.2) in D^+ and

$$u_{z\bar{z}} - \mathrm{Re}\,[A_1 u_z] - A_2 u = 0 \text{ in } D^+ \tag{2.8}$$

respectively and satisfy the boundary conditions

$$\psi(z) = 0, \; \Psi(z) = 1 \text{ on } \Gamma \cup L_0, \tag{2.9}$$

where $\psi(z)$, $\Psi(z)$ satisfies the estimates

$$C_\beta^1[\psi, \overline{D^+}] \leq M_2, \quad \|\psi\|_{W_{p_0}^2(D^+)} \leq M_2, \tag{2.10}$$

$$C_\beta^1[\Psi, \overline{D^+}] \leq M_3, \quad \|\Psi\|_{W_{p_0}^2(D^+)} \leq M_3, \Psi(z) \geq M_4 > 0, \quad z \in \overline{D^+}, \tag{2.11}$$

in which $\beta\,(0 < \beta \leq \alpha)$, $p_0\,(2 < p_0 \leq p)$, $M_j = M_j(p_0, \beta, k, D)\,(j = 2, 3, 4)$ are non-negative constants, $k = (k_0, k_1, k_2)$. Moreover $U(z)$ is a solution of the equation

$$U_{z\bar{z}} - \mathrm{Re}\,[AU_z] = 0, \quad A = -2(\ln \Psi)_{\bar{z}} + A_1 \text{ in } D^+, \tag{2.12}$$

where $\mathrm{Im}\,[\phi(z)] = 0$, $z \in L_0 = (0, 2)$ and $\phi(z)$ satisfies the estimate

$$C_\beta[\phi, \overline{D^+}] + L_{p_0}[\phi_{\bar{z}}, \overline{D^+}] \leq M_5, \tag{2.13}$$

in which $\beta(0 < \beta \leq \alpha)$, $M_5 = M_5\,(p_0, \beta, k_0, D)$ are two non-negative constants, $\Phi(z)$ is analytic in D^+. If $u(z)$ is a solution of (2.2) in D^+ satisfying the boundary conditions (1.3) and

$$\mathrm{Re}\,[\overline{\lambda(z)}u_z]|_{z=x} = s(x), \quad \lambda(x) = 1 + i \text{ or } 1 - i, \quad x \in L_0, \ C_\beta[s(x), L_0] \leq k_3, \tag{2.14}$$

then the following estimate holds:

$$C_\beta[u(z), \overline{D^+}] + C_\beta[u_z X(z), \overline{D^+}] \leq M_6(k_1 + k_2 + k_3), \tag{2.15}$$

in which k_3 is a non-negative constant, $s(x)$ can be seen as stated in the form (2.23) below, $X(z)$ is as stated in (1.14), and $M_6 = M_6(p_0, \beta, k_0, D^+)$ is a non-negative constant.

Proof According to the method in the proof of Theorem 3.1, Chapter III, the equations (2.2),(2.8) in D^+ have the solutions $\psi(z)$, $\Psi(z)$ respectively, which satisfy the boundary condition (2.9) and the estimates (2.10),(2.11). Setting that

$$U(z) = \frac{u(z) - \psi(z)}{\Psi(z)}, \tag{2.16}$$

it is clear that $U(z)$ is a solution of equation (2.12), which can be expressed the second formula in (2.7), where $\phi(z)$ satisfies the estimate (2.13) and $\Phi(z)$ is an analytic function in D^+. If $s(x)$ in (2.14) is a known function, then the boundary value problem (2.2),(1.3),(2.14) has a unique solution $u(z)$ as stated in the form (2.7), which satisfies the estimate (2.15).

Theorem 2.2 Suppose that the equation (2.2) satisfies Condition C. Then any solution of Problem P for (2.2) can be expressed as

$$u(z) = 2\mathrm{Re}\int_0^z w(z)dz + b_0, \quad w(z) = w_0(z) + W(z), \tag{2.17}$$

where $w_0(z)$ is a solution of Problem A for the complex equation (1.8) with the boundary conditions (1.3), (1.4)$(w_0(z) = u_{0z})$, and $W(z)$ possesses the form

$$W(z) = w(z) - w_0(z) \text{ in } D, \quad w(z) = \tilde{\Phi}(z)e^{\tilde{\phi}(z)} + \tilde{\psi}(z) \text{ in } D^+,$$

$$\tilde{\phi}(z) - \tilde{\phi}_0(z) + Tg = \tilde{\psi}_0(z) - \frac{1}{\pi}\int\int_{D^+}\frac{g(\zeta)}{\zeta - z}d\sigma_\zeta, \ \tilde{\psi}(z) = Tf \text{ in } D^+, \tag{2.18}$$

$$\overline{W(z)} = \Phi(z) + \Psi(z), \quad \Psi(z) = \int_2^\nu g^1(z)d\nu e_1 + \int_0^\mu g^2(z)d\mu e_2 \text{ in } D^-,$$

in which $e_1 = (1 + i)/2$, $e_2 = (1 - i)/2$, $\mu = x + y$, $l = x - y$, $\tilde{\phi}_0(z)$ *is an analytic function in* D^+, *such that* $\text{Im}\,[\tilde{\phi}(x)] = 0$ *on* L_0, *and*

$$g(z) = \begin{cases} A_1/2 + \overline{A_1}\bar{w}/(2w), & w(z) \neq 0, \\ 0, & w(z) = 0, \quad z \in D^+, \end{cases} \quad f(z) = \text{Re}\,[A_1\tilde{\phi}_z] + A_2 u + A_3 \text{ in } D^-,$$

$$g^1(z) = g^2(z) = A\xi + B\eta + Cu + D, \quad \xi = \text{Re}\,w + \text{Im}\,w, \quad \eta = \text{Re}\,w - \text{Im}\,w, \tag{2.19}$$

$$A = \frac{\text{Re}\,A_1 + \text{Im}\,A_1}{2}, \quad B = \frac{\text{Re}\,A_1 - \text{Im}\,A_1}{2}, \quad C = A_2, D = A_3 \text{ in } D^-,$$

where $\tilde{\Phi}(z)$ *and* $\Phi(z)$ *are the solutions of equation* (1.8) *in* D^+ *and* D^- *respectively satisfying the boundary conditions*

$$\text{Re}\,[\overline{\lambda(z)}(\tilde{\Phi}(z)e^{\tilde{\phi}(z)} + \tilde{\psi}(z))] = r(z), \quad z \in \Gamma,$$

$$\text{Re}\,[\overline{\lambda(x)}(\tilde{\Phi}(x)e^{\tilde{\phi}(x)} + \tilde{\psi}(x))] = s(x), \quad x \in L_0,$$

$$\text{Re}\,[\overline{\lambda(x)}\Phi(x)] = \text{Re}\,[\overline{\lambda(x)}(\overline{W(x)} - \Psi(x))], \quad z \in L_0, \tag{2.20}$$

$$\text{Re}\,[\overline{\lambda(z)}\Phi(z)] = -\text{Re}\,[\overline{\lambda(z)}\Psi(z)], \quad z \in L_1 \text{ or } L_2,$$

$$\text{Im}\,[\overline{\lambda(z_1)}\Phi(z_1)] = -\text{Im}\,[\overline{\lambda(z_1)}\Psi(z_1)],$$

where $\lambda(x) = 1 + i$ *or* $1 - i$, $x \in L_0$. *Moreover by Theorem 1.2, Chapter IV, the solution* $w_0(z)$ *of Problem A for* (1.8) *and* $u_0(z)$ *satisfy the estimate in the form*

$$C_\beta[u_0(z), \overline{D}] + C_\beta[w_0(z)X(z), \overline{D^+}] + C_\beta[w_0^\pm(z)Y^\pm(z), \overline{D^-}] \le M_7(k_1 + k_2), \tag{2.21}$$

in which $w^\pm(z) = \text{Re}\,w(z) \mp \text{Im}\,w(z)$, $X(z), Y^\pm(z)$ *are as stated in* (1.14),

$$u_0(z) = 2\text{Re}\int_0^z w_0(z)dz + b_0, \tag{2.22}$$

and $M_7 = M_7(p_0, \beta, k_0, D)$ *is a non-negative constant. From* (2.22), *it follows that*

$$C_\beta[u_0(z), \bar{D}] \le M_8\{C_\beta[w_0(z)X(z), \overline{D^+}] + C_\beta[w_0^\pm(z)Y^\pm(z), \overline{D^-}]\} + k_2,$$

where $M_8 = M_8(D)$ *is a non-negative constant.*

Proof Let $u(z)$ be a solution of Problem P for equation (2.2), and $w(z) = u_z$. $u(z)$ be substituted in the positions of w, u in (2.19), thus the functions $g(z)$, $f(z)$, $g_1(z)$, $g_2(z)$, and $\tilde{\psi}(z), \tilde{\phi}(z)$ in $\overline{D^+}$ and $\Psi(z)$ in $\overline{D^-}$ in (2.18),(2.19) can be determined. Moreover we can find the solution $\tilde{\Phi}(z)$ in D^+ and $\Phi(z)$ in $\overline{D^-}$ of (1.8) with the boundary conditions (2.20), where

$$s(x) = \begin{cases} \dfrac{2r((1-i)x/2) - 2R((1-i)x/2)}{a((1-i)x/2) - b((1-i)x/2)} + \text{Re}\,[\overline{\lambda(x)}\Psi(x)] \text{ or} \\[4mm] \dfrac{2r((1+i)x/2 + 1 - i) - 2R((1+i)x/2 + 1 - i)}{a((1+i)x/2 + 1 - i) + b((1+i)x/2 + 1 - i)} \\[4mm] + \text{Re}\,[\overline{\lambda(x)}\Psi(x)] \text{ on } L_0, \end{cases} \tag{2.23}$$

here and later $R(z) = \mathrm{Re}\,[\overline{\lambda(z)}\Psi(z)]$ on L_1 or L_2, thus

$$w(z) = w_0(z) + W(z) = \begin{cases} \tilde{\Phi}(z)^{\tilde{\phi}(z)} + \tilde{\psi}(z) \text{ in } D^+, \\ w_0(z) + \overline{\Phi(z)} + \overline{\Psi(z)} \text{ in } D^-, \end{cases}$$

is the solution of Problem A for the complex equation

$$\begin{Bmatrix} w_{\bar{z}} \\ w_{\overline{z^*}} \end{Bmatrix} = \mathrm{Re}\,[A_1 w] + A_2 u + A_3 \text{ in } \begin{Bmatrix} D^+ \\ D^- \end{Bmatrix}, \qquad (2.24)$$

which can be expressed as the second formula in (2.17), and $u(z)$ is a solution of Problem P for (2.2) as stated in the first formula in (2.17).

Theorem 2.3 *If equation (2.2) satisfies Condition C, then Problem P for (2.2) has at most one solution in D.*

Proof Let $u_1(z), u_2(z)$ be any two solutions of Problem P for (2.2). By Condition C, we see that $u(z) = u_1(z) - u_2(z)$ and $w(z) = u_z$ satisfies the homogeneous equation and boundary condition

$$\begin{Bmatrix} w_{\bar{z}} \\ w_{\overline{z^*}} \end{Bmatrix} = \mathrm{Re}\,[A_1 w] + A_2 u \text{ in } \begin{Bmatrix} D^+ \\ D^- \end{Bmatrix}, \qquad (2.25)$$

$$\mathrm{Re}\,[\overline{\lambda(z)}w(z)] = 0, \quad z \in \Gamma, \quad u(0) = 0, \quad u(2) = 0,$$
$$\mathrm{Re}\,[\lambda(z)w(z)] = 0, \quad z \in L_j (j = 1 \text{ or } 2), \quad \mathrm{Im}\,[\lambda(z_1)w(z_1)] = 0. \qquad (2.26)$$

From Theorem 2.2, the solution $w(z)$ can be expressed in the form

$$w(z) = \begin{cases} \tilde{\Phi}(z)e^{\tilde{\phi}(z)} + \tilde{\psi}(z), \quad \tilde{\psi}(z) = Tf, \quad \tilde{\phi}(z) = \tilde{\phi}_0(z) + \tilde{T}g \text{ in } D^+, \\ w_0(z) + \overline{\Phi(z)} + \overline{\Psi(z)}, \\ \Psi(z) = \int_2^\nu [A\xi + B\eta + Cu]e_1 d\nu + \int_0^\mu [A\xi + B\eta + Cu]e_2 d\mu \text{ in } D^-, \end{cases} \qquad (2.27)$$

where $g(z)$ is as stated in (2.19), $\tilde{\Phi}(z)$ in D^+ is an analytic function and $\Phi(z)$ is a solution of (1.8) in $\overline{D^-}$ satisfying the boundary condition (2.20), $\tilde{\phi}(z), \tilde{\psi}(z)$ possess the similar properties as $\phi(z), \psi_z(z)$ in Theorem 2.1. If $A_2 = 0$ in D^+, then $\tilde{\psi}(z) = 0$. Besides the functions $\tilde{\Phi}(z), \Phi(z)$ satisfy the boundary conditions

$$\begin{cases} \mathrm{Re}\,[\overline{\lambda(x)}\tilde{\Phi}(x)] = s(x), \\ \mathrm{Re}\,[\overline{\lambda(x)}\Phi(x)] = \mathrm{Re}\,[\overline{\lambda(x)}(\overline{W(x)} - \Psi(x))] \end{cases} \text{ on } L_0, \qquad (2.28)$$

where $s(x)$ is as stated in (2.23). From (2.17) with $b_0 = 0$, we can obtain

$$C[u(z), \bar{D}] \le M_8 \{C[w(z)X(z), \overline{D^+}] + C[w_0^\pm(z)Y^\pm(z), \overline{D^-}\}. \qquad (2.29)$$

By using the method of iteration, the estimate

$$C[w(z), \overline{D^-}] \leq \frac{[2M_9 M(4m+1)R']^n}{n!} \tag{2.30}$$

can be derived, where $M_9 = \max\{C[A, \overline{D^-}], C[B, \overline{D^-}], C[C, \overline{D^-}]\}$, $M = 1 + 4k_0^2(1 + k_0^2)$, and $m = C[w(z), \overline{D^-}] > 0$. Let $n \to \infty$, from (2.29), it follows that $w(z) = 0$ in $\overline{D^-}$, and $\Psi(z) = 0$, $\Phi(z) = 0$, $z \in \overline{D^-}$. Thus the solution $u(z) = 2\operatorname{Re}\int_0^z w(z)dz$ is the solution of equation (2.8) with the boundary conditions

$$\operatorname{Re}[\overline{\lambda(z)}u_z] = 0 \text{ on } \Gamma, \quad \operatorname{Re}[\overline{\lambda(x)}u_z(x)] = 0 \text{ on } L_0 = (0, 2), \quad u(0) = 0, \quad u(2) = 0, \tag{2.31}$$

in which $\lambda(x) = 1 + i$, or $1 - i$, $x \in L_0$. Similarly to the proof of Theorem 3.4, Chapter III, we can obtain $u(z) = 0$ on $\overline{D^+}$. This shows the uniqueness of solutions of Problem P for (2.2).

2.3   The solvability of the oblique derivative problem for (2.2)

Theorem 2.4   *Suppose that the mixed equation (2.2) satisfies Condition C. Then Problem P for (2.2) has a solution in D.*

Proof   It is clear that Problem P for (2.2) is equivalent to Problem A for the complex equation of first order and boundary conditions:

$$\left\{ \begin{matrix} w_{\bar{z}} \\ \bar{w}_{\bar{z}^*} \end{matrix} \right\} = F, \quad F = \operatorname{Re}[A_1 w] + A_2 u + A_3 \text{ in } \left\{ \begin{matrix} D^+ \\ D^- \end{matrix} \right\}, \tag{2.32}$$

$$\operatorname{Re}[\overline{\lambda(z)}w(z)] = r(z), \quad z \in \Gamma, \tag{2.33}$$

$$\operatorname{Re}[\lambda(z)w(z)] = r(z), \quad z \in L_j(j = 1 \text{ or } 2), \quad \operatorname{Im}[\lambda(z_1)w(z_1)] = b_1,$$

and the relation (2.17). From (2.17), it follows that

$$C[u(z), \bar{D}] \leq M_8[C(w(z)X(z), \overline{D^+}) + C(w^\pm(z)Y^\pm(z), \overline{D^-})] + k_2, \tag{2.34}$$

where $X(z), Y^\pm(z), w^\pm(z)$ are as stated in (1.14) respectively, $M_8 = M_8(D)$ is a non-negative constant. In the following, by using successive iteration, we shall find a solution of Problem A for the complex equation (2.32) in D. Firstly denoting the solution $w_0(z)(= \xi_0 e_1 + \eta_0 e_2)$ of Problem A for (1.8) and $u_0(z)$ in (2.17), and substituting them into the position of $w = (\xi e_1 + \eta e_2)$, $u(z)$ in the right-hand side of (2.32), similarly to (2.18),(2.19), we have the corresponding functions $f_1(z), g_1(z), g_2^1(z), g_2^1(z)$ and

$$w_1(z) = \tilde{\Phi}_1(z)e^{\tilde{\phi}_1(z)} + \tilde{\psi}_1(z) \text{ in } D^+,$$

$$\tilde{\phi}_1(z) = \tilde{\phi}_0(z) + Tg_1 = \tilde{\phi}_0(z) - \frac{1}{\pi}\iint_{D^+}\frac{g_1(\zeta)}{\zeta - z}d\sigma_\zeta, \tilde{\psi}_1(z) = Tf_1 \text{ in } D^+, \tag{2.35}$$

$$\overline{W_1(z)} = \Phi(z) + \Psi(z), \quad \Psi(z) = \int_2^\nu g_1^1(z)d\nu e_1 + \int_0^\mu g_1^2(z)d\mu e_2 \text{ in } D^-,$$

where $\mu = x + y$, $\nu = x - y$, where $\Phi_1(z)$ is a solution of (1.8) in D^- satisfying the boundary conditions

$$\text{Re}\,[\overline{\lambda(x)}\Phi_1(x)] = \text{Re}\,[\overline{\lambda(x)}(\overline{W_1(z)} - \Psi_1(x))], \quad z \in L_0,$$

$$\text{Re}\,[\overline{\lambda(z)}\Phi_1(z)] = -\text{Re}\,[\overline{\lambda(z)}\Psi_1(z)], \quad z \in L_1 \text{ or } L_2, \tag{2.36}$$

$$\text{Im}\,[\overline{\lambda(z_1)}\Phi_1(z_1)] = -\text{Im}\,[\overline{\lambda(z_1)}\Psi_1(z_1)],$$

and

$$w_1(z) = w_0(z) + W_1(z) = w_0(z) + \overline{\Phi_1(z)} + \overline{\Psi_1(z)} \text{ in } \overline{D^-} \tag{2.37}$$

satisfies the estimate

$$C_\beta[w(z)X(z), \overline{D^+}] + C[w_1^\pm(z)Y^\pm(z), \overline{D^-}] \le M_{10} = M_{10}(p_0, \beta, k, D^-). \tag{2.38}$$

Furthermore we substitute $w_1(z) = w_0(z) + W_1(z)$ and corresponding functions $w_1(z)$, $\xi_1(z) = w^+(z) = \text{Re}\,w_1(z) - \text{Im}\,w_1(z)$, $\eta_1(z) = w^-(z) = \text{Re}\,w_1(z) + \text{Im}\,w_1(z)$, $u_1(z)$ into the positions $w(z)$, $\xi(z)$, $\eta(z)$, $u(z)$ in (2.18),(2.19), and similarly to (2.35)–(2.37), we can find the corresponding functions $\tilde{\psi}_2(z)$, $\tilde{\phi}_2(z)$, $\tilde{\Phi}_2(z)$ in $\overline{D^+}$, $\Psi_2(z), \Phi_2(z)$ and $\overline{W_2(z)} = \Phi_2(z) + \Psi_2(z)$ in $\overline{D^-}$, and the function

$$w_2(z) = \tilde{\Phi}_2(z)e^{\tilde{\phi}_2(z)} + \tilde{\psi}_2(z) \text{ in } D^+,$$

$$w_2(z) = w_0(z) + W_2(z) = w_0(z) + \overline{\Phi_2(z)} + \overline{\Psi_2(z)} \text{ in } \overline{D^-} \tag{2.39}$$

satisfies the similar estimate in the form (2.38). Thus there exists a sequence of functions: $\{w_n(z)\}$ and

$$w_n(z) = \tilde{\Phi}_n(z)e^{\tilde{\phi}_n(z)} + \tilde{\psi}_n(z) \text{ in } D^+,$$

$$w_n(z) = w_0(z) + W_n(z) = w_0(z) + \overline{\Phi_n(z)} + \overline{\Psi_n(z)}, \tag{2.40}$$

$$\Psi_n(z) = \int_2^\nu g_n^1(z)e_1 d\nu + \int_0^\mu g_n^2(z)e_2 d\mu \text{ in } \overline{D^-},$$

and then

$$|[w_1^\pm(z) - w_0^\pm(z)]Y^\pm(z)| \le |\Phi_1^\pm(z)Y^\pm(z)|$$

$$+ \sqrt{2}[|Y^+(z)\int_2^\nu [A\xi_0 + B\eta_0 + Cu_0 + D]e_1 d\nu| \tag{2.41}$$

$$+ |Y^-(z)\int_0^\mu [A\xi_0 + B\eta_0 + Cu_0 + D]e_2 d\mu|] \le 2M_{11}M(4m+1)R' \text{ in } \overline{D^-},$$

where $M_{11} = \max_{z \in \overline{D^-}}(|A|, |B|, |C|, |D|)$, $m = C[w_0(z)X(z), \overline{D^-}]$, $R' = 2$, $M = 1 + 4k_0^2(1 + k_0^2)$. It is clear that $w_n(z) - w_{n-1}(z)$ satisfies

$$\overline{w_n(z)} - \overline{w_{n-1}(z)}$$

$$= \Phi_n(z) - \Phi_{n-1}(z) + \int_2^\nu [A(\xi_n - \xi_{n-1}) + B(\eta_n - \eta_{n-1}) + C(u_n - u_{n-1})]e_1 d\nu \tag{2.42}$$

$$+ \int_0^\mu [A(\xi_n - \xi_{n-1}) + B(\eta_n - \eta_{n-1}) + C(u_n - u_{n-1})]e_2 d\mu \text{ in } \overline{D^-},$$

where $n = 1, 2, \ldots$. From the above equality, the estimate

$$|[w_n^\pm - w_{n-1}^\pm]Y^\pm(z)| \leq [2M_{11}M(4m+1)]^n \times \int_0^{R'} \frac{R'^{n-1}}{(n-1)!}dR'$$

$$\leq \frac{[2M_{11}M(4m+1)R']^n}{n!} \quad \text{in } \overline{D^=}$$

(2.43)

can be obtained, and then we can see that the sequence of functions: $\{w_n^\pm(z)Y^\pm(z)\}$, i.e.

$$w_n^\pm(z)Y^\pm(z) = \{w_0^\pm(z) + [w_1^\pm(z) - w_0^\pm(z)] + \cdots + [w_n^\pm(z) - w_{n-1}^\pm(z)]\}Y^\pm(z) \quad (2.44)$$

$(n = 1, 2, \ldots)$ in $\overline{D^-}$ uniformly converge to $w_*^\pm(z)Y^\pm(z)$, and $w_*(z) = [w_*^+(z) + w_*^-(z) - i(w_*^+(z) - w_*^-(z))]/2$ satisfies the equality

$$\overline{w_*(z)} = \overline{w_0(z)} + \Phi_*(z) + \Psi_*(z),$$

$$\Psi_*(z) = \int_2^\nu [A\xi_* + B\eta_* + Cu_* + D]e_1 d\nu + \int_0^\mu [A\xi_* + B\eta_* + Cu_* + D]e_2 d\mu \quad \text{in } \overline{D^-},$$

(2.45)

and the corresponding function $u_*(z)$ is just a solution of Problem P for equation (2.2) in the domain D^-, and $w_*(z)$ satisfies the estimate

$$C[w_*^\pm(z)Y^\pm(z), \overline{D^-}] \leq e^{2M_{11}M(4m+1)R'}.$$

(2.46)

In the meantime we can obtain the estimate

$$C_\beta[w_n(z)X(z), \overline{D^+}] \leq M_{12} = M_{12}(p_0, \beta, k, D),$$

(2.47)

hence from the sequence $\{w_n(z)\}$, we can choose a subsequence, which uniformly converges to $w_*(z)$ in $\overline{D^+}$, and $w_*(z)$ satisfies the same estimate (2.47). Combining (2.46) and (2.47), it is obvious that the solution $w_*(z) = u_z$ of Problem A for (2.2) in $\bar{D}$ satisfies the estimate

$$C_\beta[w_*(z)X(z), \overline{D^+}] + C[w_*^\pm(z)Y^\pm(z), \overline{D^-}] \leq M_{13} = M_{13}(p_0, \beta, k, D),$$

where M_{13} is a non-negative constant. Moreover the function $u(z)$ in (2.17) is a solution of Problem P for (2.2), where $w(z) = w^*(z)$.

From Theorems 2.3 and 2.4, we see that under Condition C, Problem A for equation (2.32) has a unique solution $w(z)$, which can be found by using successive iteration and the corresponding solution $u(z)$ of Problem P satisfies the estimates

$$C_\beta[u(z), \overline{D^+}] + C_\beta[u_z X(z), \overline{D^+}] \leq M_{14},$$

$$C[u(z), \bar{D}] + C[u_z^\pm Y^\pm(z), \bar{D}] \leq M_{15},$$

(2.48)

where $X(z), Y^\pm(z)$ is as stated in (1.14), and $M_j = M_j(p_0, \beta, k, D)$ $(j = 14, 15)$ are non-negative constants, $k = (k_0, k_1, k_2)$. Moreover we can derive the following theorem.

Theorem 2.5 *Suppose that equation* (2.2) *satisfies Condition C. Then any solution* $u(z)$ *of Problem P for* (2.2) *satisfies the estimates*

$$C_\beta[u(z), \overline{D^+}] + C_\beta[u_z X(z), \overline{D^+}] \le M_{16}(k_1 + k_2),$$
$$C[u(z), \overline{D^-}] + C[u_{\bar{z}}^\pm Y^\pm(z), \overline{D^-}] \le M_{17}(k_1 + k_2),$$

(2.49)

in which $M_j = M_j(p_0, \beta, k_0, D)$ $(j = 16, 17)$ *are non-negative constants.*

From the estimates (2.48),(2.49), we can see that the regularity of solutions of Problem P for (2.2) (see [85]15)).

3 Discontinuous Oblique Derivative Problems for Second Order Linear Equations of Mixed Type

This section deals with an application of method of integral equations to second order equations of mixed type. We mainly discuss the discontinuous Poincaré boundary value problem for second order linear equation of mixed (elliptic-hyperbolic) type, i.e. the generalized Lavrent'ev-Bitsadze equation with weak conditions by the method of integral equations. We first give the representation of solutions for the above boundary value problem, and then give the solvability conditions of the above problem by the Fredholm theorem for integral equations.

3.1 Formulation of the discontinuous Poincaré problem for mixed equations of second order

Let D be a simply connected bounded domain D in the complex plane $\mathbb{C}$ with the boundary $\partial D = \Gamma \cup L$ as stated in Section 1. We consider the second order linear equation of mixed type (2.1) and its complex form (2.2) with Condition C'.

In order to introduce the discontinuous Poincaré boundary value problem for equation (2.2), let the functions $a(z)$, $b(z)$ possess the discontinuities of first kind at $m + 2$ distinct points $z_0 = 2, z_1, \ldots, z_{m+1} = 0 \in \Gamma$ and $Z = \{z_0, z_1, \ldots, z_{m+1}\}$, which are arranged according to the positive direction of Γ, where m is a positive integer, and $r(z) = O(|z - z_j|^{-\beta_j})$ in the neighborhood of $z_j (j = 0, 1, \ldots, m + 1)$ on Γ, in which $\beta_j (j = 0, 1, \ldots, m+1)$ are small positive numbers. Denote $\lambda(z) = a(x) + ib(x)$ and $|a(x)| + |b(x)| \neq 0$, there is no harm in assuming that $|\lambda(z)| = 1$, $z \in \Gamma^* = \Gamma \backslash Z$. Suppose that $\lambda(z)$, $r(z)$ satisfy the conditions

$$\lambda(z) \in C_\alpha(\Gamma_j), \quad |z - z_j|^{\beta_j} r(z) \in C_\alpha(\Gamma_j), \quad j = 0, 1, \ldots, m + 1, \quad (3.1)$$

herein Γ_j is an arc from the point z_{j-1} to z_j on Γ and $z_{m+1} = 0$, and $\Gamma_j (j = 0, 1, \ldots, m+1)$ does not include the end points, and α $(0 < \alpha < 1)$ is a constant.

Problem Q Find a continuously differentiable solution $u(z)$ of (2.2) in $D^* = \bar{D} \backslash \tilde{Z}(\tilde{Z} = \{0, x - y = 2, y \le 0\}$ or $\tilde{Z} = \{x + y = 0, y \le 0, 2\})$, which is continuous in

$\bar{D}$ and satisfies the boundary conditions

$$\frac{1}{2}\frac{\partial u}{\partial l} + \varepsilon\sigma(z)u = \text{Re}\,[\overline{\lambda(z)}u_z] + \varepsilon\sigma(z)u = r(z) + Y(z)h(z), \quad z\in\Gamma, \quad u(0)=b_0, \quad (3.2)$$

$$\frac{1}{2}\frac{\partial u}{\partial l} = \text{Re}\,[\overline{\lambda(z)}u_{\bar{z}}] = r(z), \quad z\in L_1 \text{ or } L_2, \quad \text{Im}\,[\overline{\lambda(z)}u_{\bar{z}}]|_{z=z^1} = b_1, \quad (3.3)$$

where l is a vector at every point on $\Gamma\cup L_j$ ($j=1$ or 2), $z^1=1-i$, b_0,b_1 are real constants, $\lambda(z)=a(x)+ib(x)=\cos(l,x)-i\cos(l,y)$, $z\in\Gamma$, and $\lambda(z)=a(x)+ib(x)=\cos(l,x)+i\cos(l,y)$, $z\in L_j$ ($j=1$ or 2), and $\lambda(z),r(z),b_0,b_1$ satisfy the conditions

$$C_\alpha[\lambda(z),\Gamma]\le k_0, \quad C_\alpha[\sigma(z),\Gamma]\le k_0, \quad C_\alpha[r(z),\Gamma]\le k_2, \quad |b_0|,|b_1|\le k_2,$$

$$C_\alpha[\lambda(z),L_j]\le k_0, \quad C_\alpha[\sigma(z),L_j]\le k_0, \quad C_\alpha[r(z),L_j]\le k_2, \quad j=1 \text{ or } 2, \quad (3.4)$$

$$\max_{z\in L_1}\frac{1}{|a(x)-b(x)|}, \quad \text{or} \quad \max_{z\in L_2}\frac{1}{|a(x)+b(x)|}\le k_0,$$

in which $\alpha\,(1/2<\alpha<1), k_0, k_2$ are non-negative constants, ε is a real parameter. Besides, the functions $Y(z), h(z)$ are as follows

$$Y(z)=\eta\prod_{j=0}^{m+1}|z-z_j|^{\gamma_j}|z-z_*|^l, \quad z\in\Gamma^*, \quad h(z)=\begin{cases} 0, z\in\Gamma, \text{if } K\ge -1/2, \\ h_j\eta_j(z), z\in\Gamma^j, \text{if } K<-1/2, \end{cases} \quad (3.5)$$

in which $\Gamma^j\,(j=0,1,\ldots,m)$ are arcs on $\Gamma^*=\Gamma\backslash Z$ and $\Gamma^j\cap\Gamma^k=\phi, j\ne k, h_j\in J\,(J=\phi$ if $K\ge -1/2; J=1,\ldots,2K'-1$ if $K<-1/2; K'=[|K|+1/2])$ are unknown real constants to be determined appropriately, herein $h_1=0, l=1$ if $2K$ is odd, $z_*(\notin Z)\in\Gamma^*$ is any fixed point, and $l=0$ if $2K$ is even, $\Gamma^j(j=1,\ldots,2K'-1)$ are non-degenerate, mutually disjointed arcs on Γ, and $\Gamma^j\cap Z=\phi, j=1,\ldots,2K'-1$, $\eta_j(z)$ is a positive continuous function on the interior point set of Γ^j, such that $\eta_j(z)=0$ on $\overline{\Gamma\backslash\Gamma^j}$ and

$$C_\alpha[\eta_j(z),\Gamma]\le k_0, \quad j=1,\ldots,2K'-1, \quad (3.6)$$

and $\eta=1$ or -1 on $\Gamma_j\,(0\le j\le m+1,\Gamma_{m+1}=(0,2))$ as stated in [93]. The above discontinuous Poincaré boundary value problem for (2.2) is called Problem Q. Problem Q for (2.2) with $A_3(z)=0, z\in\bar{D}, r(z)=0, z\in\Gamma\cup L_j$ ($j=1$ or 2) and $b_0=b_1=0$ will be called Problem Q_0.

Denote by $\lambda(z_j-0)$ and $\lambda(z_j+0)$ the left limit and right limit of $\lambda(z)$ as $z\to z_j\,(0\le j\le m+1)$ on $\Gamma\cup L_0$, and

$$e^{i\phi_j}=\frac{\lambda(z_j-0)}{\lambda(z_j+0)}, \quad \gamma_j=\frac{1}{\pi i}\ln\frac{\lambda(z_j-0)}{\lambda(z_j+0)}=\frac{\phi_j}{\pi}-K_j,$$

$$K_j=\left[\frac{\phi_j}{\pi}\right]+J_j, \quad J_j=0 \text{ or } 1, \quad j=0,1,\ldots,m+1, \quad (3.7)$$

in which $z_{m+1} = 0$, $z_0 = 2$, $\lambda(z) = e^{i\pi/4}$ on $L_0 = (0, 2)$ and $\lambda(z_0 - 0) = \lambda(z_{m+1} + 0) = \exp(i\pi/4)$, or $\lambda(z) = e^{-i\pi/4}$ on L_0 and $\lambda(z_0 - 0) = \lambda(z_{m+1} + 0) = \exp(-i\pi/4)$, and $0 \leq \gamma_j < 1$ when $J_j = 0$ and $-1 < J_j < 0$ when $J_j = 1$, $j = 0, 1, \ldots, m+1$, and

$$K = \frac{1}{2}(K_0 + K_2 + \cdots + K_{m+1}) = \sum_{j=0}^{m+1} \left(\frac{\phi_j}{2\pi} - \frac{\gamma_j}{2} \right) \tag{3.8}$$

is called the index of Problem Q and Problem Q_0. Let $\beta_j + \gamma_j < 1$, $j = 0, 1, \ldots, m+1$, we can require that the solution $u(z)$ satisfy the condition: $u_z = O(|z - z_j|^{-\delta_j})$ in the neighborhood of z_j $(j = 0, 1, \ldots, m + 1)$ in D^*, where

$$\tau_j = \begin{cases} \beta_j + \tau, & \text{for } \gamma_j \geq 0, \quad \text{and } \gamma_j < 0, \beta_j > |\gamma_j|, \\ |\gamma_j| + \tau, & \text{for } \gamma_j < 0, \quad \beta_j \leq |\gamma_j|, \end{cases} \qquad \delta_j = \begin{cases} 2\tau_j, & j = 0, m+1, \\ \tau_j, & j = 1, \ldots, m, \end{cases} \tag{3.9}$$

and τ, $\delta(< \tau)$ are small positive numbers. In order to ensure that the solution $u(z)$ of Problem Q is continuously differentiable in D^*, we need to choose $\gamma_1 > 0$ or $\gamma_2 > 0$ respectively.

3.2 The representation and solvability of the oblique derivative problem for (2.2)

Now we write a representation theorem of solutions for equation (2.2), which is similar to Theorem 2.2.

Theorem 3.1 *If equation (2.2) satisfies Condition C' and $\varepsilon = 0$, $A_2 \geq 0$ in D^+, then any solution of Problem Q for (2.2) can be expressed as*

$$u(z) = 2\mathrm{Re} \int_0^z w(z)dz + c_0, \quad w(z) = w_0(z) + W(z), \tag{3.10}$$

where $w_0(z)$ is a solution of Problem A for equation (1.8) with the boundary conditions

$$\mathrm{Re}\,[\overline{\lambda(z)}w(z)] = r(z) + Y(z)h(z), \quad z \in \Gamma,$$
$$\mathrm{Re}\,[\overline{\lambda(z)}w(z)] = r(z), \quad z \in L_j \ (j = 1 \text{ or } 2), \quad \mathrm{Im}\,[\overline{\lambda(z)}w(z)]|_{z=z^1} = b_1, \tag{3.11}$$

and $W(z)$ possesses the form

$$W(z) = w(z) - w_0(z), \quad W(z) = \tilde{\Phi}(z)e^{\tilde{\phi}(z)} + \tilde{\psi}(z),$$

$$\tilde{\phi}(z) = \tilde{\phi}_0(z) + Tg - \tilde{\phi}_0(z) \quad \frac{1}{\pi} \iint_{D^+} \frac{g(\zeta)}{\zeta - z} d\sigma_\zeta, \quad \tilde{\psi}(z) = Tf \text{ in } D^+, \tag{3.12}$$

$$\overline{W(z)} = \Phi(z) + \Psi(z), \quad \Psi(z) = \int_2^\nu g^1(z)dve_1 + \int_0^\mu g^2(z)d\mu e_2 \text{ in } D^-,$$

in which $e_1 = \dfrac{1+i}{2}, \ e_2 = \dfrac{1-i}{2}, \quad \mu = x+y, \quad \nu = x-y,$ *and*

$$g(z) = \begin{cases} A_1/2 + \overline{A_1}\overline{w}/(2w), & w(z) \neq 0, \\ 0, w(z) = 0, & z \in D^+, \end{cases} \quad f(z) = \mathrm{Re}\,[A_1\tilde{\phi}_z] + A_2 u + A_3 \ \text{in}\ D^+,$$

$$g^1(z) = g^2(z) = A\xi + B\eta + Cu + D, \quad \xi = \mathrm{Re}\,w - \mathrm{Im}\,w, \quad \eta = \mathrm{Re}\,w + \mathrm{Im}\,w, \tag{3.13}$$

$$A = \frac{\mathrm{Re}\,A_1 + \mathrm{Im}\,A_1}{2}, \quad B = \frac{\mathrm{Re}\,A_1 - \mathrm{Im}\,A_1}{2}, \quad C = A_2, \quad D = A_3 \ \text{in}\ \overline{D^-},$$

where $\tilde{\phi}_0(z)$ *is an analytic function in* D^+ *such that* $\mathrm{Im}\,[\tilde{\phi}(x)] = 0$ *on* $L_0 = (0,2),$ *and* $\tilde{\Phi}(z), \Phi(z)$ *are the solutions of the equation* (1.8) *in* D^+, D^- *respectively satisfying the boundary conditions*

$$\mathrm{Re}\,[\overline{\lambda(z)}e^{\tilde{\phi}(z)}\tilde{\Phi}(z)] = r(z) - \mathrm{Re}\,[\overline{\lambda(z)}\tilde{\psi}(z)], \quad z \in \Gamma,$$

$$\mathrm{Re}\,[\overline{\lambda(x)}(\tilde{\Phi}(x)e^{\tilde{\phi}(x)} + \tilde{\psi}(x))] = s(x), \quad x \in L_0,$$

$$\mathrm{Re}\,[\overline{\lambda(x)}\Phi(x)] = \mathrm{Re}\,[\overline{\lambda(x)}(\overline{W(x)} - \Psi(x))], \quad z \in L_0, \tag{3.14}$$

$$\mathrm{Re}\,[\overline{\lambda(z)}\Phi(z)] = -\mathrm{Re}\,[\overline{\lambda(z)}\Psi(z)], \quad z \in L_1 \ \text{or}\ L_2,$$

$$\mathrm{Im}\,[\overline{\lambda(z^1)}\Phi(z^1)] = -\mathrm{Im}\,[\overline{\lambda(z^1)}\Psi(z^1)],$$

where

$$s(x) = \begin{cases} \dfrac{2r((1-i)x/2) - 2R((1-i)x/2)}{a((1-i)x/2) - b((1-i)x/2)} + \mathrm{Re}\,[\overline{\lambda(x)}\Psi(x)], \ \text{or} \\[3mm] \dfrac{2r((1+i)x/2+1-i) - 2R((1+i)x/2+1-i)}{a((1+i)x/2+1-i) + b((1+i)x/2+1-i)} \\[3mm] + \mathrm{Re}\,[\overline{\lambda(x)}\Psi(x)] \ \text{on}\ L_0, \end{cases} \tag{3.15}$$

in which $s(x)$ *can be written similar to* (2.9). *Moreover from Theorem 1.1, if the index* $K \leq -1/2$, *the solution* $u_0(z)(w_0(z) = u_{0z}(z))$ *of Problem Q for* (1.2) *satisfies the estimate in the form*

$$C_\delta[u_0(z), \overline{D}] + C_\delta[w_0(z)X(z), \overline{D^+}] + C[w_0^\pm(z)Y^\pm(z), \overline{D^-}] \leq M_{18}(k_1 + k_2) \tag{3.16}$$

in which δ *is a small positive constant,* $p_0\,(2 < p_0 \leq p),$ $M_{18} = M_{18}\,(p_0, \delta, k_0, D)$ *are two non-negative constants,*

$$w_0^\pm(z) = \mathrm{Re}\,w_0(z) \mp \mathrm{Im}\,w_0(z), \ X(z) = \Pi_{j=0}^{m+1}|z - t_j|^{\eta_j}, \ Y^\pm(z) = \Pi_{j=1}^2|x \pm y - t_j|^{\eta_j},$$

$$\eta_j = 2|\gamma_j| + \delta, j = 0, m+1, \ \eta_j = |\gamma_j| + \delta, j = 1, \ldots, m,$$

and

$$u_0(z) = 2\mathrm{Re}\int_0^z w_0(z)dz + c_0. \tag{3.17}$$

In order to prove the solvability of Problem Q for (2.2), denote $w = u_z$ and consider the equivalent boundary value problem (Problem B) for the mixed complex equation

$$\begin{cases} w_{\bar{z}} - \text{Re}\,[A_1(z)w] = \varepsilon A_2(z)u + A_3(z), & z \in D^+, \\ \bar{w}_{\bar{z}^*} - \text{Re}\,[A_1(z)w] = A_3(z), & z \in D^-, \\ u(z) = 2\text{Re} \int_0^z w(z)dz + b_0 \end{cases} \tag{3.18}$$

with the boundary conditions

$$\begin{aligned} \text{Re}\,[\overline{\lambda(z)}w] &= r(z) - \varepsilon\sigma(z)u + Y(z)h(z), \quad z \in \Gamma, \\ \text{Re}\,[\overline{\lambda(z)}u_{\bar{z}}] &= r(z), \quad z \in L_j\,(j = 1 \text{ or } 2), \quad \text{Im}\,[\overline{\lambda(z)}u_{\bar{z}}]|_{z=z^1} = b_1, \end{aligned} \tag{3.19}$$

where b_0, b_1 are real constants are as stated in (3.2),(3.3). According to the method in Section 5, Chapter IV, we can find the general solution of Problem B_1 for the mixed complex equation

$$\begin{cases} w_{\bar{z}} - \text{Re}\,[A_1(z)w] = A_3(z), & z \in D^+, \\ \bar{w}_{\bar{z}^*} - \text{Re}\,[A_1(z)w] = A_3(z), & z \in D^-, \end{cases} \tag{3.20}$$

with the boundary conditions

$$\begin{aligned} \text{Re}\,[\overline{\lambda(z)}w(z)] &= r(z) + Y(z)h(z), \quad z \in \Gamma, \\ \text{Re}\,[\overline{\lambda(z)}w(z)] &= r(z), \quad z \in L_j\,(j = 1 \text{ or } 2), \quad \text{Im}\,[\overline{\lambda(z)}w(z)]|_{z=z^1} = b_1, \end{aligned} \tag{3.21}$$

which can be expressed as

$$\tilde{w}(z) = w_0(z) + \sum_{k=1}^{2K+1} c_k w_{k(z)} \tag{3.22}$$

in which $w_0(z)$ is a special solution of Problem B_1 and $w_k(z)(k = 1,\ldots,2K+1, K \geq 0)$ is the complete system of linear independent solutions for the homogeneous problem of Problem B_1. Moreover, denote by $H_2 u$ the solution of Problem B_2 for the complex equation

$$\begin{cases} w_{\bar{z}} - \text{Re}\,[A_1(z)w] = A_2(z)u, & z \in D^+, \\ \bar{w}_{\bar{z}^*} - \text{Re}\,[A_1(z)w] = A_2(z)u, & z \in D^-, \end{cases} \tag{3.23}$$

with the boundary conditions

$$\begin{aligned} \text{Re}\,[\overline{\lambda(z)}w(z)] &= -\sigma(z)u + Y(z)h(z), \quad z \in \Gamma, \\ \text{Re}\,[\overline{\lambda(z)}w(z)] &= 0, \quad z \in L_j\,(j = 1 \text{ or } 2), \quad \text{Im}\,[\overline{\lambda(z)}w(z)]|_{z=z^1} = 0, \end{aligned} \tag{3.24}$$

and the point conditions

$$\text{Im}\,[\overline{\lambda(a_j)}w(a_j)] = 0, \quad j \in J = \begin{cases} 1, \ldots, 2K+1, & K \geq 0, \\ \phi, & K < 0, \end{cases} \qquad (3.25)$$

where $a_j \in \Gamma \backslash Z$ are the distinct points. It is easy to see that H_2 is a bounded operator from $u(z) \in \tilde{C}^1(\bar{D})$ (i.e. $C(u, \bar{D}) + C(X(z)u_z, \overline{D^+}) + C(Y^\pm(z)u_z^\pm, \overline{D^-}) < \infty)$ to $w(z) \in \tilde{C}_\delta(\bar{D})$ (i.e. $C_\delta(u, \bar{D}) + C_\delta(X(z)w(z), \overline{D^+}) + C_\delta(Y^\pm(z)w^\pm(z), \overline{D^-}) < \infty)$, herein $X(z), Y^\pm(z)$ are functions as stated in (3.16). Furthermore denote

$$u(z) = H_1 w + c_0 = 2\text{Re} \int_0^z w(z)dz + c_0, \qquad (3.26)$$

where c_0 is arbitrary real constant. It is clear that H_1 is a bounded operator from $X(z)w(z) \in \tilde{C}_\delta(\bar{D})$ to $u(z) \in \tilde{C}^1(\bar{D})$. On the basis of Theorem 3.1, the function $w(z)$ can be expressed as an integral. From (3.26) and $w(z) = \tilde{w}(z) + \varepsilon H_2 u$, we can obtain a nonhomogeneous integral equation $(K \geq 0)$:

$$u - \varepsilon H_1 H_2 u = H_1 w(z) + c_0 + \sum_{k=1}^{2K+1} c_k H_1 w_k(z). \qquad (3.27)$$

Due to $H_1 H_2$ is a completely continuous operator in $\tilde{C}^1(\bar{D})$, we can use the Fredholm theorem for the integral equation (3.27). Denote by

$$\varepsilon_j (j = 1, 2, \ldots): 0 < |\varepsilon_1| \leq |\varepsilon_2| \leq \cdots \leq |\varepsilon_n| \leq |\varepsilon_{n+1}| \leq \cdots \qquad (3.28)$$

are the discrete eigenvalues for the homogeneous integral equation

$$u - \varepsilon H_1 H_2 u = 0. \qquad (3.29)$$

Noting that Problem Q for the complex equation (2.2) with $\varepsilon = 0$ is solvable, hence $|\varepsilon_1| > 0$. In the following, we first discuss the case of $K \geq 0$. If $\varepsilon \neq \varepsilon_j (j = 1, 2, \ldots)$, i.e. it is not an eigenvalue of the homogeneous integral equation (3.29), then the nonhomogeneous integral equation (3.27) has a solution $u(z)$ and the general solution of Problem Q includes $2K + 2$ arbitrary real constants. If ε is an eigenvalue of rank q as stated in (3.28), applying the Fredholm theorem, we obtain the solvability conditions for nonhomogeneous integral equation (3.27), there is a system of q algebraic equations to determine the $2K + 2$ arbitrary real constants, setting that s is the rank of the corresponding coefficients matrix and $s \leq \min(q, 2K+2)$, we can determine s equalities in the q algebraic equations, hence Problem Q for (2.2) has $q - s$ solvability conditions. When these conditions hold, then the general solution of Problem Q includes $2K + 2 + q - s$ arbitrary real constants. As for the case of $K < 0$, it can be similarly discussed. Thus we can write the above result as in the following theorem.

Theorem 3.2 *Suppose that the linear mixed equation (2.2) satisfies Condition C'. If $\varepsilon \neq \varepsilon_j$ $(j = 1, 2, \ldots)$, where $\varepsilon_j (j = 1, 2, \ldots)$ are the eigenvalues of the homogeneous integral equation (3.29). Then*

(1) *When $K \geq 0$, Problem Q for (2.2) is solvable, and the general solution $u(z)$ of Problem Q for (2.2) includes $2K + 2$ arbitrary real constants.*

(2) *When $K < 0$, Problem Q for (2.2) has $-2K - 1 - s$ solvability conditions, $s \leq 1$.*

If ε is an eigenvalue of homogeneous integral equation (3.29) with the rank q.

(3) *When $K \geq 0$, Problem Q for (2.2) has $q - s$ solvability conditions and $s \leq \min(q, 2K + 2)$.*

(4) *When $K < 0$, Problem Q for (2.2) has $-2K - 1 + q - s$ solvability conditions and $s \leq \min(-2K - 1 + q, 1 + q)$.*

Moreover we can derive the solvability result of Problem P for equation (2.2) with the boundary condition (3.2), in which $h(z) = 0$.

4 The Frankl Boundary Value Problem for Second Order Linear Equations of Mixed Type

This section deals with the Frankl boundary value problem for linear second order equations of mixed (elliptic-hyperbolic) type, i.e. for generalized Lavrent'ev-Bitsadze equations. We first give representation formula and prove uniqueness of solutions for the above boundary value problem, moreover we obtain a priori estimates of solutions, finally by the method of parameter extension, the existence of solutions is proved. In the books [12]1),3), the Frankl problem was discussed for the special mixed equations of second order: $u_{xx} + \mathrm{sgny}\, u_{yy} = 0$. In the book [73], the Frankl problem was discussed for the mixed equation with parabolic degeneracy $\mathrm{sgny}|y|^m u_{xx} + u_{yy} = 0$, which is a mathematical model of problem of gas dynamics. There the existence of solutions of Frankl problem was proved by using the method of integral equations. In this section, we will not use this method. We are proving the solvability of the Frankl problem for generalized linear Lavrent'ev-Bitsadze equations, generalizing the corresponding result from [12]1),3).

4.1 Formulation of the Frankl problem for second order equations of mixed type

Let D be a simply connected bounded domain in the complex plane $\mathbb{C}$ with the boundary $\partial D = \Gamma \cup A'A \cup A'C \cup CB$, where $\Gamma(\subset \{x > 0, y > 0\}) \in C_\mu^2(0 < \mu < 1)$ with the end points $A = i$ and $B = a$, $A'A = \{x = 0, -1 \leq y \leq 1\}$, $A'C = \{x - y = 1, x > 0, y < 0\}$ is the characteristic line and $CB = \{1 \leq x \leq a, y = 0\}$, and denote $D^+ = D \cap \{y > 0\}$, $D^- = D \cap \{y < 0\}$. Without loss of generality, we may assume that $\Gamma = \{x^2/a^2 + y^2 = 1, x > 0, y > 0\}$, otherwise, through a conformal mapping from D^+ onto the domain $D'^+ = \{x^2/a^2 + y^2 < 1, x > 0, y > 0\}$, such that three boundary points $i, 0, 1$ are not changed, then the above requirement can be realized.

Frankl Problem   Find a continuously differentiable solution $u(z)$ of equation (2.2) in $D^* = \overline{D}\backslash\{1, a, i, -i, x + y = 0\}$, which is continuous in $\overline{D}$ and satisfies the boundary conditions

$$u = \psi_1(s) \text{ on } \Gamma, \qquad (4.1)$$

$$u = \psi_2(x) \text{ on } CB, \qquad (4.2)$$

$$\frac{\partial u}{\partial x} = 0 \text{ on } A'A, \qquad (4.3)$$

$$u(iy) - u(-iy) = \phi(y), \quad -1 \le y \le 1. \qquad (4.4)$$

Here $\psi_1(s)$, $\psi_2(x)$, $\phi(y)$ are given real-valued functions satisfying the conditions

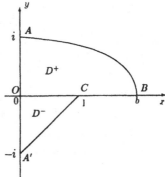

Figure 4.1

$$C_\alpha^1[\psi_1(s), S] \le k_2, \quad C_\alpha^1[\psi_2(x), CB] \le k_2,$$
$$\qquad\qquad\qquad\qquad\qquad\qquad (4.5)$$
$$C_\alpha^1[\phi(y), A'A] \le k_2, \quad \psi_1(0) = \psi_2(a),$$

in which $S = \{0 \le s \le l\}$, s is the arc length parameter on Γ normalized such that $s = 0$ at the point B, l is the length of Γ, and $\alpha \, (0 < \alpha < 1)$, k_2 are non-negative constants. The above boundary value problem is called Problem F and the corresponding homogeneous problem is called Problem F_0.

Let

$$U = \frac{1}{2}u_x, \quad V = -\frac{1}{2}u_y, \quad W = U + iV \text{ in } D, \qquad (4.6)$$

then equation (2.2) can be written as the complex equation

$$\left\{ \begin{matrix} W_{\overline{z}} \\ W_{\overline{z}^*} \end{matrix} \right\} = \text{Re}\,[A_1 W] + A_2 u + A_3 \text{ in } \left\{ \begin{matrix} D^+ \\ D^- \end{matrix} \right\},$$
$$\qquad\qquad\qquad\qquad\qquad\qquad (4.7)$$
$$u(z) = 2\text{Re} \int_a^z W(z)dz + \psi_1(0).$$

If $A_1 = A_2 = A_3 = 0$ in $\overline{D}$, then it is clear that

$$U(x, y) = \frac{1}{2}u_x = \frac{1}{2}[f(x + y) + g(x - y)],$$
$$\qquad\qquad\qquad\qquad\qquad\qquad (4.8)$$
$$-V(x, y) = \frac{1}{2}u_y = \frac{1}{2}[f(x + y) - g(x - y)],$$

in $\overline{D^-}$. From the boundary conditions (4.1)–(4.3), it follows that

$$U(0, y) = \frac{1}{2}\frac{\partial u}{\partial x} = 0, \quad -V(0, y) = \frac{1}{2}[u(0, y)]_y = \frac{1}{2}[u(0, -y)]_y + \frac{1}{2}\phi'(y)$$

$$= V(0, -y) + \frac{1}{2}\phi'(y) = -F(y) + \frac{1}{2}\phi'(y), \qquad (4.9)$$

$$F(y) = -V(0, -y), \quad -1 \le y \le 0,$$

and then

$$U(0,y) = \frac{1}{2}[f(y) + g(-y)] = 0, \quad -1 \le y \le 0,$$

$$-V(0,y) = \frac{1}{2}[f(y) - g(-y)] = -F(y) + \frac{1}{2}\phi'(y), \quad -1 \le y \le 0,$$

$$U(0,y) + V(0,y) = g(-y) = F(y) - \frac{1}{2}\phi'(y), \quad -1 \le y \le 0,$$

$$U(0,y) - V(0,y) = f(y) = -F(y) + \frac{1}{2}\phi'(y), \quad -1 \le y \le 0,$$

$$f(y) = -g(-y), \; f(y) = g(-y) - 2F(y) + \phi'(y), \quad -1 \le y \le 0, \text{ i.e.}$$

$$f(y-x) = -g(x-y), \; f(y-x) = g(x-y) - 2F(y-x) + \phi'(y-x),$$

$$U(x,y) + V(x,y) = g(x-y) = F(y-x) - \frac{1}{2}\phi'(y-x), \quad 0 \le x-y \le 1,$$

$$U(x,y) - V(x,y) = f(x+y) = -g(-x-y)$$

$$= -F(x+y) + \frac{1}{2}\phi'(x+y), \quad 0 \le -x-y \le 1.$$

(4.10)

Hence

$$U(x,y) = \frac{1}{2}[f(x+y) - f(y-x)], \quad 0 \le x-y \le 1,$$

$$-V(x,y) = \frac{1}{2}[f(x+y) - f(y-x)] - F(y-x) + \frac{1}{2}\phi'(y-x), \quad 0 \le x-y \le 1, \quad (4.11)$$

$$U(x,0) + V(x,0) = g(x) = F(-x) - \frac{1}{2}\phi'(-x), \quad 0 \le x \le 1.$$

In particular we have

$$U(x,0) = \frac{1}{2}[f(x) - f(-x)] = \frac{1}{2}[f(x) + F(-x) - \frac{1}{2}\phi'(-x)],$$

$$-V(x,0) = \frac{1}{2}[f(x) - f(-x)] - F(-x) + \frac{1}{2}\phi'(-x)$$

$$= \frac{1}{2}[f(x) - F(-x) + \frac{1}{2}\phi'(-x)] \text{ on } OC.$$

(4.12)

The boundary conditions of the Frankl problem are

$$\frac{\partial u}{\partial l} = 2\text{Re}\,[\overline{\lambda(z)}W(z)] = r(z), \quad z \in \Gamma \cup CB, \quad u(a) = b_0 = \psi_1(0),$$

$$U(0,y) = \frac{1}{2}\frac{\partial u}{\partial x} = r(0,y) = \text{Re}\,[\overline{\lambda(iy)}W(iy)] - 0, \quad 1 \le y \le 1, \quad (4.13)$$

$$\text{Re}\,[\overline{\lambda(x)}W(x)] = r(x) = \frac{1}{\sqrt{2}}[F(-x) - \frac{1}{2}\phi'(-x)], \quad x \in L_0 = (0,1), \quad (4.14)$$

in which l is the tangent vector on the boundary Γ, and

$$\lambda(z) = \begin{cases} \cos(l,x) - i\cos(l,y), \\ 1, \\ \dfrac{1+i}{\sqrt{2}}, \\ 1, \end{cases} \qquad r(z) = \begin{cases} \psi_1'(s) \text{ on } \Gamma = BA, \\ 0 \text{ on } AO, \\ \dfrac{1}{\sqrt{2}}[F(-x) - \dfrac{1}{2}\phi'(-x)] \text{ on } OC, \\ \psi_2'(x) \text{ on } CB. \end{cases}$$

We shall prove the solvability of the Frankl problem for equation (2.2) by using the methods of parameter extension and symmetry extension.

We can choose the index $K = -1/2$ of $\lambda(z)$ on the boundary ∂D^+ of D^+. In fact, due to the boundary condition

$$\text{Re}\,[\overline{\lambda(z)}W(z)] = \frac{1}{2}\text{Re}\,[\overline{\lambda(z)}(u_x - iu_y)] = r(z) \text{ on } \partial D^+ = AO \cup OB \cup BA, \quad (4.15)$$

and $\lambda(z) = 1$ on $AO \cup CB$, $\lambda(z) = \exp(i\pi/4)$ on OC, $\lambda(z) = \cos(l,x) - i\cos(l,y)$ on Γ, denote $t_1 = 0, t_2 = 1, t_3 = a, t_4 = i$, it is seen $\lambda(a + 0) = \exp(i3\pi/2)$ and $\lambda(i - 0) = \exp(i\pi)$, we have

$$K_j = \left[\frac{\phi_j}{\pi}\right] + J_j, \quad J_j = 0 \text{ or } 1, \quad e^{i\phi_j} = \frac{\lambda(t_j - 0)}{\lambda(t_j + 0)}, \quad \gamma_j = \frac{\phi_j}{\pi} - K_j, \quad j = 1,\ldots,4,$$

$$e^{i\phi_1} = \frac{\lambda(t_1 - 0)}{\lambda(t_1 + 0)} = \frac{e^{i0}}{e^{i\pi/4}} = e^{-i\pi/4}, \quad 0 < \gamma_1 = \frac{\phi_1}{\pi} - K_1 = -\frac{1}{4} - K_1 < \frac{3}{4} < 1,$$

$$e^{i\phi_2} = \frac{\lambda(t_2 - 0)}{\lambda(t_2 + 0)} = \frac{e^{i\pi/4}}{e^{i0}} = e^{i\pi/4}, \quad -1 < \gamma_2 = \frac{\phi_2}{\pi} - K_2 = \frac{1}{4} - K_2 = -\frac{3}{4} < 0, \qquad (4.16)$$

$$e^{i\phi_3} = \frac{\lambda(t_3 - 0)}{\lambda(t_3 + 0)} = \frac{e^{i0}}{e^{i3\pi/2}}, \quad 0 \le \gamma_3 = \frac{\phi_3}{\pi} - K_3 = -\frac{3}{2} - K_3 = \frac{1}{2} < 1,$$

$$e^{i\phi_4} = \frac{\lambda(t_4 - 0)}{\lambda(t_4 + 0)} = \frac{e^{i\pi}}{e^{i0}} = e^{i\pi}, \quad 0 \le \gamma_4 = \frac{\phi_4}{\pi} - K_4 = 1 - K_4 = 0 < 1,$$

here $[b]$ is the largest integer not exceeding the real number b, we choose $K_1 = -1$, $K_3 = -2$, $K_2 = K_4 = 1$. Under these conditions, the index K of $\lambda(z)$ on the boundary ∂D^+ of D^+ is just as follows:

$$K = \frac{1}{2}(K_1 + K_2 + K_3 + K_4) = -\frac{1}{2}. \qquad (4.17)$$

Noting that $U(0,y) = 0$ on $A'A$, we can extend $W(z)$ onto the reflected domain $\tilde{D}$ of D about the segment $A'A$. In fact, we introduce the function

$$\tilde{W}(z) = \begin{cases} W(z) \text{ in } D, \\ -\overline{W(-\bar{z})} \text{ on } \tilde{D}, \end{cases} \qquad (4.18)$$

this function $\tilde{W}(z)$ is a solution of the equation

$$\left\{\begin{matrix} \tilde{W}_{\bar{z}} \\ \overline{\tilde{W}_{\bar{z}^*}} \end{matrix}\right\} = \text{Re}\,[\tilde{A}_1 \tilde{W}] + \tilde{A}_2 \tilde{u} + \tilde{A}_3 \quad \text{in} \quad \left\{\begin{matrix} D \\ \tilde{D} \end{matrix}\right\}$$

$$\tilde{u}(z) = 2\text{Re} \int_1^z \tilde{W}(z)dz + \psi_1(0),$$

(4.19)

with the boundary conditions

$$2\text{Re}[\overline{\tilde{\lambda}(z)}\tilde{W}(z)]=\tilde{r}(z), \quad z\in\Gamma\cup CB\cup\tilde{\Gamma}\cup\widetilde{BC}, \quad u(a)=b_0=\psi_1(0)=u(-a),$$

$$\text{Re}[\overline{\tilde{\lambda}(x)}\tilde{W}(x)]=\tilde{r}(x), \quad x\in\tilde{L}_2=(0,1)\cup(-1,0),$$

(4.20)

in which

$$\tilde{A}_1 = \begin{cases} A_1(z), \\ -\overline{A_1(-\bar{z})}, \end{cases} \tilde{A}_2 = \begin{cases} A_2(z), \\ \overline{A_2(-\bar{z})}, \end{cases} \tilde{A}_3 = \begin{cases} A_3(z) \text{ in } D, \\ \overline{A_3(-\bar{z})} \text{ in } \tilde{D}^+ \cup \tilde{D}^-, \end{cases}$$

(4.21)

and

$$\tilde{\lambda}(z) = \begin{cases} \lambda(z), \\ \overline{\lambda(-\bar{z})}, \end{cases} \tilde{r}(z) = \begin{cases} r(z) \ \Gamma\cup CB, \\ -r(-\bar{z}) \ \tilde{\Gamma}\cup\tilde{BC}, \end{cases}$$

$$\tilde{\lambda}(z) = \begin{cases} \dfrac{1+i}{\sqrt{2}}, \\ \dfrac{1-i}{\sqrt{2}}, \end{cases} \tilde{r}(z) = \begin{cases} r(z) \text{ on } OC = (0,1), \\ -r(-\bar{z}) \text{ on } \tilde{CO} = (-1,0), \end{cases}$$

(4.22)

herein $\tilde{\Gamma}, \tilde{BC} = (-a,-1), \tilde{CO}$ and $\tilde{AB}$ are the reflected curves of Γ, CB, OC and BA about the imaginary axis, respectively. We choose the index of the function $\tilde{\lambda}(z)$ on the boundary $\partial(D^+\cup\tilde{D}^+\cup AO)$ of the elliptic domain $D^+\cup\tilde{D}^+\cup AO$ as $K = -1/2$. In fact, noting that $\lambda(z) = 1$ on $CB\cup\tilde{BC}$, $\lambda(z) = \exp(i\pi/4)$ on OC, $\tilde{\lambda}(z) = \exp(-i\pi/4)$ on $\tilde{CO}$, we denote $t_1 = 0, t_2 = 1, t_3 = a, t_4 = i, t_5 = -a, t_6 = -1$, it is seen $\lambda(a+0) = \exp(i3\pi/2), \lambda(i-0) = \lambda(i+0) = \exp(i\pi), \lambda(-a-0) = \exp(i\pi/2)$, hence we have

$$K_j=\left[\dfrac{\phi_j}{\pi}\right]+J_j, \quad J_j=0 \text{ or } 1, \quad e^{j\phi_j}=\dfrac{\tilde{\lambda}(t_j-0)}{\tilde{\lambda}(t_j+0)}, \quad \gamma_j=\dfrac{\phi_j}{\pi}-K_j, \quad j=1,\ldots,6,$$

$$e^{i\phi_1}=\dfrac{\tilde{\lambda}(t_1-0)}{\tilde{\lambda}(t_1+0)}=\dfrac{e^{-i\pi/4}}{e^{i\pi/4}}=e^{-i\pi/2}, \quad 0<\gamma_1=\dfrac{\phi_1}{\pi}-K_1=-\dfrac{1}{2}-K_1=\dfrac{1}{2}<1,$$

$$e^{i\phi_2}=\dfrac{\tilde{\lambda}(t_2-0)}{\tilde{\lambda}(t_2+0)}=\dfrac{e^{i\pi/4}}{e^{i0}}=e^{i\pi/4}, \quad -1<\gamma_2=\dfrac{\phi_2}{\pi}-K_2=\dfrac{1}{4}-K_2=-\dfrac{3}{4}<0,$$

$$e^{i\phi_3} = \frac{\tilde{\lambda}(t_3-0)}{\tilde{\lambda}(t_3+0)} = \frac{e^{i0}}{e^{i3\pi/2}} = e^{-i3\pi/2}, \quad 0 < \gamma_3 = \frac{\phi_3}{\pi} - K_3 = -\frac{3}{2} - K_3 = \frac{1}{2} < 1,$$

$$e^{i\phi_4} = \frac{\tilde{\lambda}(t_4-0)}{\tilde{\lambda}(t_4+0)} = \frac{e^{i\pi}}{e^{i\pi}} = e^{i0}, \quad 0 \le \gamma_4 = \frac{\phi_4}{\pi} - K_4 = 0 - K_4 = 0 < 1,$$

$$e^{i\phi_5} = \frac{\tilde{\lambda}(t_5-0)}{\tilde{\lambda}(t_5+0)} = \frac{e^{i\pi/2}}{e^{i0}} = e^{i\pi/2}, \quad 0 < \gamma_5 = \frac{\phi_5}{\pi} - K_5 = \frac{1}{2} - K_5 = \frac{1}{2} < 1,$$

$$e^{i\phi_6} = \frac{\tilde{\lambda}(t_6-0)}{\tilde{\lambda}(t_6+0)} = \frac{e^{i0}}{e^{-i\pi/4}} = e^{i\pi/4}, \quad -1 < \gamma_6 = \frac{\phi_6}{\pi} - K_6 = \frac{1}{4} - K_6 = -\frac{3}{4} < 0.$$

(4.23)

If we choose $K_1 = -1$, $K_2 = K_6 = 1$, $K_3 = -2$, $K_4 = K_5 = 0$, the index K of $\tilde{\lambda}(z)$ is just

$$K = \frac{1}{2}(K_1 + K_2 + \cdots + K_6) = -\frac{1}{2}. \tag{4.24}$$

We can discuss the solvability of the corresponding boundary value problem (4.19), (4.20), and then derive the existence of solutions of the Frankl problem for equation (2.2).

4.2 Representation and a priori estimates of solutions to the Frankl problem for (2.2)

First of all, similarly to Lemma 2.1, we can prove the following theorem.

Theorem 4.1 *Let equation (2.2) satisfy Condition C in D^+, and $u(z)$ be a continuous solution of (2.2) in $D_*^+ = \overline{D^+}\backslash\{0,1,a,i\}$. Then $u(z)$ can be expressed as*

$$u(z) = U(z)\Psi(z) + \psi(z) \text{ in } D^+,$$

$$U(z) = 2\mathrm{Re}\int_a^z w(z)dz + b_0, \ w(z) = \Phi(z)e^{\phi(z)} \text{ in } D^+. \tag{4.25}$$

Here $\psi(z)$, $\Psi(z)$ are the solutions of equation (2.2) in D^+ and

$$u_{z\bar{z}} - \mathrm{Re}\,[A_1 u_z] - A_2 u = 0 \text{ in } D^+, \tag{4.26}$$

respectively and satisfy the boundary conditions

$$\psi(z) = 0, \ \Psi(z) = 1 \text{ on } \Gamma \cup L,$$

$$\frac{\partial\psi(z)}{\partial x} = 0, \ \frac{\partial\Psi(z)}{\partial x} = 0 \text{ on } AO, \tag{4.27}$$

where $L = (0,a)$. They satisfy the estimates

$$C_\gamma^1[X(z)\psi(z), \overline{D^+}] \le M_{19}, \ \| X(z)\psi(z) \|_{W_{p_0}^2(D^+)} \le M_{19}, \tag{4.28}$$

$$C_\gamma^1[X(z)\Psi(z), \overline{D^+}] \le M_{20}, \| X(z)\Psi(z) \|_{W_{p_0}^2(D^+)} \le M_{20}, \Psi(z) \ge M_{21} > 0, z \in \overline{D^+}, \quad (4.29)$$

in which $X(z) = |x + y - t_1|^m \prod_{j=2}^4 |z - t_j|^{n_j}$, $\eta_j = \max\{-\gamma_j + \delta, \delta\}$, $j = 1, 2, 3, 4$, herein t_j, $\gamma_j(j = 1, 2, 3, 4)$ are as stated in (4.16), δ, $\gamma(\gamma < \delta)$ are small positive constants, $p_0 (2 < p_0 \le p)$, $M_j = M_j(p_0, \gamma, k, D)$ $(j = 19, 20, 21)$ are non-negative constants, $k = (k_0, k_1, k_2)$. Moreover, $U(z)$ is a solution of the equation

$$U_{z\bar{z}} - \mathrm{Re}\,[AU_z] = 0, \quad A = -2(\ln \Psi)_{\bar{z}} + A_1 \text{ in } D^+, \qquad (4.30)$$

where $\mathrm{Im}\,[\phi(z)] = 0$, $z \in \partial D^+$, $\mathrm{Re}\,[\phi(0)] = 0$ and $\phi(z)$ satisfies the estimate

$$C_\beta[\phi(z), \overline{D^+}] + L_{p_0}[\phi_{\bar{z}}, \overline{D^+}] \le M_{22}, \qquad (4.31)$$

in which $\beta\,(0 < \beta \le \alpha)$, $M_{22} = M_{22}(p_0, \alpha, k_0, D^+)$ are two non-negative constants, $\Phi(z)$ is analytic in D^+. If $u(z)$ is a solution of Problem F, then $W(z) = u_z$ satisfies the boundary conditions

$$\mathrm{Re}\,[\overline{\lambda(z)}W(z)] = r(z) \text{ on } \Gamma \cup AO, \quad u(a) = b_0 = \psi_1(0), \qquad (4.32)$$

$$\mathrm{Re}\,[\overline{\lambda(x)}W(x)] = r(x), \quad \lambda(x) = \begin{cases} \dfrac{1+i}{\sqrt{2}} \text{ on } L_0 = (0, 1), \\ 1, \text{ on } L_1 = (1, a). \end{cases} \qquad (4.33)$$

Theorem 4.2 *Suppose that equation (2.2) satisfies Condition C. Then any solution of the Frankl problem for (2.2) can be expressed as*

$$u(z) = 2\mathrm{Re} \int_a^z W(z)dz + b_0, \quad b_0 = \psi_1(0). \qquad (4.34)$$

Here $W(z)$ is a solution of the equation

$$\begin{Bmatrix} W_{\bar{z}} \\ \overline{W_{z^*}} \end{Bmatrix} = \mathrm{Re}\,[A_1 W] + A_2 u + A_3 \text{ in } \begin{Bmatrix} D^+ \\ D^- \end{Bmatrix}, \qquad (4.35)$$

satisfying the boundary conditions (4.13) − (4.14) $(W(z) = u_z)$, and $W(z)$ possesses the form

$$W(z) = \tilde{\Phi}(z)e^{\tilde{\phi}(z)} + \tilde{\psi}(z),$$

$$\tilde{\phi}(z) = \tilde{\phi}_0(z) + Tg = \tilde{\phi}_0(z) - \frac{1}{\pi} \iint_{D^+} \frac{q(\zeta)}{\zeta - z} d\sigma_\zeta, \quad \tilde{\psi}(z) = Tf \text{ in } D^+, \qquad (4.36)$$

$$\overline{W(z)} = \Phi(z) + \Psi(z), \quad \Psi(z) = \int_1^\nu g_1(z)d\nu e_1 + \int_0^\mu g_2(z)d\mu e_2 \text{ in } D^-,$$

in which $e_1 = (1+i)/2$, $e_2 = (1-i)/2$, $\mu = x+y$, $\nu = x-y$, $\tilde{\phi}_0(z)$ *is an analytic function in* D^+, *such that* $\operatorname{Im}\tilde{\phi}(x) = 0$ *on* $\Gamma \cup AO \cup L$, *and*

$$g(z) = \begin{cases} A_1/2 + \overline{A_1 W}/(2W), & W(z) \neq 0 \\ 0, & W(z) = 0 \end{cases} \quad \text{in } D^+$$

$$f(z) = \operatorname{Re}\left[A_1 u_z\right] + A_2 u + A_3, \quad f(z) = \operatorname{Re}\left[A_1 \tilde{\phi}_z\right] + A_2 u + A_3 \text{ in } D^+,$$

$$g^1(z) = g^2(z) = A\xi + B\eta + Cu + D, \quad \xi = \operatorname{Re}W + \operatorname{Im}W, \quad \eta = \operatorname{Re}W - \operatorname{Im}W, \tag{4.37}$$

$$A = \frac{\operatorname{Re}A_1 + \operatorname{Im}A_1}{2}, \quad B = \frac{\operatorname{Re}A_1 - \operatorname{Im}A_1}{2}, \quad C = A_2, \quad D = A_3 \text{ in } D^-,$$

where $\tilde{\Phi}(z)$ *in* D^+ *and* $\Phi(z)$ *in* D^- *are solutions of the equation*

$$\begin{Bmatrix} W_{\bar{z}} \\ W_{\overline{z^*}} \end{Bmatrix} = 0 \text{ in } \begin{Bmatrix} D^+ \\ D^- \end{Bmatrix}, \tag{4.38}$$

satisfying the boundary conditions

$$\operatorname{Re}\left[\overline{\lambda(z)}e^{\tilde{\phi}(z)}\tilde{\Phi}(z)\right] = r(z) - \operatorname{Re}\left[\overline{\lambda(z)}\tilde{\psi}(z)\right], \quad z \in \Gamma,$$

$$\operatorname{Re}\left[e^{\tilde{\phi}(z)}\tilde{\Phi}(z)\right] = -\operatorname{Re}\left[\tilde{\psi}(z)\right], \quad z = iy \in AO,$$

$$\operatorname{Re}\left[\overline{\lambda(x)}\tilde{\Phi}(x)e^{\tilde{\phi}(x)}\right] = r(x) - \operatorname{Re}\left[\overline{\lambda(x)}\tilde{\psi}(x)\right], \quad x \in L_0 = (0,1), \tag{4.39}$$

$$\operatorname{Re}\left[\overline{\lambda(x)}(\Phi(x) + \Psi(x))\right] = r(x) = \operatorname{Re}\left[\overline{\lambda(x)}W(x)\right], \quad x \in L_0 = (0,1),$$

$$\operatorname{Re}\left[\Phi(x)\right] = -\operatorname{Re}\left[\Psi(x)\right], \quad x \in OA', \quad u(a) = b_0 = \psi_1(0),$$

where $\lambda(x)$ *on* $L = (0,a)$ *is as stated in* (4.15).

Proof Let $u(z)$ be a solution of the Frankl problem for equation (2.2), and $W(z) = u_z$, $u(z)$ be substituted in the positions of w, u in (4.37). Thus the functions $g(z)$, $f(z)$, $g_1(z)$, $g_2(z)$, and $\tilde{\psi}(z)$, $\tilde{\phi}(z)$ in $\overline{D^+}$ and $\Psi(z)$ in $\overline{D^-}$ in (4.36),(4.37) are determined. Moreover, we can find the solution $\tilde{\Phi}(z)$ in D^+ and $\Phi(z)$ in $\overline{D^-}$ of (4.38) with the boundary condition (4.39), where $r(z)$ as stated in (4.15), namely

$$r(z) = H(F, \phi), \quad z \in \Gamma \cup AO \cup L, \tag{4.40}$$

thus

$$W(z) = \begin{cases} \tilde{\Phi}(z)e^{\tilde{\phi}(z)} + \tilde{\psi}(z) \text{ in } D^+, \\ \overline{\Phi(z)} + \overline{\Psi(z)} \text{ in } D^-, \end{cases} \tag{4.41}$$

is the solution of Problem A for the complex equation (4.35) with the boundary conditions (4.13),(4.14), which can be expressed as in (4.36), and $u(z)$ is a solution of the Frankl problem for (2.2) as stated in (4.34).

Next, we discuss the uniqueness of solutions of the Frankl problem for (2.2).

Theorem 4.3 *Suppose that the mixed equation (2.2) satisfies Condition C. Then the Frankl problem for (2.2) has at most one solution $u(z) \in C(\bar{D}) \cap C^1(D)$.*

Proof We consider equation (2.2) in D^+. As stated before, if $u_1(z), u_2(z)$ are two solutions of the Frankl problem for (2.2), then $u(z) = u_1(z) - u_2(z)$ is a solution of the homogeneous equation

$$\begin{Bmatrix} u_{z\bar{z}} \\ u_{\bar{z}\bar{z}^*} \end{Bmatrix} = \mathrm{Re}\,[A_1 u_z] + A_2 u \text{ in } \begin{Bmatrix} D^+ \\ D^- \end{Bmatrix},$$

$$u(z) = 2\,\mathrm{Re}\int_a^z W(z)dz, \quad W(z) = u_z \text{ in } D, \qquad (4.42)$$

$$u(z) = U(z)\tilde{\Psi}(z), \quad U_z = \tilde{\Phi}(z)e^{\tilde{\phi}(z)} \text{ in } D^+,$$

$$W(z) = \Phi_0(z) + \overline{\Phi(z)} + \overline{\Psi(z)} \text{ in } D^-,$$

in which

$$\Psi(z) = \int_1^\nu g^1(z)d\nu e_1 + \int_0^\mu g^2(z)d\mu e_2, \quad g^1(z) = g^2(z) = A\xi + B\eta + Cu \text{ in } D^-,$$

and $\tilde{\Psi}(z), \tilde{\Phi}(z), \tilde{\phi}(z), \Phi(z)$ are similar to those in Theorem 4.2, and $\tilde{\Phi}(z), \Phi(z), \Phi_0(z)$ are solutions of equation (4.38) in $\overline{D^+}$ and $\overline{D^-}$ respectively satisfying the conditions

$$2\,\mathrm{Re}\,[\overline{\lambda(z)}\tilde{\Phi}(z)] = r(z) = 0 \text{ on } \Gamma \cup AO \cup CB, \quad U(a) = 0,$$

$$\mathrm{Re}\,[\overline{\lambda(x)}\tilde{\Phi}(x)] = \frac{1}{2}[f(x) + F(-x)] \text{ on } L_0 = (0,1),$$

$$\tilde{\Phi}(x) = \Phi(x), \quad \Phi_0(x) = u_z(x) - \Psi(x) - U_z(x)e^{-\phi(x)} \text{ on } L_0, \qquad (4.43)$$

$$\mathrm{Re}\,[\Phi_0(z)] = \mathrm{Re}\,[\Psi(z)], \quad \mathrm{Im}\,[\Phi_0(z)] = \mathrm{Im}\,[u_z - \Psi(z) - U_z(z)e^{-\phi(z)}] \text{ on } A'O.$$

According to Theorem 4.1, the solution $\tilde{U}(z) = 2\,\mathrm{Re}\int_a^z \tilde{\Phi}(z)dz$ of equation (1.2) satisfies the boundary conditions

$$\tilde{U}(z) = S(z) = 0 \text{ on } \Gamma \cup CB,$$

$$\tilde{U}(x) = 2\int_a^x \tilde{\Phi}(x)dx = 2\int_a^x U_z(x)e^{-\tilde{\phi}(x)}dx = \int_0^x \tilde{\Phi}(x)dx - \int_0^a \tilde{\Phi}(x)dx \qquad (4.44)$$

$$= S(x) = \int_0^x [f(x) + F(-x)]dx = g(x) + \frac{1}{2}\tilde{U}(ix) \text{ on } (0,1),$$

where $g(x) = \int_0^x f(x)dx$, $\int_0^x F(-x)dx = \tilde{U}(ix)/2$. Besides the harmonic function $\tilde{U}(z)$ in D^+ satisfies the boundary condition

$$\frac{\partial \tilde{U}(z)}{\partial x} = 0 \text{ on } AO \qquad (4.45)$$

Moreover, there exists a conjugate harmonic function $\tilde{V}(z)$ in D^+ such that $\tilde{V}(0) = 0$. From the above last formula, we can derive that $\tilde{V}(iy) = \int_0^y \tilde{V}_y dy = \int_0^y \tilde{U}_x dy = 0$ on

AO. By the Cauchy theorem, we have

$$\int_{\partial D^+} [\tilde{U}(z) + i\tilde{V}(z)]^2 dz = 0$$

$$= -\int_1^a [\tilde{V}(x)]^2 dx - \int_\Gamma [\tilde{V}(z)]^2 \left(\frac{dx}{ds} + i\frac{dy}{ds}\right) ds \qquad (4.46)$$

$$+i\int_1^0 [\tilde{U}(iy)]^2 dy + \int_0^1 [\tilde{U}^2(x) - \tilde{V}^2(x) + 2i\,\tilde{U}(x)\tilde{V}(x)] dx.$$

Due to the continuity of $\tilde{U}_y$ on $(0,1)$, $\tilde{V}(x) = \int_0^x \tilde{V}_x dx = -\int_0^x \tilde{U}_y dx = 2\int_0^x \tilde{V}(x) dx = -\int_0^x [f(x) - F(-x)] dx = -g(x) + \tilde{U}(ix)/2$ is obtained. From the imaginary part in (4.46) and the above formula, it is clear that

$$\int_\Gamma [\tilde{V}(z)]^2 \frac{\partial y}{\partial s} ds + \int_0^1 [\tilde{U}(iy)]^2 dy + 2\int_0^1 \{[g(x)]^2 - \frac{1}{4}[\tilde{U}(ix)]^2\} dx = 0. \qquad (4.47)$$

Hence, we get

$$\tilde{U}(iy) = 0 \text{ on } AO, \quad g(x) = 0 \text{ on } OC,$$

and then $f(x) = g'(x) = 0$, $F(-x) = [\tilde{U}(ix)]_x/2 = 0$ on OC. Due to the function $r(z) = S(z) = 0$ on ∂D^+ in (4.43),(4.44) and the index $K = -1/2$, hence $\tilde{\Phi}(z) = 0$ in $\overline{D^+}$, and then the solution $u(z)$ of the homogeneous Frankl problem for (4.42) in D^+ satisfies $u(z) = u_1(z) - u_2(z) = 0$. Moreover, we can derive $u(z) = u_1(z) - u_2(z) = 0$ in D^-. This proves the uniqueness of solutions for the Frankl problem for (2.2) in D.

Finally we give an a priori estimate of solutions to the Frankl problem for equation (2.2). From the estimate we can see the singular behavior of u_z at the discontinuity set $Z = \{1, a, i, -i, x + y = 0\}$. It becomes infinity of an order not exceeding $3/4$ at $z = 1$, infinite of order not exceeding a small positive number δ at the points $\{i, -i\}$ and u_z is bounded at the point set $\{a, x+y = 0\}$. In fact, we can prove that $z = i, -i$ are removable singular points. In [12]3), the author pointed out that u_z can become infinity of an order less than 1.

Theorem 4.4 *Suppose that equation (2.2) satisfies Condition C in $\bar{D}$ and the function $r(z)$ in (4.14) is $H(F, \phi)$, especially*

$$r(x) = H(F, \phi) = \frac{1}{\sqrt{2}}[F(-x) - \frac{1}{2}\phi'(-x)], \quad x \in L_0 = (0, 1). \qquad (4.48)$$

Then any solution $u(z)$ of the Frankl problem for equation (2.2) in D^+ satisfies the estimate

$$\tilde{C}_\gamma^1[u, \overline{D^+}] = C_\gamma[u(z), \overline{D^+}] + C[u_z X(z), \overline{D^+}] \le M_{23}(k_1 + k_2), \qquad (4.49)$$

where $X(z)$ is as stated in (4.29), i.e.

$$X(z) = |x + y - t_1|^m \prod_{j=2}^4 |z - t_j|^{\eta_j}, \quad \eta_j = \max\{-\gamma_j + \delta, \delta\}, \quad j = 1, 2, 3, 4, \qquad (4.50)$$

and $M_{23} = M_{23}(p_0, \gamma, \delta, k_0, D)$ *is a non-negative constant.*

Proof On the basis of the uniqueness of solutions of the Frankl problem for (2.2) in Theorem 4.3 and the results in [12]3) by using reductio ad absurdum we can derive the estimate (4.49). In fact, from (4.83),(4.84) in the proof of Theorem 4.6 below, we see that the function $[\tilde{W}_{n+1}, \tilde{u}_{n+1}]$ $(\tilde{W}_{n+1}(z) = W_{n+1}(z) - W_n(z), \tilde{u}_{n+1}(z) = u_{n+1}(z) - u_n(z))$ is a solution of the boundary value problem

$$[W_{n+1}]_{\bar{z}} - t_0 G(z, \tilde{u}_{n+1}, \tilde{W}_{n+1}) = (t - t_0)G(z, \tilde{u}_n, \tilde{W}_n), \quad z \in D^+,$$

$$\mathrm{Re}\,[\overline{\lambda(z)}\tilde{W}_{n+1}(z)] = 0, \quad z \in \Gamma \cup AO \cup CB,$$

$$\mathrm{Re}\,[\overline{\lambda(z)}\tilde{W}_{n+1}(z)] - t_0 H(\tilde{F}_{n+1}, 0) = (t - t_0)H(\tilde{F}_n, 0)] \text{ on } L_0 = (0,1), \tag{4.51}$$

$$\tilde{u}_{n+1}(z) = \int_a^z \tilde{W}_{n+1}(z)dz, \quad z \in D^+,$$

where $G(z, u, W) = \mathrm{Re}\,[A_1 W] + A_2 u + A_3$, $\tilde{G}(z, \tilde{u}_n, \tilde{w}_n) = G(z, u_{n+1}, w_{n+1}) - G(z, u_n, w_n)$, $\tilde{F}_{n+1} = F_{n+1} - F_n$. On the basis of Theorem 4.1, the solution $\tilde{W}_{n+1}$ of the boundary value problem (4.51) can be expressed as

$$\tilde{u}_{n+1}(z) = U_{n+1}(z)\tilde{\Psi}_{n+1}(z) + \tilde{\psi}_{n+1}(z) \text{ in } D^+,$$

$$\tilde{U}_{n+1}(z) = 2\,\mathrm{Re}\int_1^z \tilde{\Phi}_{n+1}(z)dz, U_{n+1z} = \tilde{\Phi}_{n+1}(z)e^{\tilde{\phi}_{n+1}(z)} \text{ in } D^+, \tag{4.52}$$

$$\tilde{W}_{n+1}(z) = \Phi_{n+1}^0(z) + \overline{\Phi_{n+1}(z)} + \overline{\Psi_{n+1}(z)} \text{ in } D^-,$$

where $\Phi_{n+1}(z), \Phi_{n+1}^0(z)$ are the solutions of equation (4.38) in D^-, $\Psi_{n+1}(z)$ is a solution of the equation in (4.51) in D^-, $\tilde{\psi}_{n+1}(z), \tilde{\Psi}_{n+1}(z)$ are the solutions of the equation in (4.51) and its homogeneous complex equation in $\overline{D^+}$ satisfying the boundary conditions

$$\tilde{\psi}_{n+1}(z) = 1, \ \tilde{\Psi}_{n+1}(z) = 1 \text{ on } \Gamma \cup L,$$

$$\frac{\partial \tilde{\psi}_{n+1}(z)}{\partial x} = 0, \ \frac{\partial \tilde{\Psi}_{n+1}(z)}{\partial x} = 0 \text{ on } AO. \tag{4.53}$$

According to the proof of Theorem 4.3, we see that the function $\tilde{U}_{n+1}(z)$ satisfies the boundary conditions

$$\tilde{U}_{n+1}(z) = S(z) = 0 \text{ on } \Gamma \cup CB, \quad \tilde{U}_{n+1}(a) = 0,$$

$$\tilde{U}_{n+1}(x) = \mathrm{Re}\int_a^x \tilde{\Phi}_{n+1x}dx = S(x),$$

$$S(x) = \int_0^x [f(x) + \tilde{F}_{n+1}(-x)]dx - \frac{1}{2}\int_0^x \tilde{\phi}_{n+1}(-x)dx \tag{4.54}$$

$$= g(x) + \frac{1}{2}\tilde{U}_{n+1}(ix) + \frac{1}{2}\tilde{\phi}_{n+1}(-x) \text{ on } (0,1),$$

where

$$g(x) = \int_0^x f(x)dx, \quad \int_0^x \tilde{F}_{n+1}(-x)dx = \frac{\tilde{U}_{n+1}(ix)}{2}, \quad \tilde{U}_{n+1}(-x) = -\int_0^x \tilde{\phi}'_{n+1}(-x)dx.$$

Besides we can see that the harmonic function $\tilde{U}_{n+1}(z)$ in D^+ satisfies the boundary condition

$$\frac{\partial \tilde{U}_{n+1}(z)}{\partial x} = 0 \text{ on } AO. \tag{4.55}$$

Moreover there exists a conjugate harmonic function $\tilde{V}_{n+1}(z)$ in D^+ such that $\tilde{V}_{n+1}(0) = 0$. We shall verify that

$$\lim_{n\to\infty} \max_{\overline{D}^+} |X(x)\tilde{U}_{n+1x}| = 0, \quad \lim_{n\to\infty} \int_0^1 [\tilde{U}_{n+1}(iy)]^2 dy = 0. \tag{4.56}$$

Suppose that $\lim_{n\to\infty} \int_0^1 |\tilde{U}_{n+1}(iy)|dy = C > 0$, due to

$$\int_\Gamma [\tilde{V}_{n+1}(z)]^2 \frac{\partial y}{\partial s} ds + \int_0^1 [\tilde{U}_{n+1}(iy)]^2 dy$$

$$+ 2\int_0^1 \left\{ [g(x)]^2 - \frac{1}{4}[t_0\tilde{U}_{n+1}(ix) + (t-t_0)\tilde{U}_n(ix)]^2 \right\} dx = 0, \tag{4.57}$$

provided that $|t - t_0|$ is sufficiently small, such that $|t - t_0|^2 \int_0^1 |\tilde{U}_n(iy)|^2 dy \leq \int_0^1 |\tilde{U}_{n+1}(iy)|^2 dy/2$ for $n = n_k \to \infty$, then similarly to (4.47), from (4.57) we can derive that

$$\tilde{U}_{n+1}(iy) = 0 \text{ on } AO, \quad g(x) = 0 \text{ on } OC. \tag{4.58}$$

This contradiction proves that $\int_0^1 [\tilde{U}_{n+1}(ix)]^2 dx = 0$, $\int_0^1 [\tilde{U}_{n+1}(ix)]^2 dx = 0$ and $\tilde{U}_{n+1} = U_{n+1}(z) - U_n(z) = 0$ in D^+ for $n \geq N_0$, where N_0 is a sufficiently large positive number. Hence $\tilde{u}_{n+1} = u_{n+1}(z) - u_n(z) = \psi_{n+1}(z) - \psi_n(z)$ in D^+ for $n \geq N_0$. Similarly to the proof of the first estimate in (4.28), we can obtain

$$C_\gamma^1[X(z)\tilde{u}_{n+1}(z), \overline{D}^+] \leq M_{24}|t - t_0|C_\gamma^1[X(z)\tilde{u}_n(z), \overline{D}^+], \tag{4.59}$$

in which $M_{24} = M_{24}(p, \gamma, \delta, k_0, D^+)$ is a non-negative constant. Choosing the constant ε so small that $\varepsilon M_{24} \leq 1/2$ and $|t - t_0| \leq \varepsilon$, it follows that

$$\tilde{C}_\gamma^1[\tilde{u}_{n+1}, \overline{D}^+] \leq \varepsilon M_{24}|t - t_0|C_\gamma^1[\tilde{u}_n, \overline{D}^+] \leq \frac{1}{2}C_\gamma^1[\tilde{u}_n, \overline{D}^+],$$

$$\tilde{C}_\gamma^1[\tilde{u}_{n+1}, \overline{D}^+] \leq 2^{-n+N_0} \sum_{j=N_0}^{\infty} 2^{-j}\tilde{C}_\gamma^1[u_1 - u_0, \overline{D}^+] \leq 2^{-n+N_0+1}\tilde{C}_\gamma^1[u_1 - u_0, \overline{D}^+]$$

for $n > N_0$. Therefore there exists a continuous function $u_*(z)$ on $\overline{D}^+$, such that

$$u_*(z) = \sum_{j=0}^{\infty} \tilde{u}_{n+1} = \sum_{j=0}^{\infty} [u_{n+1} - u_n(z)].$$

From the estimate of $\sum_{j=0}^{n}[u_{j+1}(z) - u_j(z)] = u_{n+1}(z) - u_0(z)$ in D^+, the estimate

$$\tilde{C}_\gamma^1[u_{n+1}, \bar{D}] = C_\gamma[u_{n+1}, \overline{D^+}] + C[u_{n+1z}X, \overline{D^+}] \le M_{25} \qquad (4.60)$$

can be derived, where $M_{25} = M_{25}(p_0, \gamma, \delta, k_0, D^+)$ is a non-negative constant. Moreover we can derive a similar estimate of $u_*(z)$ in $\overline{D^+}$ and $\overline{D^-}$, which gives the estimate (4.49).

4.3   The solvability of the Frankl problem for (2.2)

Theorem 4.5   *Suppose that the mixed equation (2.2) satisfies Condition C and $A_1(z) = A_2(z) = 0$ in $\bar{D}$, i.e.*

$$\begin{cases} u_{z\bar{z}} = A_3(z), & z \in D^+, \\ u_{\bar{z}\bar{z}^*} = A_3(z), & z \in D^-. \end{cases} \qquad (4.61)$$

Then the Frankl problem for (4.61) has a solution in D.

Proof   It is clear that the Frankl problem for (4.61) is equivalent to the following Problem A for the complex equation of first order and boundary conditions:

$$\begin{Bmatrix} W_{\bar{z}} \\ W_{\overline{z^*}} \end{Bmatrix} = A_3(z) \text{ in } \begin{Bmatrix} D^+ \\ D^- \end{Bmatrix}, \qquad (4.62)$$

$$\begin{aligned} \text{Re}\,[\overline{\lambda(z)}W(z)] = r(z) = H(F, \phi), & \quad z \in \Gamma \cup AO \cup L, \\ \text{Re}\,[\overline{\lambda(z)}W(z)] = r(z) = 0, \quad \lambda(z) = 1, & \quad z \in OA', \end{aligned} \qquad (4.63)$$

and the relation

$$u(z) = 2\text{Re}\int_a^z W(z)dz + b_0 \text{ in } D, \qquad (4.64)$$

in which $\lambda(z)$, $r(z)$ are as stated in (4.15) and (4.40).

In order to find a solution $W(z)$ of Problem A in D, we can express $W(z)$ in the form (4.34)–(4.37). In the following, by using the method of parameter extension, we shall find a solution of Problem A for the complex equation (4.62). We consider equation (4.62) and the boundary conditions with the parameter $t \in [0, 1]$:

$$\text{Re}\,[\overline{\lambda(z)}W(z)] = tH(F, \phi) + R(z) \text{ on } \partial D^+ = \Gamma \cup AO \cup L, \qquad (4.65)$$

in which $H(F, \phi)$ on $\partial D^+ = \Gamma \cup AO \cup L$ is as stated in (4.48), and $R(z)X(z) \in C_\gamma(\partial D^+)$, this problem is called Problem F_t.

When $t = 0$, the unique solution of Problem F_0 for the complex equation (4.61) can be found by a method given in Section 1, and its solution $[W_0(z), u_0(z)]$ can be expressed as

$$u_0(z) = 2\mathrm{Re}\int_a^z W_0(z)dz + b_0, \quad W_0(z) = \tilde{W}(z) \text{ in } D, \quad b_0 = \psi_1(0),$$

$$\tilde{W}(z) = \tilde{\Phi}(z) + \tilde{\psi}(z), \quad \tilde{\psi}(z) = TA_3 = -\frac{1}{\pi}\iint_{D^+}\frac{A_3(\zeta)}{\zeta - z}d\sigma_\zeta \text{ in } D^+, \qquad (4.66)$$

$$\overline{W(z)} = \Phi(z) + \Psi(z), \quad \Psi(z) = \int_1^\nu A_3(z)e_1 d\nu + \int_0^\mu A_3(z)e_2 d\mu \text{ in } \overline{D^-},$$

where $\tilde{\Phi}(z)$ is an analytic function in D^+ and $\Phi(z)$ is a solution of (4.38) in $\overline{D_-}$ satisfying the boundary conditions

$$\mathrm{Re}\,[\overline{\lambda(z)}\tilde{W}(z)] = R(z), \quad z \in \Gamma \cup L,$$

$$\mathrm{Re}\,[\overline{\lambda(z)}(\tilde{\Phi}(z) + \tilde{\Psi}(z))] = R(z), \quad \lambda(z) = 1, \quad z = iy \in AO,$$

$$\mathrm{Re}\,[\overline{\lambda(z)}(\Phi(z) + \Psi(z))] = R(z), \quad \lambda(z) = 1, \quad z = iy \in OA', \qquad (4.67)$$

$$\mathrm{Re}\,[\overline{\lambda(x)}(\Phi(x) + \Psi(x))] = R(x) = \mathrm{Re}\,[\lambda(x)\tilde{W}(x)], \quad z = x \in OC, \quad u(a) = b_0.$$

Suppose that when $t = t_0$ ($0 \le t_0 < 1$), Problem F_{t_0} is solvable, i.e. Problem F_{t_0} for (4.62) has a solution $[W_0(z), u_0(z)]$ ($u_0(z) \in \tilde{C}_\gamma^1(\bar{D})$). We can find a neighborhood $T_\varepsilon = \{|t - t_0| \le \varepsilon, 0 \le t \le 1\}$ ($0 < \varepsilon < 1$) of t_0 such that for every $t \in T_\varepsilon$, Problem F_t is solvable. In fact, Problem F_t can be written in the form

$$\begin{Bmatrix} W_{\bar{z}} \\ \bar{W}_{\bar{z}^*} \end{Bmatrix} = A_3(z) \text{ in } \begin{Bmatrix} D^+ \\ D^- \end{Bmatrix}, \qquad (4.68)$$

$$\mathrm{Re}\,[\overline{\lambda(z)}W(z)] - t_0 H(F, \phi) = (t - t_0)H(F, \phi) + R(z) \text{ on } \partial D^+.$$

Replacing $W(z), u(z)$ in the right-hand sides of (4.68) by a function $W_0(z) \in C_\gamma(\partial D^+)$ and the corresponding function $u_0(z)$ in (4.66), i.e. $u_0(z) \in \tilde{C}_\gamma^1(\bar{D})$, especially, by the solution $[W_0(z), u_0(z)]$ of Problem F_0, it is obvious that the boundary value problem for such an equation in (4.68) then has a solution $[W_1(z), u_1(z)], u_1(z) \in \tilde{C}_\gamma^1(\partial D)$. Using successive iteration, we obtain a sequence of solutions $[W_n(z), u_n(z)], u_n(z) \in \tilde{C}_\gamma^1(\bar{D})$, $n = 1, 2, \ldots$, which satisfy the equations and boundary conditions

$$\begin{Bmatrix} W_{n+1\bar{z}} \\ \bar{W}_{n+1\bar{z}^*} \end{Bmatrix} = A_3(z) \text{ in } \begin{Bmatrix} D^+ \\ D^- \end{Bmatrix},$$

$$\mathrm{Re}\,[\overline{\lambda(z)}W_{n+1}(z)] - t_0 H(F_{n+1}, \phi) = (t - t_0)H(F_n, \phi) + R(z) \text{ on } \partial D^+, \qquad (4.69)$$

$$\mathrm{Re}\,[\overline{\lambda(z)}W_{n+1}(z)] = 0, \quad z \in OA'.$$

From the above formulas, it follows that

$$[W_{n+1} - W_n]_{\bar{z}} = 0, \quad z \in D,$$

$$\mathrm{Re}\,[\overline{\lambda(z)}(W_{n+1}(z) - W_n(z))] - t_0[H(F_{n+1} - F_n, 0)] = (t - t_0)[H(F_n - F_{n-1}, 0)]. \qquad (4.70)$$

Noting that

$$|t - t_0|C_\gamma[XH(\Phi_n - \Phi_{n-1}, 0), L_0] \le |t - t_0|C_\gamma[X(\Phi_n - \Phi_{n-1}), L_0], \qquad (4.71)$$

and applying Theorem 4.4, we have

$$\tilde{C}_\gamma^1[u_{n+1} - u_n, \overline{D^+}] \le M_{26}\tilde{C}_\gamma^1[\Phi_n - \Phi_{n-1}, \overline{D^+}], \qquad (4.72)$$

where $M_{26} = M_{26}(p_0, \gamma, \delta, k_0, D^+)$. Choosing the constant ε so small that $\varepsilon M_{24} \le 1/2$ and $|t - t_0| \le \varepsilon$, it follows that

$$\tilde{C}_\gamma^1[u_{n+1} - u_n, \overline{D^+}] \le \varepsilon M_{26}\tilde{C}_\gamma^1[u_n - u_{n-1}, \overline{D^+}] \le \frac{1}{2}\tilde{C}_\gamma^1[u_n - u_{n-1}, \overline{D^+}], \qquad (4.73)$$

and when $n, m \ge N_0 + 1$ (N_0 is a positive integer),

$$\tilde{C}_\gamma^1[u_{n+1} - u_n, \overline{D^+}] \le 2^{-N_0}\sum_{j=0}^{\infty} 2^{-j}\tilde{C}_\gamma^1[u_1 - u_0, \overline{D^+}] \le 2^{-N_0+1}\tilde{C}_\gamma^1[u_1 - u_0, \overline{D^+}]. \qquad (4.74)$$

Hence $\{u_n(z)\}$ is a Cauchy sequence. According to the completeness of the Banach space $\tilde{C}_\gamma^1(\overline{D^+})$, there exists a function $u_*(z) \in C_\gamma^1(\overline{D^+})$, and $W_*(z) = u_{*z}(z)$, so that $\tilde{C}_\gamma^1[u_n - u_*, \overline{D^+}] = C_\gamma[u_n - u_*, \overline{D^+}] + C[X(W_n - W_*), \overline{D^+}] \to 0$ as $n \to \infty$. We can see that $[W_*(z), u_*(z)]$ is a solution of Problem F_t for every $t \in T_\varepsilon = \{|t - t_0| \le \varepsilon\}$. Because the constant ε is independent of t_0 ($0 \le t_0 < 1$), therefore from the solvability of Problem F_0 when $t_0 = 0$, we can derive the solvability of Problem F_t when $t = \varepsilon, 2\varepsilon, \ldots, [1/\varepsilon]\varepsilon, 1$. In particular, when $t = 1$ and $R(z) = 0$, Problem F_1 i.e. the Frankl problem for (4.61) in $\overline{D^+}$ is solvable.

As for the solution $[W(z), u(z)]$ in $\overline{D^-}$, it can be obtained by (4.10),(4.11) and the method in Chapters I and II, namely

$$u(z) = 2\mathrm{Re}\int_a^z W(z)dz + b_0 \text{ on } \overline{D^-}, \quad b_0 = \psi_1(0),$$

$$\overline{W(z)} = \Phi(z) + \Psi(z), \quad \Psi(z) = \int_1^\nu A_3(z)e_1 d\nu + \int_0^\mu A_3(z)e_2 d\mu,$$

$$\Phi(z) = \frac{1}{2}[(1+i)f(x+y) + (1-i)g(x-y)], \qquad (4.75)$$

$$f(x+y) = \mathrm{Re}\,[(1-i)(\overline{W(x+y)} - \Psi(x+y))],$$

$$g(x-y) = \mathrm{Re}\,[(1+i)(\overline{W(x-y)} - \Psi(x-y))], \quad z \in \overline{D^-} \cap \{x+y \ge 0\},$$

where $W(x+y)$, $W(x-y)$ are the values on $0 \le z = x+y \le 1$, $0 \le x-y \le 1$ of the solution $W(z)$ of Problem F for (4.61) in $\overline{D^+}$ and $\Psi(x+y)$, $\Psi(x-y)$ are the values on $0 \le z = x+y \le 1$, $0 \le x-y \le 1$ of $\Psi(z)$ respectively. Moreover, the function $W(z)$ in $D^- \cap \{x+y \le 0\}$ can be obtained by (4.75),(4.18). In fact, from (4.75) we have found the function $W(z)$ on $OC' = \{x+y = 0, 0 \le x \le 1/2\}$, by (4.18), we obtain the function $\tilde{W}(z) = -\overline{W(-\bar{z})}$ on $OC'' = \{x-y = 0, -1/2 \le x \le 0\}$, and denote $\sigma(x) = \operatorname{Re}\left[(1-i)\tilde{\Psi}(z)\right]$ on OC', $\tau(x) = \operatorname{Re}\left[(1+i)\tilde{\Psi}(z)\right]$ on OC''. Hence the solution $u(z)$ in $D^- \cap \{x+y \le 0\}$ is as follows:

$$u(z) = 2\operatorname{Re}\int_0^z W(z)dz + u(0), \quad \tilde{\Psi}(z) = \int_0^\nu \tilde{A}_3(z)e_1 d\nu + \int_0^\mu \tilde{A}_3(z)e_2 d\mu,$$

$$W(z) = \frac{1}{2}[(1-i)f(x+y) + (1+i)g(x-y)] + \overline{\tilde{\Psi}(z)}, \tag{4.76}$$

$$f(x+y) = \tau((x+y)/2) + \operatorname{Re}W(0) + \operatorname{Im}W(0),$$
$$g(x-y) = \sigma((x-y)/2) + \operatorname{Re}W(0) - \operatorname{Im}W(0), \quad z \in D^- \cap \{x+y \le 0\},$$

in which $\Phi(z)$ and $\Psi(z)$ are the functions from (4.75). Furthermore, we can prove that the solution $u(z)$ satisfies the boundary conditions (4.1)–(4.4). This completes the proof.

Theorem 4.6 *Suppose that the mixed equation* (2.2) *satisfies Condition C. Then the Frankl problem for* (2.2) *has a solution in D.*

Proof Similarly to the proof of Theorem 4.5, we see that the Frankl problem for (2.2) is equivalent to Problem A for first order complex equation and boundary conditions:

$$\left\{ \begin{array}{c} W_{\bar{z}} \\ W_{\overline{z^*}} \end{array} \right\} = G, \quad G = G(z, u, W) = \operatorname{Re}\left[A_1 W\right] + A_2 u + A_3 \text{ in } \left\{ \begin{array}{c} D^+ \\ D^- \end{array} \right\}, \tag{4.77}$$

$$\operatorname{Re}\left[\overline{\lambda(z)}W(z)\right] = r(z) = H(F,\phi), \quad z \in \Gamma \cup AO \cup L, \tag{4.78}$$

and the relation (4.64), in which $r(z) = H(F,\phi)$ on $z \in \partial D^+ = \Gamma \cup AO \cup L$ is as stated in (4.63).

In order to find a solution $W(z)$ of Problem A in D, we can express $W(z)$ in the form (4.34)–(4.37). In the following, by using the method of parameter extension, a solution of Problem A for the complex equation (4.77) will be found. We consider the equation and boundary conditions with the parameter $t \in [0,1]$:

$$W_{\bar{z}} = tG + K(z), \quad G = G(z, u, W) = \operatorname{Re}\left[A_1 W\right] + A_2 u + A_3 \text{ in } D^+, \tag{4.79}$$

$$\operatorname{Re}\left[\overline{\lambda(z)}W(z)\right] = tH(F,\phi) + R(z) \text{ on } \partial D^+ = \Gamma \cup AO \cup L, \tag{4.80}$$

where $K(z) \in L_p(\overline{D^+})$ and $R(z)X(z) \in C_\gamma(\partial D^+)$. This problem is called Problem F_t.

When $t = 0$, the complex equation (4.79) becomes the equation

$$W_{\bar{z}} = K(z), \quad \text{in } D^+. \tag{4.81}$$

From Theorem 4.5, we can find the unique solution of Problem F_0 for (4.79). Suppose that when $t = t_0$ ($0 \le t_0 < 1$), Problem F_{t_0} is solvable, i.e. Problem F_{t_0} for (4.79) has a solution $[W_0(z), u_0(z)]$ ($u_0 \in \tilde{C}_\gamma^1(\bar{D})$). We can find a neighborhood $T_\epsilon = \{|t - t_0| \le$

$\varepsilon, 0 \leq t \leq 1\}(0 < \varepsilon < 1)$ of t_0 such that for every $t \in T_\varepsilon$, Problem F_t is solvable. In fact, Problem F_t can be written in the form

$$W_{\bar{z}} - t_0 G(z, u, W) = (t - t_0)G(z, u, W) + K(z) \text{ in } D^+,$$

$$\text{Re}\left[\overline{\lambda(z)}W(z)\right] - t_0 H(F, \phi) = (t - t_0)H(F, \phi) + R(z) \text{ on } \partial D^+. \tag{4.82}$$

Replacing $W(z), u(z)$ in the right-hand sides of (4.82) by a function $W_0(z) \in C_\gamma(\partial D^+)$ and the corresponding function $u_0(z)$ in (4.66), i.e. $u_0(z) \in \tilde{C}_\gamma^1(\bar{D})$, especially, by the solution $[W_0(z), u_0(z)]$ of Problem F_0, it is obvious that the boundary value problem for such equation in (4.82) then has a solution $[W_1(z), u_1(z)], u(z) \in \tilde{C}_\gamma^1(\partial D)$. Using successive iteration, we obtain a sequence of solutions $[W_n(z), u_n(z)], u_n(z) \in \tilde{C}_\gamma^1(\bar{D})$, $n = 1, 2, \ldots$, which satisfy the equations and boundary conditions

$$W_{n+1\bar{z}} - t_0 G(z, u_{n+1}, W_{n+1}) = (t - t_0)G(z, u_n, W_n) + K(z), \quad \text{in } D^+, \tag{4.83}$$

$$\text{Re}\left[\overline{\lambda(z)}W_{n+1}(z)\right] - t_0 H(F_{n+1}, \phi) = (t - t_0)H(F_n, \phi) + R(z) \text{ on } \partial D^+. \tag{4.84}$$

From the above formulas, it follows that

$$[W_{n+1} - W_n]_{\bar{z}} - t_0[G(z, u_{n+1}, W_{n+1}) - G(z, u_n, W_n)]$$

$$= (t - t_0)[G(z, u_n, W_n) - G(z, u_{n-1}, W_{n-1})], \quad z \in D^+,$$

$$\text{Re}\left[\overline{\lambda(z)}(W_{n+1}(z) - W_n(z))\right] - t_0[H(F_{n+1} - F_n, 0)]$$

$$= (t - t_0)[H(F_n - F_{n-1}, 0)], \quad z \in L_0. \tag{4.85}$$

Noting that

$$L_p[(t-t_0)(G(z, u_n, W_n) - G(z, u_{n-1}, W_{n-1})), \overline{D^+}] \leq 2k_0|t - t_0|\tilde{C}_\gamma^1[u_n - u_{n-1}, \overline{D^+}],$$

$$|t - t_0|C_\gamma[XH(F_n - F_{n-1}, 0), L_0] \leq |t - t_0|C_\gamma[X(F_n - F_{n-1}), L_0], \tag{4.86}$$

and according to the method in the proof of Theorem 4.4, we can obtain

$$\tilde{C}_\gamma^1[u_{n+1} - u_n, \overline{D^+}] \leq M_{27}[2k_0 + 1]\tilde{C}_\gamma^1[u_n - u_{n-1}, \overline{D^+}], \tag{4.87}$$

where $M_{27} = M_{27}(p_0, \gamma, \delta, k_0, D^+)$. Choosing the constant ε so small that εM_{27} $(2k_0 + 1) \leq 1/2$ and $|t - t_0| \leq \varepsilon$, it follows that

$$\tilde{C}_\gamma^1[u_{n+1} - u_n, \overline{D^+}] \leq \varepsilon M_{27}(2k_0 + 1)\tilde{C}_\gamma^1[u_n - u_{n-1}, \overline{D^+}] \leq \frac{1}{2}\tilde{C}_\gamma^1[u_n - u_{n-1}, \overline{D^+}], \tag{4.88}$$

and when $n, m \geq N_0 + 1$ (N_0 is a positive integer),

$$\tilde{C}_\gamma^1[u_{n+1} - u_n, \overline{D^+}] \leq 2^{-N_0} \sum_{j=0}^{\infty} 2^{-j}\tilde{C}_\gamma^1[u_1 - u_0, \overline{D^+}] \leq 2^{-N_0+1}\tilde{C}_\gamma^1[u_1 - u_0, \overline{D^+}].$$

Hence $\{u_n(z)\}$ is a Cauchy sequence. According to the completeness of the Banach space $\tilde{C}_\gamma^1(\overline{D^+})$, there exists a function $u_*(z) \in C_\gamma^1(\overline{D^+})$, and $W_*(z) = u_{*z}(z)$, so that

$\tilde{C}^1_\gamma[u_n - u_*, \overline{D^+}] = C_\gamma[u_n - u_*, \overline{D^+}] + C[X(W_n - W_*), \overline{D^+}] \to 0$ as $n \to \infty$. We can see that $[W_*(z), u_*(z)]$ is a solution of Problem F_t for every $t \in T_\varepsilon = \{|t - t_0| \le \varepsilon\}$. Because the constant ε is independent of t_0 $(0 \le t_0 < 1)$, therefore from the solvability of Problem F_{t_0} when $t_0 = 0$, we can derive the solvability of Problem F_t when $t = \varepsilon, 2\varepsilon, \ldots, [1/\varepsilon]\varepsilon, 1$. In particular, when $t = 1$ and $K(z) = 0$, $R(z) = 0$, Problem F_1, i.e. the Frankl problem for (2.2) in D^+ is solvable.

The existence of the solution $[W(z), u(z)]$ of Problem F for (2.2) in D^- can be obtained by the method in Chapters I and II.

5   Oblique Derivative Problems for Second Order Degenerate Equations of Mixed Type

In this section we discuss the oblique derivative problem for second order degenerate equations of mixed type in a simply connected domain. We first give the representation of solutions of the boundary value problem for the equations, and then prove the uniqueness of solutions for the problem. Moreover we introduce the possibility to prove the existence of the above oblique derivative problem.

5.1   Formulation of oblique derivative problems for degenerate equations of mixed type

Let D be a simply connected bounded domain in the complex plane $\mathbb{C}$ with the boundary $\partial D = \Gamma \cup L$, where $\Gamma(\subset \{y > 0\}) \in C^2_\alpha (0 < \alpha < 1)$ with the end points $z = 0, 2$ and $L = L_1 \cup L_2$, $L_1 = \{x + \int_0^y \sqrt{-K(t)}dt = 0, x \in (0,1)\}$, $L_2 = \{x - \int_0^y \sqrt{-K(t)}dt = 2, x \in (1,2)\}$, and $z_1 = x_1 + jy_1 = 1 + jy_1$ is the intersection point of L_1 and L_2. In this section, we use the hyperbolic numbers. Denote $D^+ = D \cap \{y > 0\}$, $D^- = D \cap \{y < 0\}$. We may assume that $\Gamma = \{|z - 1| = 1, y \ge 0\}$, and consider the linear degenerate mixed equation of second order

$$Lu = K(y)u_{xx} + u_{yy} = du_x + eu_y + fu + g \text{ in } D, \tag{5.1}$$

where $K(y)$ possesses the first order continuous derivatives $K'(y)$, and $K'(y) > 0$ on $y \ne 0$, $K(0) = 0$. The following degenerate mixed equation is a special case

$$Lu = \text{sgn}y|y|^m u_{xx} + u_{yy} = du_x + eu_y + fu + g \text{ in } D, \tag{5.2}$$

where m is a positive constant, d, e, f, g are functions of $z(\in D)$. Similarly to (5.43), Chapter II, we denote $\tilde{W}(z) = \tilde{U} - i\tilde{V} = y^{m/2}U + iV = [y^{m/2}u_x - iu_y]/2$, $\tilde{W}_{\bar{z}} = [y^{m/2}\tilde{W}_x + i\tilde{W}_y]/2$ in D^+ and $\tilde{W}(z) = \tilde{U} + j\tilde{V} = |y|^{m/2}U - jV = [|y|^{m/2}u_x + ju_y]/2$, $\tilde{W}_{\bar{z}} = [|y|^{m/2}\tilde{W}_x - j\tilde{W}_y]/2$ in D^-, then equation (5.2) in D can be reduced to the form

$$\left\{\begin{matrix}\tilde{W}_{\bar{z}}\\ \tilde{W}_{\bar{z}}\end{matrix}\right\} = A_1(z)\tilde{W} + A_2(z)\overline{\tilde{W}} + A_3(z)u + A_4(z) \text{ in } \left\{\begin{matrix}D^+\\ D^-\end{matrix}\right\},$$

$$A_1 = \begin{cases} \dfrac{im}{8y} + \dfrac{d}{4y^{m/2}} + \dfrac{ie}{4} = \dfrac{d}{4y^{m/2}} + i\left(\dfrac{m}{8y} + \dfrac{e}{4}\right), \\[2mm] \dfrac{jm}{8|y|} + \dfrac{-d}{4|y|^{m/2}} - \dfrac{je}{4} = \dfrac{-d}{4|y|^{m/2}} + j\left(\dfrac{m}{8|y|} - \dfrac{e}{4}\right), \end{cases} \text{ in } \left\{\begin{matrix}D^+\\ D^-\end{matrix}\right\} \qquad (5.3)$$

$$A_2 = \begin{cases} \dfrac{d}{4y^{m/2}} + i\left(\dfrac{m}{8y} - \dfrac{e}{4}\right), \\[2mm] \dfrac{-d}{4|y|^{m/2}} + j\left(\dfrac{m}{8|y|} + \dfrac{e}{4}\right), \end{cases} \quad A_3 = \left\{\begin{matrix}\dfrac{f}{4},\\[2mm] -\dfrac{f}{4},\end{matrix}\right. \quad A_4 = \left\{\begin{matrix}\dfrac{g}{4},\\[2mm] -\dfrac{g}{4}\end{matrix}\right. \text{ in } \left\{\begin{matrix}D^+\\ D^-\end{matrix}\right\}$$

and

$$u(z) = \begin{cases} 2\,\mathrm{Re}\displaystyle\int_0^z u_z dz + u(0) \text{ in } D^+, \\[3mm] 2\,\mathrm{Re}\displaystyle\int_0^z (U - jV)d(x + jy) + u(0) \text{ in } D^- \end{cases}$$

is a solution of equation (5.2).

Suppose that equation (5.2) satisfies the following conditions: **Condition** C

The coefficients $A_j(z)\,(j = 1, 2, 3)$ in (5.2) are continuous in $\overline{D^+}$ and continuous in $\overline{D^-}$ and satisfy

$$C[A_j, \overline{D^+}] \le k_0, \quad j = 1, 2, \quad C[A_3, \overline{D^+}] \le k_1, \quad A_2 \ge 0 \text{ in } D^+,$$
$$C[A_j, \overline{D^-}] \le k_0, \quad j = 1, 2, \quad C[A_3, \overline{D^-}] \le k_1. \qquad (5.4)$$

where $p\,(>2)$, k_0, k_1 are non-negative constants. If the above conditions is replaced by

$$C_\alpha^1[A_j, \overline{D^\pm}] \le k_0, \quad j = 1, 2, \quad C_\alpha^1[A_3, \overline{D^\pm}] \le k_1, \qquad (5.5)$$

in which $\alpha\,(0 < \alpha < 1)$ is a real constant, then the conditions will be called Condition C'.

Now we formulate the oblique derivative boundary value problem as follows:

Problem P Find a continuously differentiable solution $u(z)$ of (5.2) in $D^* = \bar{D}\backslash\{0, L_2\}$, which is continuous in $\bar{D}$ and satisfies the boundary conditions

$$lu = \frac{\partial u}{\partial l} = 2\mathrm{Re}\,[\overline{\lambda(z)}u_z] = r(z), \quad z \in \Gamma, \quad u(0) = b_0, \quad u(2) = b_2, \qquad (5.6)$$

$$\mathrm{Re}\,[\overline{\lambda(z)}u_{\bar{z}}] = r(z), \quad z \in L_1, \quad \mathrm{Im}\,[\overline{\lambda(z)}u_{\bar{z}}]|_{z=z_1} = b_1, \qquad (5.7)$$

where $u_{\bar{z}} = [\sqrt{-K}u_x + iu_y]/2$, $\lambda(z) = a(x) + ib(x) = \cos(l, x) - i\cos(l, y)$, if $z \in \Gamma$ and $\lambda(z) = a(z) + jb(z)$, if $z \in L_1$, b_0, b_1, b_2 are real constants, and $\lambda(z)(|\lambda(z)| = 1)$, $r(z)$, b_0, b_1, b_2 satisfy the conditions

$$C_\alpha[\lambda(z),\Gamma]\le k_0, \quad C_\alpha[r(z),\Gamma]\le k_2, \quad C_\alpha[\lambda(z),L_1]\le k_0, \quad C_\alpha[r(z),L_1]\le k_2,$$

$$\cos(l,n)\ge 0 \text{ on } \Gamma, \quad |b_0|,|b_1|,|b_2|\le k_2, \quad \max_{z\in L_1}\frac{1}{|a(z)-b(z)|}\le k_0, \tag{5.8}$$

in which n is the outward normal vector at every point on Γ, $\alpha\,(1/2<\alpha<1)$, k_0,k_2 are non-negative constants. For convenience, we may assume that $u_{\bar z}(z_1)=0$, otherwise through a transformation of function $U_{\bar z}(z)=u_{\bar z}(z)-\lambda(z_1)\,[r(z_1)+jb_1]/[a^2(z_1)+b^2(z_1)]$, the requirement can be realized. If $\cos(l,n)=0$ on Γ, where n is the outward normal vector on Γ, then the problem is called Problem D, in which $u(z)=2\mathrm{Re}\int_0^z u_z dz+b_0=\phi(z)$ on Γ.

Problem P for (5.2) with $A_3(z)=0$, $z\in\bar D$, $r(z)=0$, $z\in\Gamma\cup L_j\,(j=1$ or $2)$ and $b_0=b_1=0$ will be called Problem P_0. The number

$$K=\frac{1}{2}(K_1+K_2)$$

is called the index of Problem P and Problem P_0, where

$$K_j=\left[\frac{\phi_j}{\pi}\right]+J_j, \quad J_j=0\text{ or }1, \quad e^{i\phi_j}=\frac{\lambda(t_j-0)}{\lambda(t_j+0)}, \quad \gamma_j=\frac{\phi_j}{\pi}-K_j, \quad j=1,2, \tag{5.9}$$

in which $t_1=2$, $t_2=0$, $\lambda(t)=e^{i\pi/4}$ on $L_0=(0,2)$ and $\lambda(t_1-0)=\lambda(t_2+0)=\exp(i\pi/4)$. Here we choose $K=0$, or $K=-1/2$ on the boundary ∂D^+ of D^+ if $\cos(\nu,n)\equiv 0$ on Γ and the condition $u(2)=b_2$ can be canceled. In fact, if $\cos(l,n)\equiv 0$ on Γ, from the boundary condition (5.6), we can determine the value $u(2)$ by the value $u(0)$, namely

$$u(2)=2\mathrm{Re}\int_0^2 u_z dz+u(0)=2\int_0^2\mathrm{Re}\,[i(z-1)u_z]d\theta+b_0=2\int_\pi^0 r(z)d\theta+b_0, \tag{5.10}$$

in which $\overline{\lambda(z)}=i(z-1)$, $\theta=\arg(z-1)$ on Γ. In order to ensure that the solution $u(z)$ of Problem P is continuously differentiable in D^*, we need to choose $\gamma_1>0$. If we require that the solution is only continuous, it suffices to choose $-2\gamma_2<1$, $-2\gamma_1<1$ respectively. In the following, we shall only discuss the case: $K=0$, and the case: $K=-1/2$ can be similarly discussed. Problem P in this case still includes the Dirichlet problem (Problem D) as a special case.

5.2 Representation and uniqueness of solutions of oblique derivative problem for degenerate equations of mixed type

Now we give the representation theorem of solutions for equation (5.2).

Theorem 5.1 *Suppose that the equation* (5.2) *satisfies Condition* C'. *Then any solution of Problem P for* (5.2) *can be expressed as*

$$u(z)=2\mathrm{Re}\int_0^z w(z)dz+b_0, \quad w(z)=w_0(z)+W(z), \tag{5.11}$$

where $w_0(z)$ is a solution of Problem A for the complex equation

$$W_{\bar{z}} = 0 \text{ in } D \tag{5.12}$$

with the boundary conditions (5.6), (5.7) $(w_0(z) = u_{0z}$ on Γ, $w_0(z) = u_{0\bar{z}}$ on $L_1)$, and $W(z)$ in D^- possesses the form

$$\overline{W(z)} = \Phi(z) + \Psi(z), \quad \Psi(z) = \int_2^\nu g_1(z) d\nu e_1 + \int_0^\mu g_2(z) d\mu e_2 \text{ in } D^-, \tag{5.13}$$

in which $e_1 = (1+j)/2$, $e_2 = (1-j)/2$, $\mu = x - 2|y|^{m/2+1}/(m+2)$, $\nu = x + 2|y|^{m/2+1}/(m+2)$,

$$g_1(z) = \tilde{A}_1 \xi + \tilde{B}_1 \eta + \tilde{C} u + \tilde{D}, \quad \xi = \text{Re}\, w + \text{Im}\, w, \quad \eta = \text{Re}\, w - \text{Im}\, w,$$

$$g_2(z) = \tilde{A}_2 \xi + \tilde{B}_2 \eta + \tilde{C} u + \tilde{D}, \quad \tilde{C} = -\frac{f}{4|y|^{m/2}}, \quad \tilde{D} = -\frac{g}{4|y|^{m/2}},$$

$$\tilde{A}_1 = \frac{1}{4|y|^{m/2}} \left[\frac{m}{2y} - \frac{d}{|y|^{m/2}} - e \right], \quad \tilde{B}_1 = \frac{1}{4|y|^{m/2}} \left[\frac{m}{2y} - \frac{d}{|y|^{m/2}} + e \right], \tag{5.14}$$

$$\tilde{A}_2 = \frac{1}{4|y|^{m/2}} \left[\frac{m}{2|y|} - \frac{d}{|y|^{m/2}} - e \right], \quad \tilde{B}_2 = \frac{1}{4|y|^{m/2}} \left[\frac{m}{2|y|} - \frac{d}{|y|^{m/2}} + e \right] \text{ in } D^-,$$

$\Phi(z)$ *is the solutions of equation (5.2), and $w(z)$ in D^+ and $\Phi(z)$ in D^- satisfy the boundary conditions*

$$\text{Re}\,[\overline{\lambda(z)} w(z)] = r(z), \quad z \in \Gamma, \quad \text{Re}\,[\overline{\lambda(x)} w(x)] = s(x), \quad x \in L_0,$$

$$\text{Re}\,[\overline{\lambda(x)} \Phi(x)] = \text{Re}\,[\overline{\lambda(x)}(\overline{W(z)} - \Phi(x))], \quad x \in L_0 = (0, 2), \tag{5.15}$$

$$\text{Re}\,[\overline{\lambda(z)}(\Phi(z) + \Psi(z))] = 0, \quad z \in L_1, \quad \text{Im}\,[\overline{\lambda(z_1)}(\Phi(z_1) + \Psi(z_1))] = 0,$$

where $\lambda(x) = 1 + i, x \in L_0$. Moreover by Section 5, Chapter II, we see that $w_0(z)$ is a solution of Problem A for equation (5.12), and

$$u_0(z) = 2\text{Re} \int_0^z w_0(z) dz + b_0 \text{ in } D. \tag{5.16}$$

Proof Let $u(z)$ be a solution of Problem P for equation (5.2), and $w(z) = u_z$, $u(z)$ be substituted in the positions of w, u in (5.13), (5.14), thus the functions $g_1(z)$, $g_2(z)$, and $\Psi(z)$ in $\overline{D^-}$ in (5.13),(5.14) can be determined. Moreover we can find the solution $\Phi(z)$ in $\overline{D^-}$ of (5.12) with the boundary condition (5.15), where $s(x)$ on L_0 is a function of $\lambda(z), r(z), \Psi(z)$, thus

$$w(z) = w_0(z) + \overline{\Phi(z)} + \overline{\Psi(z)} \text{ in } D^- \tag{5.17}$$

is the solution of Problem A in D^- for equation (5.2), which can be expressed as the second formula in (5.11), and $u(z)$ is a solution of Problem P for (5.2) as stated in the first formula in (5.11).

Theorem 5.2 *Suppose that equation* (5.2) *satisfies Condition* C'. *Then Problem P for* (5.2) *has at most one solution in* D.

Proof Let $u_1(z), u_2(z)$ be any two solutions of Problem P for (5.2). It is easy to see that $u(z) = u_1(z) - u_2(z)$ and $W(z) = u_{\bar{z}}$ satisfy the homogeneous equation and boundary conditions

$$\left\{ \begin{matrix} W_{\bar{z}} \\ W_{\bar{z}} \end{matrix} \right\} = A_1 W + A_2 \overline{W} + A_3 u \text{ in } \left\{ \begin{matrix} D^+ \\ D^- \end{matrix} \right\}, \tag{5.18}$$

$$\begin{aligned} \text{Re}\,[\overline{\lambda(z)}W(z)] &= 0, \quad z \in \Gamma, \quad u(0) = 0, \quad u(2) = 0, \\ \text{Re}\,[\lambda(z)W(z)] &= 0, \quad z \in L_1, \quad \text{Im}\,[\lambda(z_1)W(z_1)] = 0, \end{aligned} \tag{5.19}$$

where $W(z) = u_z$ in $\overline{D^+}$. According to the method as stated in Section 5, Chapter II, the solution $W(z)$ in the hyperbolic domain D^- can be expressed in the form

$$\begin{aligned} W(z) &= \overline{\Phi(z)} + \overline{\Psi(z)}, \\ \Psi(z) &= \int_2^\nu [\tilde{A}\xi + \tilde{B}\eta + \tilde{C}u]e_1 d\nu + \int_0^\mu [\tilde{A}\xi + \tilde{B}\eta + \tilde{C}u]e_2 d\mu \text{ in } D^-, \end{aligned} \tag{5.20}$$

where $\Phi(z)$ is a solution of (5.12) in $\overline{D^-}$ satisfying the boundary condition (5.15). Similarly to the method in Section 5, Chapter II, $\Psi(z) = 0$, $\Phi(z) = 0$, $W(z) = 0$, $z \in \overline{D^-}$ can be derived. Thus the solution $u(z) = 2\,\text{Re}\int_0^z w(z)dz$ is the solution of the homogeneous equation of (5.2) with homogeneous boundary conditions of (5.6) and (5.7):

$$2\text{Re}\,[\overline{\lambda(z)}u_z] = 0 \text{ on } \Gamma, \quad \text{Re}\,[\lambda(x)u_{\bar{z}}(x)] = 0 \text{ on } L_0, \quad u(0) = 0, \quad u(2) = 0, \tag{5.21}$$

in which $\lambda(x) = 1 + i$, $x \in L_0 = (0, 2)$.

Now we verify that the above solution $u(z) \equiv 0$ in D^+. If $u(z) \not\equiv 0$ in D^+, noting that $u(z)$ satisfies the boundary condition (5.21), and similarly to the proof of Theorem 3.4, Chapter III, we see that its maximum and minimum cannot attain in $D^+ \cup \Gamma$. Hence $u(z)$ attains its maximum and minimum at a point $z^* = x^* \in L_0 = (0, 2)$. By using Lemma 4.1, Chapter III, we can derive that $u_x(x_*) = 0$, $u_y(x^*) \neq 0$, and then

$$\text{Re}\,[\lambda(x)u_{\bar{z}}(x^*)] = \frac{1}{2}\left[\sqrt{-K(y)}\,u_x(x^*) + u_y(x^*)\right] \neq 0,$$

this contradicts the second equality in (5.21). Thus $u(z) \equiv 0$ in $\overline{D^+}$. This completes the proof.

5.3   Solvabilty problem of oblique derivative problems for degenerate equations of mixed type

From the above discussion, we see in order to prove the existence of solutions of Problem P for equation (5.2), the main problem is to find a solution of the oblique derivative problem for the degenerate elliptic equation of second order, i.e. equation (5.2) in elliptic domain D^+, and the oblique derivative boundary conditions is (5.6) and

$$\text{Re}\,[\lambda(x)u_{\bar{z}}(x)] = s(x) \text{ on } L_0 = (0, 2), \quad \text{i.e.}$$

$$\frac{1}{2}\left[\sqrt{-K(y)}\,u_x + u_y\right] = s(x) \text{ on } L_0, \tag{5.22}$$

which is more general than the case as stated in Section 4, Chapter III. We try to solve the problem by using the method of integral equations or the method of auxiliary functions, which will be discussed in detail in our other publishers.

The references for this chapter are [1],[10],[12],[17],[21],[22],[28],[37],[43],[47],[49], [57],[62],[66],[69],[70],[73],[77],[85],[91],[93].

CHAPTER VI

SECOND ORDER QUASILINEAR EQUATIONS OF MIXED TYPE

This chapter deals with several oblique derivative boundary value problems for second order quasilinear equations of mixed (elliptic-hyperbolic) type. We shall discuss oblique derivative boundary value problems and discontinuous oblique derivative boundary value problems for second order quasilinear equations of mixed (elliptic-hyperbolic) type. Moreover we shall discuss oblique derivative boundary value problems for general second order quasilinear equations of mixed (elliptic-hyperbolic) type and the boundary value problems in multiply connected domains. In the meantime, we shall give a priori estimates of solutions for above oblique derivative boundary value problems.

1 Oblique Derivative Problems for Second Order Quasilinear Equations of Mixed Type

In this section, we first give the representation of solutions for the oblique derivative boundary value problem, and then prove the uniqueness and existence of solutions of the problem and give a priori estimates of solutions of the above problem. Finally we prove the solvability of oblique derivative problems for general quasilinear second order equations of mixed type.

1.1 Formulation of the oblique derivative problem for second order equations of mixed type

Let D be a simply connected bounded domain D in the complex plane $\mathbb{C}$ as stated in Chapter V. We consider the second order quasilinear equation of mixed type

$$u_{xx} + \text{sgn}y\, u_{yy} = au_x + bu_y + cu + d \quad \text{in} \quad D, \tag{1.1}$$

where a, b, c, d are functions of $z(\in D), u, u_x, u_y\,(\in \mathbb{R})$, its complex form is the following complex equation of second order

$$Lu_z = \begin{Bmatrix} u_{z\bar{z}} \\ u_{\bar{z}\bar{z}^*} \end{Bmatrix} = F(z, u, u_z), \quad F = \text{Re}\,[A_1 u_z] + A_2 u + A_3 \quad \text{in} \quad \begin{Bmatrix} D^+ \\ D^- \end{Bmatrix}, \tag{1.2}$$

where $A_j = A_j(z, u, u_z)$, $j = 1, 2, 3$, and

$$u_{z\bar{z}} = \frac{1}{4}[u_{xx} + u_{yy}], \quad u_{\bar{z}\bar{z}} = \frac{1}{2}[(u_{\bar{z}})_x - i(\overline{u_{\bar{z}}})_y] = \frac{1}{4}[u_{xx} - u_{yy}],$$

$$A_1 = \frac{a + ib}{2}, \quad A_2 = \frac{c}{4}, \quad A_3 = \frac{d}{4} \text{ in } D.$$

Suppose that the equation (1.2) satisfies the following conditions, namely

Condition C

1) $A_j(z, u, u_z)$ $(j = 1, 2, 3)$ are continuous in $u \in \mathbb{R}, u_z \in \mathbb{C}$ for almost every point $z \in D^+$, and measurable in $z \in D^+$ and continuous in $\overline{D^-}$ for all continuously differentiable functions $u(z)$ in $D^* = \bar{D}\backslash\{0, x - y = 2\}$ or $D^* = \bar{D}\backslash\{x + y = 0, 2\}$ and satisfy

$$L_p[A_j, \overline{D^+}] \leq k_0, \quad j = 1, 2, \quad L_p[A_3, \overline{D^+}] \leq k_1, \quad A_2 \geq 0 \text{ in } D^+,$$
$$C[A_j, \overline{D^-}] \leq k_0, \quad j = 1, 2, \quad C[A_3, \overline{D^-}] \leq k_1. \tag{1.3}$$

2) For any continuously differentiable functions $u_1(z), u_2(z)$ in D^*, the equality

$$F(z, u_1, u_{1z}) - F(z, u_2, u_{2z}) = \text{Re}\,[\tilde{A}_1(u_1 - u_2)_z] + \tilde{A}_2(u_1 - u_2) \text{ in } D \tag{1.4}$$

holds, where $\tilde{A}_j = \tilde{A}_j(z, u_1, u_2)$ $(j = 1, 2)$ satisfy the conditions

$$L_p[\tilde{A}_j, \overline{D^+}] \leq k_0, \quad C[\tilde{A}_j, \overline{D^-}] \leq k_0, \quad j = 1, 2 \tag{1.5}$$

in (1.3),(1.5), $p (> 2), k_0, k_1$ are non-negative constants. In particular, when (1.2) is a linear equation, the condition (1.4) obviously holds.

Problem P The oblique derivative boundary value problem for equation (1.2) is to find a continuously differentiable solution $u(z)$ of (1.2) in $D^* = \overline{D}\backslash\{0, x - y = 2\}$, which is continuous in $\overline{D}$ and satisfies the boundary conditions

$$\frac{1}{2}\frac{\partial u}{\partial l} = \text{Re}\,[\overline{\lambda(z)}u_z] = r(z), \quad z \in \Gamma, \quad \frac{1}{2}\frac{\partial u}{\partial l} = \text{Re}\,[\overline{\lambda(z)}u_{\bar{z}}] = r(z), \quad z \in L_1,$$
$$\text{Im}\,[\overline{\lambda(z)}u_{\bar{z}}]_{z=z_1} = b_1, \quad u(0) = b_0, \quad u(2) = b_2, \tag{1.6}$$

where l is a given vector at every point on $\Gamma \cup L_1$, $\lambda(z) = a(x) + ib(x) = \cos(l, x) \mp i\cos(l, y)$, and $\mp$ are determined by $z \in \Gamma$ and $z \in L_1$ respectively, b_0, b_1, b_2 are real constants, and $\lambda(z), r(z), b_0, b_1, b_2$ satisfy the conditions

$$C_\alpha[\lambda(z), \Gamma] \leq k_0, \quad C_\alpha[r(z), \Gamma] \leq k_2, \quad \cos(l, n) \geq 0 \text{ on } \Gamma, \quad |b_0|, |b_1|, |b_2| \leq k_2,$$
$$C_\alpha[\lambda(z), L_1] \leq k_0, \quad C_\alpha[r(z), L_1] \leq k_2, \quad \max_{z \in L_1} |1/|a(x) - b(x)|| \leq k_0, \tag{1.7}$$

in which n is the outward normal vector at every point on Γ, $\alpha (1/2 < \alpha < 1), k_0, k_2$ are non-negative constants. For convenience, we may assume that $w(z_1) = 0$, otherwise through a transformation of function $W(z) = w(z) - \overline{\lambda(z_1)}[r(z_1) - ib_1]$, the requirement can be realized. Here we mention that if $A_2(z) = 0$ in D_1, we can cancel the assumption $\cos(l, n) \geq 0$ on Γ, and if the boundary condition $\text{Re}\,[\overline{\lambda(z)}u_{\bar{z}}] = r(z)$, $z \in L_1$ is replaced by $\text{Re}\,[\overline{\lambda(z)}u_z] = r(z)$, $z \in L_1$, then Problem P does not include the Dirichlet problem (Tricomi problem) as a special case.

The boundary value problem for (1.2) with $A_3(z, u, u_z) = 0$, $z \in D, u \in \mathbb{R}$, $u_z \in \mathbb{C}$, $r(z) = 0$, $z \in \partial D$ and $b_0 = b_2 = b_1 = 0$ will be called Problem P_0. The number

$$K = \frac{1}{2}(K_1 + K_2) \tag{1.8}$$

is called the index of Problem P and Problem P_0, where

$$K_j = \left[\frac{\phi_j}{\pi}\right] + J_j, \quad J_j = 0 \text{ or } 1, \quad e^{i\phi_j} = \frac{\lambda(t_j - 0)}{\lambda(t_j + 0)}, \quad \gamma_j = \frac{\phi_j}{\pi} - K_j, \quad j = 1, 2, \tag{1.9}$$

in which $[a]$ is the largest integer not exceeding the real number a, $t_1 = 2$, $t_2 = 0$, $\lambda(t) = \exp(i\pi/4)$ on L_0 and $\lambda(t_1 - 0) = \lambda(t_2 + 0) = \exp(i\pi/4)$, here we only discuss the case of $K = 0$ on ∂D^+ if $\cos(l, n) \not\equiv 0$ on Γ, or $K = -1/2$ if $\cos(l, n) \equiv 0$ on Γ, because in this case the last point condition in (1.6) can be eliminated, and the solution of Problem P is unique. In order to ensure that the solution $u(z)$ of Problem P in D^* is continuously differentiable, we need to choose $\gamma_1 > 0$. If we require that the solution of Problem P in $\bar{D}$ is only continuous, it suffices to choose $-2\gamma_1 < 1$, $-2\gamma_2 < 1$.

Besides, if $A_2 = 0$ in D, the last condition in (1.6) is replaced by

$$\text{Im}\,[\overline{\lambda(z)}u_{\bar{z}}]|_{z=z_2} = b_2, \tag{1.10}$$

where the integral path is along two family of characteristic lines similar to that in (2.10), Chapter II, $z_2(\neq 0, 2) \in \Gamma$, and b_2 is a real constant with the condition $|b_2| \leq k_2$ and here the condition $\cos(l, n) \geq 0$ is canceled, then the boundary value problem for (1.2) will be called Problem Q.

1.2   The existence and uniqueness of solutions for the oblique derivative problem for (1.2)

Similarly to Section 2, Chapter V, we can prove the following results.

Lemma 1.1   *Let equation* (1.2) *satisfy Condition C. Then any solution of Problem P for* (1.2) *can be expressed as*

$$u(z) = 2\text{Re} \int_0^z w(z)dz + b_0, \quad w(z) = w_0(z) + W(z) \text{ in } D, \tag{1.11}$$

where the integral path in $\overline{D^-}$ is the same as in Chapter II, and $w_0(z)$ is a solution of Problem A for the equation

$$Lw = \left\{ \begin{matrix} w_{\bar{z}} \\ \overline{w_{\bar{z}^*}} \end{matrix} \right\} = 0 \text{ in } \left\{ \begin{matrix} D^+ \\ D^- \end{matrix} \right\}, \tag{1.12}$$

with the boundary condition (1.6) $(w_0(z) = u_{0z})$, *and* $W(z)$ *possesses the form*

$$W(z) = w(z) - w_0(z) \text{ in } D, \quad w(z) = \tilde{\Phi}(z)e^{\tilde{\phi}(z)} + \tilde{\psi}(z) \text{ in } D^+,$$

$$\tilde{\phi}(z) = \tilde{\phi}_0(z) + Tg = \tilde{\phi}_0(z) - \frac{1}{\pi}\int\int_{D^+}\frac{g(\zeta)}{\zeta - z}d\sigma_\zeta, \quad \tilde{\psi}(z) = Tf \text{ in } D^+, \quad (1.13)$$

$$\overline{W(z)} = \Phi(z) + \Psi(z), \quad \Psi(z) = \int_2^\nu g_1(z)d\nu e_1 + \int_0^\mu g_2(z)d\mu e_2 \text{ in } D^-,$$

in which $\text{Im}[\tilde{\phi}(z)] = 0$ *on* $L_0 = (0,2)$, $e_1 = (1+i)/2$, $e_2 = (1-i)/2$, $\mu = x + y$, $\nu = x - y$, $\tilde{\phi}_0(z)$ *is an analytic function in* D^+ *and continuous in* $\overline{D^+}$, *such that* $\text{Im}[\phi(x)] = 0$ *on* L_0, *and*

$$g(z) = \begin{cases} A_1/2 + \overline{A_1}\bar{w}/(2w), w(z) \neq 0, \\ 0, w(z) = 0, z \in D^+, \end{cases} \quad f(z) = \text{Re}[A_1\tilde{\phi}_z] + A_2u + A_3 \text{ in } D^+,$$

$$g_1(z) = g_2(z) = A\xi + B\eta + Cu + D, \quad \xi = \text{Re}w + \text{Im}w, \quad \eta = \text{Re}w - \text{Im}w, \quad (1.14)$$

$$A = \frac{\text{Re}A_1 + \text{Im}A_1}{2}, \quad B = \frac{\text{Re}A_1 - \text{Im}A_1}{2}, \quad C = A_2, \quad D = A_3 \text{ in } D^-,$$

where $\tilde{\Phi}(z)$ *is an analytic function in* D^+ *and* $\Phi(z)$ *is a solution of equation* (1.12) *in* D^- *satisfying the boundary conditions*

$$\text{Re}[\overline{\lambda(z)}(\tilde{\Phi}(z)e^{\tilde{\phi}(z)} + \tilde{\psi}(z))] = r(z), \quad z \in \Gamma,$$

$$\text{Re}[\overline{\lambda(x)}(\tilde{\Phi}(x)e^{\tilde{\phi}(x)} + \tilde{\psi}(x))] = s(x), \quad x \in L_0,$$

$$\text{Re}[\overline{\lambda(x)}\Phi(x)] = \text{Re}[\overline{\lambda(x)}(\overline{W(x)} - \Psi(x))], \quad z \in L_0, \quad (1.15)$$

$$\text{Re}[\overline{\lambda(z)}\Phi(z)] = -\text{Re}[\overline{\lambda(z)}\Psi(z)], \quad z \in L_1,$$

$$\text{Im}[\overline{\lambda(z_1)}\Phi(z_1)] = -\text{Im}[\overline{\lambda(z_1)}\Psi(z_1)],$$

in which $\lambda(x) = 1 + i$, $x \in L_0 = (0,2)$ *and* $s(x)$ *is as stated in* (2.23), *Chapter* V. *Moreover by Theorem 1.1, Chapter* V, *the solution* $w_0(z)$ *of Problem A for* (1.12) *and* $u_0(z)$ *satisfy the estimate in the form*

$$C_\beta[u_0(z), \overline{D}] + C_\beta[w_0(z)X(z), \overline{D^+}] + C_\beta[w_0^\pm(\mu,\nu)Y^\pm(\mu,\nu), \overline{D^-}] \leq M_1(k_1+k_2), \quad (1.16)$$

where

$$X(z) = \prod_{j=1}^2 |z - t_j|^{\eta_j}, \quad Y^\pm(z) = Y^\pm(\mu,\nu) = [|\nu - 2||\mu - 2|]^{\eta_j},$$

$$\eta_j = \begin{cases} 2|\gamma_j| + \delta, \gamma_j < 0, \\ \delta, \gamma_j \geq 0, \end{cases} \quad j = 1, 2, \quad (1.17)$$

herein $w_0^\pm(\mu,\nu) = \text{Re}w_0(z) \mp \text{Im}w_0(z)$, $w_0(z) = w_0(\mu,\nu)$, $\mu = x + y$, $\nu = x - y$, *and* γ_1, γ_2 *are the real constants in* (1.9), $\beta(< \delta)$, δ *are sufficiently small positive constants, and*

$$u_0(z) = 2\text{Re}\int_0^z w_0(z)dz + b_0 \text{ in } \overline{D} \quad (1.18)$$

where $p_0(2 < p_0 \leq p)$, $M_1 = M_1(p_0, \beta, k_0, D)$ *are non-negative constants.*

Theorem 1.2 *Suppose that equation (1.2) satisfies Condition C. Then Problem P for (1.2) has a unique solution $u(z)$ in D.*

Theorem 1.3 *Suppose that the equation (1.2) satisfies Condition C. Then any solution $u(z)$ of Problem P for (1.2) satisfies the estimates*

$$\tilde{C}^1_\beta[u, \overline{D^+}] = C_\beta[u(z), \overline{D^+}] + C_\beta[u_z X(z), \overline{D^+}] \le M_2,$$

$$\tilde{C}^1[u, \overline{D^-}] = C_\beta[u(z), \overline{D^-}] + C[u_z^\pm(\mu, \nu)Y^\pm(\mu, \nu), \overline{D^-}] \le M_3, \qquad (1.19)$$

$$\tilde{C}^1_\beta[u, \overline{D^+}] \le M_4(k_1 + k_2), \quad \tilde{C}^1[u, \overline{D^-}] \le M_4(k_1 + k_2),$$

where $X(z), Y^\pm(\mu, \nu)$ are stated in (1.17), and $M_j = M_j(p_0, \beta, k_0, D)\,(j = 2, 3, 4)$ are non-negative constants.

1.3 $C^1_\alpha(\bar{D})$-estimate of solutions of Problem P for second order equations of mixed type

Now, we give the $C^1_\alpha(\bar{D})$-estimate of solutions $u(z)$ for Problem P for (1.2), but it needs to assume the following conditions: For any real numbers u_1, u_2 and complex numbers w_1, w_2, we have

$$|A_j(z_1, u_1, w_1) - A_j(z_2, u_2, w_2)| \le k_0[|z_1 - z_2|^\alpha + |u_1 - u_2|^\alpha + |w_1 - w_2|], \quad j = 1, 2,$$

$$|A_3(z_1, u_1, w_1) - A_3(z_2, u_2, w_2)| \le k_1[|z_1 - z_2|^\alpha + |u_1 - u_2|^\alpha + |w_1 - w_2|], \quad z_1, z_2 \in \overline{D^-},$$
$$(1.20)$$

where $\alpha\,(0 < \alpha < 1)$, k_0, k_1 are non-negative constants.

Theorem 1.4 *If Condition C and (1.20) hold, then any solution $u(z)$ of Problem P for equation (1.2) in $\overline{D^-}$ satisfies the estimates*

$$\tilde{C}^1_\beta[u, \overline{D^-}] = C_\beta[u, \overline{D^-}] + C_\beta[u_z^\pm(\mu, \nu)Y^\pm(\mu, \nu), \overline{D^-}]$$
$$\le M_5, \quad \tilde{C}^1_\beta[u, \overline{D^-}] \le M_6(k_1 + k_2), \qquad (1.21)$$

in which $u_z^\pm(\mu, \nu) = \mathrm{Re}\,u_z \mp \mathrm{Im}\,u_z$, $\beta\,(0 < \beta \le \alpha)$, $M_5 = M_5(p_0, \beta, k, D)$, $M_6 = M_6(p_0, \beta, k_0, D)$ are non-negative constants, $k = (k_0, k_1.k_2)$.

Proof Similarly to Theorem 1.3, it suffices to prove the first estimate in (1.21). Due to the solution $u(z)$ of Problem P for (1.2) is found by the successive iteration through the integral expressions (1.11), (1.13) and (1.14), we first choose the solution of Problem A of (1.12) in the form (1.18), i.e.

$$u_0(z) = 2\mathrm{Re} \int_0^z w_0(z)dz + b_0, \quad w_0(z) = \xi_0(z)e_1 + \eta_0(z)e_2 \text{ in } D, \qquad (1.22)$$

and substitute them into the positions of u_0, w_0 in the right-hand side of (1.14), we can obtain $\Psi_1(z), w_1(z), u_1(z)$ as stated in (1.11)–(1.14). Denote

$$u_1(z) = 2\mathrm{Re} \int_0^z w_1(z)dz + b_0, \quad w_1(z) = w_0(z) + \overline{\Phi_1(z)} + \overline{\Psi_1(z)},$$

$$\Psi_1^1(z) = \int_2^\nu G_1(z)d\nu, \quad G_1(z) = A\xi_0 + B\eta_0 + Cu_0 + D, \qquad (1.23)$$

$$\Psi_1^2(z) = \int_0^\mu G_2(z)d\mu, \quad G_2(z) = A\xi_0 + B\eta_0 + Cu_0 + D,$$

from the last two equalities in (1.23), it is not difficult to see that $G_1(z) = G_1(\mu, \nu)$, $\Psi_1^1(z) = \Psi_1^1(\mu, \nu)$ and $G_2(z) = G_2(\mu, \nu)$, $\Psi_1^2(z) = \Psi_1^2(\mu, \nu)$ satisfy the Hölder estimates about ν, μ respectively, namely

$$C_\beta[G_1(\cdot, \nu), \overline{D^-}] \le M_7, \quad C_\beta[\Psi_1^1(\cdot, \nu), \overline{D^-}] \le M_7 R,$$
$$C_\beta[G_2(\mu, \cdot), \overline{D^-}] \le M_8, \quad C_\beta[\Psi_1^2(\mu, \cdot), \overline{D^-}] \le M_8 R, \qquad (1.24)$$

where $M_j = M_j(p_0, \beta, k, D)$ $(j = 7, 8)$ and $R = 2$. Moreover, from (1.23), we can derive that $\Psi_1^1(\mu, \nu)$, $\Psi_1^2(\mu, \nu)$ about μ, ν satisfy the Hölder conditions respectively, namely

$$C_\beta[\Psi_1^1(\mu, \cdot), \overline{D^-}] \le M_9 R, \quad C_\beta[\Psi_1^2(\cdot, \nu), \overline{D^-}] \le M_9 R, \qquad (1.25)$$

where $M_0 = M_0(p_0, \beta, k, D)$. Besides we can obtain the estimate of $\Phi_1(z)$, i.e.

$$C_\beta[\Phi_1(z), \overline{D^-}] \le M_{10} R = M_{10}(p_0, \beta, k, D)R, \qquad (1.26)$$

in which $\Phi_1(z)$ satisfies equation (1.12) and boundary condition of Problem P, but in which the function $\Psi(z)$ is replaced by $\Psi_1(z)$. Setting $w_1(z) = w_0(z) + \overline{\Phi_1(z)} + \overline{\Psi_1(z)}$ and by the first formula in (1.23), we can find the function $u_1(z)$ from $w_1(z)$. Furthermore from (1.25),(1.26), we can derive that the functions $\tilde{w}_1^\pm(z) = \tilde{w}_1^\pm(\mu, \nu) = \mathrm{Re}\,\tilde{w}_1(z) \mp \mathrm{Im}\,\tilde{w}_1(z)$ $(\tilde{w}_1(z) = w_1(z) - w_0(z))$ and $\tilde{u}_1(z) = u_1(z) - u_0(z)$ satisfy the estimates

$$C_\beta[\tilde{w}_1^\pm(\mu, \nu)Y^\pm(\mu, \nu), \overline{D^-}] \le M_{11} R, C_\beta[\tilde{u}_1(z), \overline{D^-}] \le M_{11} R, \qquad (1.27)$$

where $M_{11} = M_{11}(p_0, \beta, k, D)$. Thus, according to the successive iteration, we can obtain the estimates of functions $\tilde{w}_n^\pm(z) = \tilde{w}_n^\pm(\mu, \nu) = \mathrm{Re}\,\tilde{w}_n(z) \mp \mathrm{Im}\,\tilde{w}_n(z)$ $(\tilde{w}_n(z) = w_n(z) - w_{n-1}(z))$ and the corresponding function $\tilde{u}_n(z) = u_n(z) - u_{n-1}(z)$ satisfy the estimates

$$C_\beta[\tilde{w}_n^\pm(\mu, \nu)Y^\pm(\mu, \nu), \overline{D^-}] \le \frac{(M_{11} R)^n}{n!}, \quad C_\beta[\tilde{u}_n(z), \overline{D^-}] \le \frac{(M_{11} R)^n}{n!}. \qquad (1.28)$$

Therefore the sequences of functions

$$w_n(z) = \sum_{m=1}^n \tilde{w}_m(z) + w_0(z), \ u_n(z) = \sum_{m=1}^n \tilde{u}_m(z) + u_0(z), \ n = 1, 2, \ldots, \qquad (1.29)$$

uniformly converge to $w(z)$, $u(z)$ in any close subset of D^* respectively, and $w(z)$, $u(z)$ satisfy the estimates

$$C_\beta[w^\pm(\mu,\nu)Y^\pm(\mu,\nu),\overline{D^-}] \le e^{M_{11}R}, \quad C_\beta[u(z),\overline{D^-}] \le M_5, \qquad (1.30)$$

this is just the first estimate in (1.21).

From the estimates (1.19) and (1.21), we can see the regularity of solutions of Problem P for (1.2). Moreover it is easy to see that the derivatives $[w^+(\mu,\nu)]_\nu$, $[w^-(\mu,\nu)]_\mu$ satisfy the estimates similar to those in (1.30).

As for Problem Q for (1.2), we can similarly discuss its unique solvability.

1.4   The solvability for the oblique derivative problem for general second order quasilinear equations of mixed type

Now, we consider the general quasilinear equation of second order

$$Lu_z = \left\{ \begin{matrix} u_{z\bar{z}} \\ u_{\bar{z}\bar{z}^*} \end{matrix} \right\} = F(z,u,u_z) + G(z,u,u_z), \quad z \in \left\{ \begin{matrix} D^+ \\ D^- \end{matrix} \right\}, \qquad (1.31)$$

$$F = \mathrm{Re}\,[A_1 u_z] + A_2 u + A_3, \quad G = A_4|u_z|^\sigma + A_5|u|^\tau, \quad z \in D,$$

where $F(z,u,u_z)$ satisfies Condition C, σ, τ are positive constants, and $A_j(z,u,u_z)$ $(j=4,5)$ satisfy the conditions in Condition C, the main conditions of which are

$$L_p[A_j(\,\cdot\,,u,u_z),\overline{D^+}] \le k_0, \quad C[A_j(z,u,u_z),\overline{D^-}] \le k_0, \quad j=4,5,$$

and denote the above conditions by Condition C'.

Theorem 1.5   *Let the complex equation* (1.31) *satisfy Condition C'.*

(1) *When* $0 < \max(\sigma,\tau) < 1$, *Problem P for* (1.31) *has a solution* $u(z) \in C(\bar{D})$.

(2) *When* $\min(\sigma,\tau) > 1$, *Problem P for* (1.31) *has a solution* $u(z) \in C(\bar{D})$, *provided that*

$$M_{12} = k_1 + k_2 + |b_0| + |b_1| \qquad (1.32)$$

is sufficiently small.

(3) *When* $\min(\sigma,\tau) > 1$, *Problem P for the equation*

$$Lu_z = \left\{ \begin{matrix} u_{z\bar{z}} \\ u_{\bar{z}\bar{z}^*} \end{matrix} \right\} = F(z,u,u_z) + \varepsilon G(z,u,u_z), \quad z \in \left\{ \begin{matrix} D^+ \\ D^- \end{matrix} \right\}. \qquad (1.33)$$

has a solution $u(z) \in C(\bar{D})$, *provided that the positive number ε in* (1.33) *is appropriately small, where the functions* $F(z,u,u_z)$, $G(z,u,u_z)$ *are as stated in* (1.31).

Proof   (1) Consider the algebraic equation

$$M_4\{k_1 + k_2 + 2k_0 t^\sigma + 2k_0 t^\tau + |b_0| + |b_1|\} = t \qquad (1.34)$$

for t, where M_4 is the constant stated in (1.19). It is not difficult to see that equation (1.34) has a unique solution $t = M_{13} \geq 0$. Now, we introduce a bounded, closed and convex subset B^* of the Banach space $B = \tilde{C}^1(\bar{D})$, whose elements are the functions $u(z)$ satisfying the condition

$$\tilde{C}^1[u(z), \bar{D}] = C_\beta[u(z), \bar{D}] + C_\beta[u_z X(z), \overline{D^+}] + C[u_z^\pm(\mu, \nu) Y^\pm(\mu, \nu), \overline{D^-}] \leq M_{13}. \quad (1.35)$$

We arbitrarily choose a function $u_0(z) \in B$ for instance $u_0(z) = 0$ and substitute it into the position of u in the coefficients of (1.31) and $G(z, u, u_z)$. From Theorem 1.2, it is clear that problem P for

$$Lu_z - \text{Re}\,[A_1(z, u_0, u_{0z})u_z] - A_2(z, u_0, u_{0z})u - A_3(z, u_0, u_{0z}) = G(z, u_0, u_{0z}), \quad (1.36)$$

has a unique solution $u_1(z)$. From Theorem 1.4, we see that the solution $u_1(z)$ satisfies the estimate in (1.35). By using successive iteration, we obtain a sequence of solutions $u_m(z)\ (m = 1, 2, ...) \in B^*$ of Problem P, which satisfy the equations

$$Lu_{m+1z} - \text{Re}\,[A_1(z, u_m, u_{mz})u_{m+1z}] - A_2(z, u_m, u_{mz})u_{m+1}$$
$$+A_3(z, u_m, u_{mz}) = G(z, u_m, u_{mz}) \text{ in } D, \quad m = 1, 2, \dots \quad (1.37)$$

and $u_{m+1}(z) \in B^*$. From (1.37), we see that and $\tilde{u}_{m+1}(z) = u_{m+1}(z) - u_m(z)$ satisfies the equations and boundary conditions

$$L\tilde{u}_{m+1z} - \text{Re}[\tilde{A}_1\tilde{u}_{m+1z}] - \tilde{A}_2\tilde{u}_{m+1} = G(z, u_m, u_{mz}) - G(z, u_{m-1}, u_{m-1z}) \text{ in } D,$$
$$\text{Re}[\overline{\lambda(z)}\tilde{u}_{m+1z}] = 0 \text{ on } \Gamma, \ \text{Re}[\overline{\lambda(z)}\tilde{u}_{m+1\bar{z}}] = 0 \text{ on } L_1, \ \text{Im}[\overline{\lambda(z)}\tilde{u}_{m+1\bar{z}}]|_{z=z_1} = 0, \quad (1.38)$$

in which $m = 1, 2, \dots$. Noting that $C[G(z, u_m, u_{mz}) - G(z, u_{m-1}, u_{m-1z}), \bar{D}] \leq 2k_0 M_{13}$, M_{13} is a solution of the algebraic equation (1.34), and according to Theorem 1.3, the estimate

$$\| \tilde{u}_{m+1} \| = \tilde{C}^1[\tilde{u}_{m+1}, \bar{D}] \leq M_{14} = M_{14}(p_0, \beta, k_0, D) \quad (1.39)$$

can be obtained. Due to $\tilde{u}_{m+1}$ can be expressed as

$$\tilde{u}_{m+1}(z) = 2\text{Re} \int_0^z w_{m+1}(z)dz, \text{ in } D, \quad w_{m+1}(z) = \overline{\Phi_{m+1}(z)} + \overline{\Psi_{m+1}(z)},$$

$$\Psi_{m+1}(z) = \int_2^{x-y} [\tilde{A}\xi_{m+1} + \tilde{B}\eta_{m+1} + \tilde{C}u_{m+1} + \tilde{D}]e_1 d(x - y) \quad (1.40)$$

$$+ \int_0^{x+y} [\tilde{A}\xi_{m+1} + \tilde{B}\eta_{m+1} + \tilde{C}u_{m+1} + \tilde{D}]e_2 d(x + y) \text{ in } D^-,$$

in which $\Phi_{m+1}(z), \Psi_{m+1}(z)$ are similar to the functions $\Phi(z), \Psi(z)$ in (1.13), the relation between $\tilde{A}_1, \tilde{A}_2, \tilde{G}$ and $\tilde{A}, \tilde{B}, \tilde{C}, \tilde{D}$ is the same as that of A_1, A_2, A_3 and A, B, C, D in (1.14) and $\tilde{G} = G(z, u_m, u_{mz}) - G(z, u_{m-1}, u_{m-1z})$. By using the method from the proof of Theorem 1.3, we can obtain

$$\| u_{m+1} - u_m \| = \tilde{C}^1[\tilde{u}_{m+1}, \bar{D}] \leq \frac{(M_{14}R')^m}{m!},$$

where $M_{14} = 2M_4(M_{15} + 1)MR(4m_0 + 1)$, $R = 2$, $m_0 = \| w_0(z)X(z) \|_{C(\bar{D})}$, herein $M_{15} = \max\{C[\tilde{A}, Q], C[\tilde{B}, Q], C[\tilde{C}, Q]\}, C[\tilde{D}, Q]\}$, $M = 1 + 4k_0^2(1 + k_0^2)$. From the above inequality, we see that the sequence of functions: $\{u_m(z)\}$, i.e.

$$u_m(z) = u_0(z) + [u_1(z) - u_0(z)] + \cdots + [u_m(z) - u_{m-1}(z)], \quad m = 1, 2, \ldots \quad (1.41)$$

uniformly converges to a function $u_*(z)$, and $w_*(z) = u_{*z}$ satisfies the equality

$$w_*(z) = w_0(z) + \overline{\Phi_*(z)} + \overline{\Psi_*(z)} \text{ in } D^-,$$

$$\Psi_*(z) = + \int_2^{x-y} [A\xi_* + B\eta_* + Cu_* + D]e_1 d(x - y) \quad (1.42)$$

$$+ \int_0^{x+y} [A\xi_* + B\eta_* + Cu_* + D]e_2 d(x + y) \text{ in } D^-,$$

and the function

$$u_*(z) = 2\mathrm{Re} \int_0^z w_*(z)dz + b_0 \text{ in } \bar{D} \quad (1.43)$$

is just a solution of Problem P for the general quasilinear equation (1.31) in $\bar{D}$.

(2) Consider the algebraic equation

$$M_4\{k_1 + k_2 + 2k_0t^\sigma + 2k_0t^\tau + |b_0| + |b_1|\} = t \quad (1.44)$$

for t. It is not difficult to see that the equation (1.44) has a solution $t = M_{13} \geq 0$, provided that the positive constant M_{12} in (1.32) is small enough. Now, we introduce a bounded, closed and convex subset B_* of the Banach space $\tilde{C}^1(\bar{D})$, whose elements are the functions $u(z)$ satisfying the condition

$$\tilde{C}^1[u(z), \overline{D}] = C_\beta[u(z), \overline{D}] + C_\beta[u_z X(z), \overline{D^+}] + C[u_z^\pm(\mu, \nu)Y^\pm(\mu, \nu), \overline{D^-}] \leq M_{13}. \quad (1.45)$$

By using the same method as in (1), we can find a solution $u(z) \in B_*$ of Problem P for equation (1.31) with $\min(\sigma, \tau) > 1$.

(3) There is no harm in assuming that k_1, k_2 in (1.3),(1.7) are positive constants, we introduce a bounded, closed and convex subset B' of the Banach space $\tilde{C}^1(\bar{D})$, whose elements are the functions $u(z)$ satisfying the condition

$$\tilde{C}^1[u(z), \overline{D}] \leq (M_4 + 1)(2k_1 + k_2) \quad (1.46)$$

where M_4 is a constant as stated in (1.19) and we can choose an appropriately small positive number ε such that $C[\varepsilon G(z, u, u_z), \bar{D}] \leq k_1$. Moreover, we are free to choose a function $u_0(z) \in B'$ for instance $u_0(z) = 0$ and substitute it into the position of u in the coefficients of (1.33) and $G(z, u, u_z)$. From Theorem 1.2, it is seen that there exists a unique solution of Problem P for

$$Lu - \mathrm{Re}\,[A_1(z, u_0, u_{0z})u_z] - A_2(z, u_0, u_{0z})u - A_3(z, u_0, u_{0z}) = G(z, u_0, u_{0z}),$$

and $u_1(z) \in B'$. Thus similarly to the proof in (1), by the successive iteration, a solution of Problem P for equation (1.33) can be obtained.

By using a similar method as before, we can discuss the solvability of Problem $\tilde{P}$ and the corresponding Problem $\tilde{Q}$ for equation (1.2) or (1.31) with the boundary conditions

$$\text{Re}\,[\overline{\lambda(z)}u_z] = r(z), \quad z \in \Gamma, \quad u(0) = b_0,\, u(2) = b_2,$$

$$\text{Re}\,[\overline{\lambda(z)}u_{\bar{z}}] = r(z), \quad z \in L_2, \quad \text{Im}\,[\overline{\lambda(z)}u_{\bar{z}}]|_{z=z_1} = b_1$$

in which the coefficients $\lambda(z)$, $r(z)$, b_0, b_1, b_2 satisfy the condition (1.7), but where the conditions $C_\alpha[\lambda(z), L_1] \leq k_0$, $C_\alpha[r(z), L_1] \leq k_2$, $\max_{z \in L_1}[1/|a(x) - b(x)|] \leq k_0$ are replaced by $C_\alpha[\lambda(z), L_2] \leq k_0$, $C_\alpha[r(z), L_2] \leq k_2$, $\max_{z \in L_2}[1/|a(x) + b(x)|] \leq k_0$ and in (1.9) the condition $\lambda(t) = e^{i\pi/4}$ on $L_0 = (0, 2)$ and $\lambda(t_1 - 0) = \lambda(t_2 + 0) = \exp(i\pi/4)$ is replaced by $\lambda(t) = e^{-i\pi/4}$ on $L_0 = (0, 2)$ and $\lambda(t_1 - 0) = \lambda(t_2 + 0) = \exp(-i\pi/4)$. Besides the set $D_* = \bar{D}\backslash\{0, x-y = 2\}$ in Condition C is replaced by $D_* = \bar{D}\backslash\{x+y = 0, 2\}$, if the constant $\gamma_2 > 0$ in (1.9).

2 Oblique Derivative Problems for Second Order Equations of Mixed Type in General Domains

This section deals with oblique derivative boundary value problem for second order quasilinear equations of mixed (elliptic-hyperbolic) type in general domains. We prove the uniqueness and existence of solutions of the above problem. In refs. [12]1),3), the author discussed the Dirichlet problem (Tricomi problem) for second order equations of mixed type: $u_{xx} + \text{sgn}y\, u_{yy} = 0$ by using the method of integral equations and a complicated functional relation. In the present section, by using a new method, the solvability result of oblique derivative problem for more general domains is obtained.

2.1 Oblique derivative problem for second order equations of mixed type in another domain.

Let D be a simply connected bounded domain D in the complex plane $\mathcal{C}$ with the boundary $\partial D = \Gamma \cup L$, where $\Gamma(\subset \{y > 0\}) \in C^2_\mu\,(0 < \mu < 1)$ with the end points $z = 0, 2$ and $L = L_1 \cup L_2 \cup L_3 \cup L_4$, and

$$L_1 = \{x + y = 0,\, 0 \leq x \leq a/2\}, \quad L_2 = \{x - y = a,\, a/2 \leq x \leq a\},$$

$$L_3 = \{x + y = a,\, a \leq x \leq 1 + a/2\}, \quad L_4 = \{x - y = 2,\, 1 + a/2 \leq x \leq 2\}, \tag{2.1}$$

where $a\,(0 < a < 2)$ is a constant. Denote $D^+ = D \cap \{y > 0\}$ and $D^- = D \cap \{y < 0\}$, $D^-_1 = D^- \cap \{x - y < a\}$ and $D^-_2 = D^- \cap \{x + y > a\}$. Without loss of generality, we may assume that $\Gamma = \{|z - 1| = 1,\, y > 0\}$, otherwise through a conformal mapping, this requirement can be realized.

We consider the quasilinear second order mixed equation (1.2) and assume that (1.2) satisfies Condition C in $\bar{D}$, here D^- is as stated before. Problem P for (1.2)

in D is to find a continuously differentiable
solution of (1.2) in $D^* = \overline{D} \backslash \{0, a, 2\}$ satisfying the boundary conditions

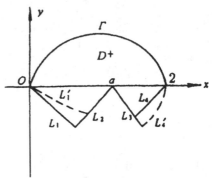

$$\frac{1}{2} \frac{\partial u}{\partial l} = \mathrm{Re}[\overline{\lambda(z)} u_z] = r(z), \quad z \in \Gamma,$$

$$u(0) = b_0, \quad u(a) = b_1, \quad u(2) = b_2, \quad (2.2)$$

$$\frac{1}{2} \frac{\partial u}{\partial l} = \mathrm{Re}[\overline{\lambda(z)} u_{\bar{z}}] = r(z), \quad z \in L_1 \cup L_4,$$

$$\mathrm{Im}[\overline{\lambda(z)} u_{\bar{z}}]|_{z=z_j} = b_{j+2}, \quad j = 1, 2,$$

Figure 2.1

where l is a given vector at every point on $\Gamma \cup L_1 \cup L_4$, $z_1 = (1 - i)a/2$, $z_2 = (1 + a/2) + i(1 - a/2)$, $\lambda(z) = a(x) + ib(x) = \cos(l, x) - i \cos(l, y)$, $z \in \Gamma$, and $\lambda(z) = a(x) + ib(x) = \cos(l, x) + i \cos(l, y)$, $z \in L_1 \cup L_4$, $b_j (j = 0, 1, \ldots, 4)$ are real constants, and $\lambda(z), r(z), b_j (j = 0, 1, \ldots, 4)$ satisfy the conditions

$$C_\alpha[\lambda(z), \Gamma] \le k_0, \quad C_\alpha[r(z), \Gamma] \le k_2, \quad C_\alpha[\lambda(z), L_j] \le k_0,$$

$$C_\alpha[r(z), L_j] \le k_2, \quad j = 1, 4, \quad |b_j| \le k_2, \quad j = 0, 1, \ldots, 4, \qquad (2.3)$$

$$\cos(l, n) \ge 0 \text{ on } \Gamma, \quad \max_{z \in L_1} \frac{1}{|a(x) - b(x)|}, \quad \max_{z \in L_4} \frac{1}{|a(x) + b(x)|} \le k_0,$$

in which n is the outward normal vector at every point on Γ, $\alpha \, (1/2 < \alpha < 1)$, k_0, k_2 are non-negative constants. The number

$$K = \frac{1}{2}(K_1 + K_2 + K_3), \qquad (2.4)$$

is called the index of Problem P on the boundary ∂D^+ of D^+, where

$$K_j = \left[\frac{\phi_j}{\pi}\right] + J_j, \quad J_j = 0 \text{ or } 1, \quad e^{i\phi_j} = \frac{\lambda(t_j - 0)}{\lambda(t_j + 0)}, \quad \gamma_j = \frac{\phi_j}{\pi} - K_j, \quad j = 1, 2, 3, \qquad (2.5)$$

in which $[b]$ is the largest integer not exceeding the real number b, $t_1 = 2$, $t_2 = 0$, $t_3 = a$, $\lambda(t) = e^{i\pi/4}$ on $(0, a)$ and $\lambda(t_3 - 0) = \lambda(t_2 + 0) = \exp(i\pi/4)$, and $\lambda(t) = e^{-i\pi/4}$ on $(a, 2)$ and $\lambda(t_1 - 0) = \lambda(t_3 + 0) = \exp(-i\pi/4)$. Here we only discuss the case $K = 1/2$, or $K = 0$ if $\cos(l, n) = 0$ on Γ, because in this case the solution of Problem P is unique and includes the Dirichlet problem (Tricomi problem) as a special case. We mention that if the boundary condition $\mathrm{Re}[\overline{\lambda(z)} u_{\bar{z}}] = r(z)$, $z \in L_j (j = 1, 4)$ is replaced by

$$\mathrm{Re}[\overline{\lambda(z)} u_z] = r(z), \quad z \in L_j \, (j = 1, 4),$$

then Problem P does not include the Dirichlet problem (Tricomi problem) as a special case. In order to ensure that the solution $u(z)$ of Problem P is continuously differentiable in D^*, we need to choose $\gamma_1 > 0$, $\gamma_2 > 0$ and can select $\gamma_3 = 1/2$. If we only require that the solution $u(z)$ in $\overline{D}$ is continuous, it is sufficient to choose $-2\gamma_1 < 1$, $-2\gamma_2 < 1$, $-\gamma_3 < 1$.

211

Besides, we consider the oblique derivative problem (Problem Q) for equation (1.2) with $A_2 = 0$ and the boundary condition (2.2), but the last point conditions $u(a) = b_1$, $u(2) = b_2$ in (2.2) is replaced by

$$\text{Im}\left[\overline{\lambda(z)}u_z\right]\big|_{z_j'} = c_j, \quad j = 1, 2, \tag{2.6}$$

in which $z_j'(j = 1, 2) \in \Gamma^* = \Gamma\backslash\{0, 2\}$ are two points and c_1, c_2 are real constants and $|c_1|, |c_2| \leq k_2$.

Similarly to Section 1, we can give a representation theorem of solutions of Problem P for equation (1.2), in which the functions $\Psi(z)$ in (1.13), $\lambda(x)$, $s(x)$ on L_0 in (1.15), $X(z), Y^\pm(\mu, \nu)$ in (1.17) are replaced by

$$\Psi(z) = \begin{cases} \int_a^\nu g_1(z)d\nu e_1 + \int_0^\mu g_2(z)d\mu e_2, & z \in D_1^-, \\ \int_2^\nu g_1(z)d\nu e_1 + \int_a^\mu g_2(z)d\mu e_2, & z \in D_2^-, \end{cases} \tag{2.7}$$

$$\lambda(x) = \begin{cases} 1+i, & x \in L_0' = (0, a), \\ 1-i, & x \in L_0'' = (a, 2), \end{cases} \tag{2.8}$$

$$s(x) = \begin{cases} \dfrac{2r((1-i)x/2) - 2\text{Re}\overline{[\lambda((1-i)x/2)}\Psi((1-i)x/2)}{a((1-i)x/2) - b((1-i)x/2)} + \text{Re}[\overline{\lambda(x)}\Psi(x)] \\[2mm] \quad - \dfrac{[a((1-i)x/2) + b((1-i)x/2)]f(0)}{a((1-i)x/2) - b((1-i)x/2)}, \quad x \in (0, a), \\[4mm] \dfrac{2r((1+i)x/2+1-i) - 2\text{Re}[\overline{\lambda((1+j)x/2+1-i)}\Phi((1+j)x/2+1-i)]}{a((1+i)x/2+1-i) + b((1+i)x/2+1-i)} \\[2mm] \quad - \dfrac{a((1+i)x/2+1-i) - b((1+i)x/2+1-i)g(2) - h(x)}{a((1+i)x/2+1-i) + b((1+i)x/2+1-i)}, \\[2mm] h(x) = \text{Re}[\overline{\lambda(x)}\Psi(x)] \\ \quad \times [a((1+i)x/2+1-i) + b((1+i)x/2+1-i)]/2, x \in (a, 2), \\[2mm] f(0) = [a(z_1) + b(z_1)]r(z_1) + [a(z_1) - b(z_1)]b_3, \\[2mm] g(2) = [a(z_2) - b(z_2)]r(z_2) - [a(z_2) + b(z_2)]b_4, \end{cases} \tag{2.9}$$

$$X(z) = \prod_{j=1}^3 |z - t_j|^{\eta_j}, \quad Y^\pm(z) = Y^\pm(\mu, \nu) = \prod_{j=1}^3 [|\mu - t_j||\nu - t_j|]^{\eta_j}, \tag{2.10}$$

$$\eta_j = \begin{cases} 2\max(-\gamma_j, 0) + \delta, & j = 1, 2, \\ \max(-\gamma_j, 0) + \delta, & j = 3, \end{cases}$$

respectively, δ is a sufficiently small positive constant, besides L_1 and the point z_1 in (1.15) should be replaced by $L_1 \cup L_4$ and $z_1 = (1-i)a/2, z_2 = 1 + a/2 + (1-a/2)i$. Now we first prove the unique solvability of Problem Q for equation (1.2).

Theorem 2.1   *If the mixed equation* (1.2) *in the domain* D *satisfies Condition* C, *then Problem* Q *for* (1.2) *has a unique solution* $u(z)$ *as stated in the form*

$$u(z) = 2\mathrm{Re}\int_0^z w(z)dz + b_0 \text{ in } \overline{D}, \tag{2.11}$$

where

$$w(z) = \tilde{\Phi}(z)e^{\tilde{\phi}(z)} + \tilde{\psi}(z) \text{ in } D^+,$$

$$\tilde{\phi}(z) = \tilde{\phi}_0(z) + Tg = \tilde{\phi}_0(z) - \frac{1}{\pi}\iint_{D^+}\frac{g(\zeta)}{\zeta - z}d\sigma_\zeta, \quad \tilde{\psi}(z) = Tf \text{ in } D^+,$$

$$w(z) = w_0(z) + \overline{\Phi(z)} + \overline{\Psi(z)} \text{ in } D^-, \tag{2.12}$$

$$\Psi(z) = \begin{cases} \int_a^\nu g_1(z)d\nu e_1 + \int_0^\mu g_2(z)d\mu e_2, & z \in D_1^-, \\[2mm] \int_2^\nu g_1(z)d\nu e_1 + \int_a^\mu g_2(z)d\mu e_2, & z \in D_2^-, \end{cases}$$

where $\mathrm{Im}\,[\tilde{\phi}(z)] = 0$ *on* $L_0 = (0,2)$, $e_1 = (1+i)/2$, $e_2 = (1-i)/2$, $\mu = x+y$, $\nu = x - y$, $\phi_0(z)$ *is an analytic function in* D^+ *and continuous in* $\overline{D^+}$, *such that* $\mathrm{Im}\,[\phi(z)] = 0$ *on* L_0, *and* $f(z), g(z), g_1(z), g_2(z)$ *are as stated in* (1.14), *and* $\tilde{\Phi}(z)$ *is an analytic function in* D^+ *and* $\Phi(z)$ *is a solution of equation* (1.12) *satisfying the boundary conditions: the first conditions and*

$$\mathrm{Re}\,[\overline{\lambda(z)}\Phi(z)] = -\mathrm{Re}\,[\overline{\lambda(z)}\Psi(z)], \quad z \in L_1 \cup L_4,$$

$$\mathrm{Im}\,[\overline{\lambda(z_j)}\Phi(z_j)] = -\mathrm{Im}\,[\overline{\lambda(z_j)}\Psi(z_j)], \quad j = 1,2, \tag{2.13}$$

$$\mathrm{Im}\,[\overline{\lambda(z_j')}\Phi(z_j')] = c_j, \quad j = 1,2,$$

in which $\lambda(x)$, $s(x)$ *are as stated in* (2.8), (2.9).

Proof   By using a similar method as stated in Section 2, Chapter V, we can prove Theorem 2.1, provided that L_1 or L_2 in the boundary conditions, Section 2, Chapter V is replaced by $L_1 \cup L_4$; the point conditions $\mathrm{Im}\,[\lambda(z_1)w(z_1)] = b_1$ in Section 2, Chapter V is replaced by $\mathrm{Im}\,[\lambda(z_j)w(z_j)] = b_j, j = 1,2$ and so on; the formula (2.14), Chapter V is replaced by

$$\mathrm{Re}\,[\overline{\lambda(z)}u_z] = \mathrm{Re}\,[\overline{\lambda(z)}w(z)] = s(x),$$

$$\lambda(x) = \begin{cases} 1+i, & x \in L_0' = (0,a), \\ 1-i, & x \in L_0'' = (a,2), \end{cases} \quad C_\beta[s(x), L_0'] + C_\beta[s(x), L_0''] \le k_3.$$

Theorem 2.2   *Suppose that the equation* (1.2) *satisfies Condition* C. *Then Problem* P *for* (1.2) *has a unique solution* $u(z)$, *and the solution* $u(z)$ *satisfies the estimates*

$$\tilde{C}_\beta^1[u(z), \overline{D^+}] = C_\beta[u(z), \overline{D^+}] + C_\beta[u_z X(z), \overline{D^+}] \le M_{15},$$

$$\hat{C}^1[u(z), \overline{D^-}] = C_\beta[u(z), \overline{D^-}] + C[u_z^\pm Y^\pm(z), \overline{D^-}] \le M_{15}, \tag{2.14}$$

$$\tilde{C}_\beta^1[u(z), \overline{D^+}] \le M_{16}(k_1+k_2), \quad \hat{C}^1[u(z), \overline{D^-}] \le M_{16}(k_1+k_2),$$

where

$$X(z)=\prod_{j=1}^{3}|z-t_j|^{\eta_j}, \quad Y^{\pm}(z)=\prod_{j=1}^{3}|x\pm y-t_j|^{\eta_j}, \quad \eta_j=\begin{cases} 2\max(-\gamma_j,0)+\delta, & j=1,2, \\ \max(-\gamma_j,0)+\delta, & j=3, \end{cases}$$

(2.15)

herein $t_1 = 2, t_2 = 0, t_3 = a, \gamma_1, \gamma_2, \gamma_3$ *are real constants in* (2.5), $\beta(< \delta), \delta$ *are sufficiently small positive constants, and* $M_{15} = M_{15}(p_0, \beta, k, D)$, $M_{16} = M_{16}(p_0, \beta, k_0, D)$ *are non-negative constants,* $k = (k_0, k_1, k_2)$.

Proof First of all, we prove the uniqueness of solutions of Problem P for (1.2). Suppose that there exist two solutions $u_1(z), u_2(z)$ of Problem P for (1.2). By Condition C, we can see that $u(z) = u_1(z) - u_2(z)$ and $w(z) = u_z$ satisfies the homogeneous equation and boundary conditions

$$\left\{\begin{matrix} w_{\bar{z}} \\ \bar{w}_{\bar{z}^*} \end{matrix}\right\} = f, \quad f = \mathrm{Re}\,[A_1 w] + A_2 u \text{ in } \left\{\begin{matrix} D^+ \\ D^- \end{matrix}\right\},$$

(2.16)

$$\mathrm{Re}\,[\overline{\lambda(z)}w(z)] = r(z), \quad z \in \Gamma, \quad u(0) = 0, \quad u(a) = 0,$$

$$u(2) = 0, \quad \mathrm{Re}\,[\lambda(z)w(z)] = 0, \quad z \in L_1 \cup L_4,$$

(2.17)

and (2.11). By using the method of proofs in Theorems 2.3 and 2.4, Chapter V, $w(z) = u_z = 0, u(z) = 0$ in $\overline{D^-}$ can be derived. Thus we have

$$2\mathrm{Re}\left[\frac{1-i}{\sqrt{2}}u_z\right] = \frac{\partial u}{\partial l} = 0 \text{ on } (0,a), \quad 2\mathrm{Re}\left[\frac{1+i}{\sqrt{2}}u_z\right] = \frac{\partial u}{\partial l} = 0 \text{ on } (a,2), \quad (2.18)$$

it is clear that $(1-i)/\sqrt{2} = \cos(l,x) - i\cos(l,y) = \exp(-i\pi/4)$ on $(0,a)$ and $(1+i)/\sqrt{2} = \cos(l,x) + i\cos(l,y) = \exp(i\pi/4)$ on $(a,2)$. On the basis of the maximum principle of solutions for (1.2) with $A_3 = 0$ in D^+, if $\max_{\overline{D^+}} u(z) > 0$, then its maximum M attains at a point $z^* \in \Gamma \cup L_0$, obviously $z^* \neq 0, a$ and 2, and we can prove $z^* \notin \Gamma$ by the method as stated in the proof of Theorem 2.3, Chapter V. Moreover, it is not difficult to prove that if $z^* \in L_0$ then $\partial u/\partial l \neq 0$ at z^*. This contradicts (2.17). Thus $\max_{\overline{D^+}} u(z) = 0$. By the similar method, we can prove $\min_{\overline{D^+}} u(z) = 0$. Hence $u(z) = 0, u_1(z) = u_2(z)$ in $\overline{D^+}$.

Secondly, we first prove the existence of solutions of Problem P for the linear equation (1.2) with $A_2 = 0$, i.e.

$$\begin{cases} u_{z\bar{z}} = \mathrm{Re}\,[A_1 u_z] + A_3, & z \in D^+, \\ u_{\bar{z}^*} = \mathrm{Re}\,[A_1 u_\tau] + A_3, & z \in \overline{D^-}, \end{cases}$$

(2.19)

By Theorem 2.1, we can prove the solvability of Problem P for (2.19). In fact if $u_0(a) = b_1, u_0(2) = b_2$, then the solution $u_0(z)$ is just a solution of Problem P for (1.2). Otherwise, $u_0(a) = c_1' \neq b_1$, or $u_0(2) = c_2' \neq b_2$, we find a solution $u_2(z)$ of

Problem P for (2.19) with the boundary conditions

$$\mathrm{Re}\,\overline{[\lambda(z)u_{kz}]} = 0, \quad z \in \Gamma, \quad u_k(0) = 0,$$

$$\mathrm{Re}\,\overline{[\lambda(z)u_{k\bar{z}}]} = 0, \quad z \in L_1 \cup L_4, \quad k = 1, 2, \qquad (2.20)$$

$$\mathrm{Im}\,\overline{[\lambda(z)u_{k\bar{z}}]}|_{z=z_j'} = \delta_{kj}, \quad k, j = 1, 2,$$

in which $\delta_{11} = \delta_{22} = 1$, $\delta_{12} = \delta_{21} = 0$. It is clear that

$$J = \begin{vmatrix} u_1(a) & u_2(a) \\ u_1(2) & u_2(2) \end{vmatrix} \neq 0. \qquad (2.21)$$

Because otherwise there exist two real constants d_1, d_2 $(|d_1| + |d_2| \neq 0)$, such that $d_1 u_1(z) + d_2 u_2(z) \not\equiv 0$ in D and

$$d_1 u_1(a) + d_2 u_2(a) = 0, \, d_1 u_1(2) + d_2 u_2(2) = 0. \qquad (2.22)$$

According to the proof of uniqueness as before, we can derive $d_1 u_1(z) + d_2 u_2(z) \equiv 0$ in $\overline{D}$, the contradiction proves $J \neq 0$. Hence there exist two real constants d_1, d_2, such that

$$d_1 u_1(a) + d_2 u_2(a) = c_1' - b_1, \, d_1 u_1(2) + d_2 u_2(2) = c_2' - b_2, \qquad (2.23)$$

thus the function

$$u(z) = u_0(z) - d_1 u_1(z) - d_2 u_2(z) \text{ in } D \qquad (2.24)$$

is just a solution of Problem P for equation (1.2) in the linear case. Moreover, we can obtain that the solution $u(z)$ of Problem P for (1.2) satisfies the estimates in (2.14), we can rewrite in the form

$$\hat{C}^1[u, \overline{D}] = C_\beta[u(z), \overline{D}] + C_\beta[u_z X(z), \overline{D^+}]$$

$$+ C[u_{\bar{z}}^{\pm}(\mu, \nu) Y^{\pm}(\mu, \nu), \overline{D^-}] \leq M_{17}, \qquad (2.25)$$

$$\hat{C}^1[u, \overline{D}] \leq M_{18}(k_1 + k_2),$$

where $X(z)$, $Y^{\pm}(\mu, \nu)$ are as stated in (2.10), and $M_{17} = M_{17}(p_0, \beta, k, D)$, $M_{18} = M_{18}(p_0, \beta, k_0, D)$ are non-negative constants, $k = (k_0, k_1, k_2)$. By using the estimate and method of parameter extension, the existence of solutions of Problem P for quasilinear equation (1.2) can be proved.

2.2 Oblique derivative problem for second order equations of mixed type in general domains

Now, we consider the domain D' with the boundary $\partial D' = \Gamma \cup L'$, where $L' = L_1' \cup L_2' \cup L_3' \cup L_4'$ and the parameter equations of the four curves L_1', L_2', L_3', L_4' are

$$L_1' = \{\gamma_1(x) + y = 0, \, 0 \leq x \leq l_1\}, \quad L_2' = \{x - y = a, \, l_1 \leq x \leq a\},$$

$$L_3' = \{x + y = a, \, a \leq x \leq l_2\}, \quad L_4' = \{\gamma_2(x) + y = 0, \, l_2 \leq x \leq 2\}, \qquad (2.26)$$

in which $\gamma_1(0) = 0$, $\gamma_2(2) = 0$; $\gamma_1(x) > 0$ on $0 \le x \le l_1 = \gamma(l_1) + a$; $\gamma_2(x) > 0$ on $l_2 = -\gamma(l_2) + a \le x \le 2$; $\gamma_1(x)$ on $0 \le x \le l_1$, $\gamma_2(x)$ on $l_2 \le x \le 2$ are continuous, and $\gamma_1(x)$, $\gamma_2(x)$ are differentiable on $0 \le x \le l_1$, $l_2 \le x \le 2$ except some isolated points, and $1 + \gamma_1'(x) > 0$ on $0 \le x \le l_1$, $1 - \gamma_2'(x) > 0$ on $l_2 \le x \le 2$. Denote $D'^+ = D' \cap \{y > 0\} = D^+$, $D'^- = D' \cap \{y < 0\}$, $D_1'^- = D'^- \cap \{x < a\}$ and $D_2'^- = D'^- \cap \{x > a\}$. Here we mention that in [12]1),3), the author assumed $0 < -\gamma_1'(x) < 1$ on $0 \le x \le l_1$ and some other conditions.

We consider the quasilinear second order equation of mixed (elliptic-hyperbolic) type: (1.2) in D'. Assume that equation (1.2) satisfies Condition C, but the hyperbolic domain D^- is replaced by D'^-.

Problem P' The oblique derivative problem for equation (1.2) is to find a continuously differentiable solution of (1.2) in $D^* = \overline{D'} \backslash \{0, a, 2\}$ for (1.2) satisfying the boundary conditions

$$\frac{1}{2} \frac{\partial u}{\partial l} = \mathrm{Re}\,[\overline{\lambda(z)} u_z] = r(z), \quad z \in \Gamma, \quad u(0) = b_0, \quad u(a) = b_1, \quad u(2) = b_2,$$

$$\frac{1}{2} \frac{\partial u}{\partial l} = \mathrm{Re}\,[\overline{\lambda(z)} u_{\bar{z}}] = r(z), \quad z \in L_1' \cup L_4', \quad \mathrm{Im}\,[\overline{\lambda(z)} u_{\bar{z}}]|_{z=z_j} = b_{j+2}, \quad j = 1, 2. \tag{2.27}$$

Here l is a given vector at every point on $\Gamma \cup L_1' \cup L_4'$, $z_1 = l_1 - i\gamma_1(l_1)$, $z_2 = l_2 - i\gamma_2(l_2)$, $\lambda(z) = a(x) + ib(x) = \cos(l, x) - i\cos(l, y)$, $z \in \Gamma$, and $\lambda(z) = a(x) + ib(x) = \cos(l, x) + i\cos(l, y)$, $z \in L_1' \cup L_4'$, $b_j(j = 0, 1, \ldots, 4)$ are real constants, and $\lambda(z)$, $r(z)$, $b_j(j = 0, 1, \ldots, 4)$ satisfy the conditions

$$C_\alpha[\lambda(z), \Gamma] \le k_0, \quad C_\alpha[r(z), \Gamma] \le k_2,$$

$$C_\alpha[\lambda(z), L_j'] \le k_0, \quad C_\alpha[r(z), L_j'] \le k_2, \quad j = 1, 4,$$

$$\cos(l, n) \ge 0 \text{ on } \Gamma, \quad |b_j| \le k_2, \quad 0 \le j \le 4, \tag{2.28}$$

$$\max_{z \in L_1'} \frac{1}{|a(x) - b(x)|}, \quad \max_{z \in L_4'} \frac{1}{|a(x) + b(x)|} \le k_0,$$

in which n is the outward normal vector at every point on Γ, $\alpha\,(1/2 < \alpha < 1)$, k_0, k_2 are non-negative constants. In particular, if $L_j' = L_j(j = 1, 2, 3, 4)$, then Problem P' in this case is called Problem P.

In the following, we discuss the domain D' with the boundary $\Gamma \cup L_1' \cup L_2' \cup L_3' \cup L_4'$, where L_1', L_2', L_3', L_4' are as stated in (2.26), and $\gamma_1(x), \gamma_2(x)$ satisfy the conditions $1 + \gamma_1'(x) > 0$ on $0 \le x \le l_1$ and $1 - \gamma_2'(x) > 0$ on $l_2 \le x \le 2$. By the conditions, the inverse functions $x = \sigma(\nu)$, $x = \tau(\mu)$ of $x + \gamma_1(x) = \nu = x - y$, $x - \gamma_2(x) = \mu = x + y$ can be found respectively, namely

$$\mu = 2\sigma(\nu) - \nu, \quad 0 \leq \nu \leq a, \quad \nu = 2\tau(\mu) - \mu, \quad a \leq \mu \leq 2. \tag{2.29}$$

They are other expressions for the curves L_1', L_4'. Now, we make a transformation in D'^-:

$$\tilde{\mu} = \frac{a[\mu - 2\sigma(\nu) + \nu]}{a - 2\sigma(\nu) + \nu}, \quad \tilde{\nu} = \nu, \quad 2\sigma(\nu) - \nu \leq \mu \leq a, \quad 0 \leq \nu \leq a,$$

$$\tilde{\mu} = \mu, \quad \tilde{\nu} = \frac{a[2\tau(\mu) - \mu - 2] + (2 - a)\nu}{2\tau(\mu) - \mu - a}, \quad a \leq \mu \leq 2, \quad a \leq \nu \leq 2\tau(\mu) - \mu, \tag{2.30}$$

in which μ, ν are variables. If $(\mu, \nu) \in L_1'$, L_2', L_3', L_4', then $(\tilde{\mu}, \tilde{\nu}) \in L_1$, L_2, L_3, L_4 respectively. The inverse transformation of (2.30) is

$$\mu = \frac{1}{a}[a - 2\sigma(\nu) + \nu]\tilde{\mu} + 2\sigma(\nu) - \nu$$

$$= \frac{1}{a}[a - 2\sigma(x + \gamma_1(x)) + x + \gamma_1(x)](\tilde{x} + \tilde{y}) + 2\sigma(x + \gamma_1(x)) - x - \gamma_1(x),$$

$$\nu = \tilde{\nu} = \tilde{x} - \tilde{y}, \quad 0 \leq \tilde{\mu} \leq a, \ 0 \leq \tilde{\nu} \leq a, \ \mu = \tilde{\mu} = \tilde{x} + \tilde{y},$$

$$\nu = \frac{1}{2-a}[(2\tau(\mu) - \mu)(\tilde{\nu} - a) - a(\tilde{\nu} - 2)] \tag{2.31}$$

$$= \frac{1}{2-a}[(2\tau(x - \gamma_2(x)) - x + \gamma_2(x))(\tilde{x} - \tilde{y} - a) - a(\tilde{x} - \tilde{y} - 2)],$$

$$a \leq \tilde{\mu} \leq 2, \quad a \leq \tilde{\nu} \leq 2.$$

It is not difficult to see that the transformations in (2.31) map the domains $D_1'^-$, $D_2'^-$ onto the domains D_1^-, D_2^- respectively. Moreover, we have

$$\tilde{x} = \frac{1}{2}(\tilde{\mu} + \tilde{\nu}) = \frac{2ax - (a + x - y)[2\sigma(x + \gamma_1(x)) - x - \gamma_1(x)]}{2a - 4\sigma(x + \gamma_1(x)) + 2x + 2\gamma_1(x)},$$

$$\tilde{y} = \frac{1}{2}(\tilde{\mu} - \tilde{\nu}) = \frac{2ay - (a - x + y)[2\sigma(x + \gamma_1(x)) - x - \gamma_1(x)]}{2a - 4\sigma(x + \gamma_1(x)) + 2x + 2\gamma_1(x)},$$

$$x = \frac{1}{2}(\mu + \nu) = \frac{1}{2a}[a - 2\sigma(x + \gamma_1(x)) + x + \gamma_1(x)](\tilde{x} + \tilde{y}) \tag{2.32}$$

$$\qquad + \sigma(x + \gamma_1(x)) - \frac{1}{2}(x + \gamma_1(x) - \tilde{x} + \tilde{y}),$$

$$y = \frac{1}{2}(\mu - \nu) = \frac{1}{2a}[a - 2\sigma(x + \gamma_1(x)) + x + \gamma_1(x)](\tilde{x} + \tilde{y})$$

$$\qquad + \sigma(x + \gamma_1(x)) - \frac{1}{2}(x + \gamma_1(x) + \tilde{x} - \tilde{y}),$$

and

$$\tilde{x}=\frac{1}{2}(\tilde{\mu}+\tilde{\nu})=\frac{[2\tau(x-\gamma_2(x))-x+\gamma_2(x)](x+y+a)+2(x-y)-2a(1+x)}{2[2\tau(x-\gamma_2(x))-x+\gamma_2(x)-a]},$$

$$\tilde{y}=\frac{1}{2}(\tilde{\mu}-\tilde{\nu})=\frac{[2\tau(x-\gamma_2(x))-x+\gamma_2(x)](x+y-a)-2(x-y)+2a(1-y)}{2[2\tau(x-\gamma_2(x))-x+\gamma_2(x)-a]},$$

$$x=\frac{1}{2}(\mu+\nu)=\frac{1}{2(2-a)}[2\tau(x-\gamma_2(x))-x+\gamma_2(x))(\tilde{x}-\tilde{y}-a) \tag{2.33}$$
$$+2(\tilde{x}+\tilde{y}+a)-2a\tilde{x}],$$

$$y=\frac{1}{2}(\mu-\nu)=\frac{1}{2(2-a)}[2\tau(x-\gamma_2(x))-x+\gamma_2(x))(-\tilde{x}+\tilde{y}+a)$$
$$+2(\tilde{x}+\tilde{y}-a)-2a\tilde{y}].$$

Denote by $\tilde{z}=\tilde{x}+i\tilde{y}=f(z)$, $\tilde{z}=\tilde{x}+i\tilde{y}=g(z)$ and $z=x+iy=f^{-1}(\tilde{z})$, $z=x+iy=g^{-1}(\tilde{z})$ the transformations and their inverse transformations in (2.32), (2.33) respectively. Through the transformation (2.30), we have

$$(U+V)_{\tilde{\nu}}=(U+V)_\nu, \quad (U-V)_{\tilde{\mu}}=\frac{1}{a}[a-2\sigma(\nu)+\nu](U-V)_\mu \text{ in } D_1'^-,$$
$$\tag{2.34}$$
$$(U+V)_{\tilde{\nu}}=\frac{2\tau(\mu)-\mu-a}{2-a}(U+V)_\nu, \quad (U-V)_{\tilde{\mu}}=(U-V)_\mu \text{ in } D_2'^-.$$

Equation (1.2) in D'^- can be rewritten in the form

$$\xi_\nu=A\xi+B\eta+Cu+D, \quad \eta_\mu=A\xi+B\eta+Cu+D, \quad z\in\overline{D'^-}, \tag{2.35}$$

where $\xi=U+V=(u_x-u_y)/2$, $\eta=U-V=(u_x+u_y)/2$, under the transformation (2.34), it is clear that system (2.35) in $\overline{D'^-}$ is reduced to

$$\xi_{\tilde{\nu}}=A\xi+B\eta+Cu+D, \quad \eta_{\tilde{\mu}}=\frac{1}{a}[a-2\sigma(\nu)+\nu][A\xi+B\eta+Cu+D] \text{ in } D_1^-,$$
$$\tag{2.36}$$
$$\xi_{\tilde{\nu}}=\frac{2\tau(\mu)-\mu-a}{2-a}[A\xi+B\eta+Cu+D], \quad \eta_{\tilde{\mu}}=A\xi+B\eta+Cu+D \text{ in } D_2^-.$$

Moreover, through the transformations (2.32),(2.33), the boundary condition (2.23) on $L_1'\cup L_4'$ is reduced to

$$\text{Re}\,[\lambda(f^{-1}(\tilde{z}))w(f^{-1}(\tilde{z}))]=r[f^{-1}(\tilde{z})], \quad \tilde{z}\in L_1, \quad \text{Im}\,[\lambda(f^{-1}(\tilde{z}_3))w(f^{-1}(\tilde{z}_3))]=b_1,$$
$$\tag{2.37}$$
$$\text{Re}\,[\lambda(g^{-1}(\tilde{z}))w(g^{-1}(\tilde{z}))]=r[g^{-1}(\tilde{z})], \quad \tilde{z}\in L_4, \quad \text{Im}\,[\lambda(g^{-1}(\tilde{z}_4))w(g^{-1}(\tilde{z}_4))]=b_2,$$

in which $\tilde{z}_3=f(z_3)$, $\tilde{z}_4=g(z_4)$. Therefore the boundary value problem (1.2) (in $\overline{D^+}$), (2.35), (2.27), (2.6) is transformed into the boundary value problem (1.2), (2.36), (2.27), (2.37). According to the proof of Theorem 2.1, we see that the boundary value problem (1.2), (2.36), (2.27), (2.37) has a unique solution $w(\tilde{z})$, and then $w[\tilde{z}(z)]$ is a solution of the boundary value problem (1.2),(2.2) ($w = u_z$) in $\overline{D'^-}$, and the function

$$
u(z) = \begin{cases}
2\mathrm{Re} \displaystyle\int_0^z w(z)dz + b_0 \text{ in } D^+, \\[3mm]
2\mathrm{Re} \displaystyle\int_a^z w[f(z)]dz + u(a) \text{ in } \overline{D_1'^-}, \\[3mm]
2\mathrm{Re} \displaystyle\int_2^z w[g(z)]dz + u(2) \text{ in } \overline{D_2'^-},
\end{cases}
\tag{2.38}
$$

is just a solution of Problem P for (1.2) in D', where $u(a) = b_1$, $u(2) = b_2$.

Theorem 2.3 *If the mixed equation* (1.2) *in the domain* D' *satisfies Condition C, then Problem P' for* (1.2) *with the boundary condition* (2.2) *has a unique solution* $u(z)$ *as stated in* (2.38)*, where* $z_1 = l_1 - i\gamma_1(l_1)$*,* $z_2 = l_2 - i\gamma_2(l_2)$*.*

By using the above method and the method in Section 4, Chapter IV, we can discuss the unique solvability of Problem P' for equation (1.2) in some more general domains D'' including the domain $D'' = \{|z - 1| < 1, \mathrm{Im}\, z \geq 0\} \cup \{|z - a/2| < a^2/4, \mathrm{Im}\, z < 0\} \cup \{|z - 1 - a/2| < (2 - a)^2/4, \mathrm{Im}\, z < 0\}$.

3 Discontinuous Oblique Derivative Problems for Second Order Quasilinear Equations of Mixed Type

This section deals with discontinuous oblique derivative problems for quasilinear second order equations of mixed (elliptic-hyperbolic) type in a simply connected domain. Firstly, we give a representation theorem and prove the uniqueness of the solution for the above boundary value problem, and then by using the method of successive iteration, the existence of solutions for the above problem is proved.

3.1 Formulation of discontinuous oblique derivative problems for second order equations of mixed type

Let D be a simply connected domain with the boundary $\Gamma \cup L_1 \cup L_2$ as stated before, where $D^+ = \{|z-1| < 1, \mathrm{Im}\, z > 0\}$. We discuss the second order quasilinear complex equations of mixed type as stated in (1.2) with Condition C. In order to introduce the discontinuous oblique derivative boundary value problem for equation (1.2), let the functions $a(z), b(z)$ possess discontinuities of first kind at $m - 1$ distinct points $z_1, z_2, \ldots, z_{m-1} \in \Gamma$, which are arranged according to the positive direction of Γ, and $Z = \{z_0 = 2, z_1, \ldots, z_m = 0\} \cup \{x + y = 0, x - y = 2, \mathrm{Im}\, z \leq 0\}$, where m is a positive integer, and $r(z) = O(|z - z_j|^{-\beta_j})$ in the neighborhood of $z_j(j = 0, 1, \ldots, m)$ on $\overline{\Gamma}$, in which $\beta_j(j = 0, 1, \ldots, m)$ are sufficiently small positive numbers. Denote $\lambda(z) = a(z) + ib(z)$ and $|a(z)| + |b(z)| \neq 0$. There is no harm in assuming that

$|\lambda(z)| = 1$, $z \in \Gamma^* = \Gamma \backslash Z$. Suppose that $\lambda(z)$, $r(z)$ satisfy the conditions

$$\lambda(z) \in C_\alpha(\Gamma_j), \quad |z - z_j|^{\beta_j} r(z) \in C_\alpha(\Gamma_j), \quad j = 1, \ldots, m, \tag{3.1}$$

herein Γ_j is the arc from the point z_{j-1} to z_j on Γ and $z_0 = 2$, and $\Gamma_j(j = 1, \ldots, m)$ does not include the end points, α $(0 < \alpha < 1)$ is a constant. Besides, there exist n points $E_1 = a_1, E_2 = a_2, \ldots, E_n = a_n$ on the segment $AB = L_0 = (0, 2)$ and $E_0 = 0, E_{n+1} = 2$, where $a_0 = 0 < a_1 < a_2 < \cdots < a_n < a_{n+1} = 2$. Denote by $A = A_0 = 0, A_1 = (1 - i)a_1/2, A_2 = (1 - i)a_2/2, \ldots, A_n = (1 - i)a_n/2, A_{n+1} = C = 1 - i$ and $B_1 = 1 - i + (1 + i)a_1/2, B_2 = 1 - i + (1 + i)a_2/2, \ldots, B_n = 1 - i + (1 + i)a_n/2, B_{n+1} = B = 2$ on the segments AC, CB respectively. Moreover, we denote $D_1^- = \overline{D^-} \cap \{\cup_{j=0}^{[n/2]}(a_{2j} \leq x - y \leq a_{2j+1})\}$, $D_2^- = \overline{D^-} \cap \{\cup_{j=1}^{[(n+1)/2]}(a_{2j-1} \leq x+y \leq a_{2j})\}$ and $\tilde{D}_{2j+1}^- = \overline{D^-} \cap \{a_{2j} \leq x - y \leq a_{2j+1}\}$, $j = 0, 1, \ldots, [n/2]$, $\tilde{D}_{2j}^- = \overline{D^-} \cap \{a_{2j-1} \leq x + y \leq a_{2j}\}$, $j = 1, \ldots, [(n + 1)/2]$, and $D_*^- = \overline{D^-} \backslash \{\cup_{j=0}^{n+1}(x \pm y = a_j, y \leq 0)\}$, $D_* = D^+ \cup D_*^-$.

The discontinuous oblique derivative boundary value problem for equation (1.2) may be formulated as follows:

Problem P' Find a continuous solution $u(z)$ of (1.2) in $\bar{D}$, which is continuously differentiable in $D_* = D^+ \cup D_*^-$ and satisfies the boundary conditions

$$\frac{1}{2}\frac{\partial u}{\partial l} = \text{Re}\left[\overline{\lambda(z)}u_z\right] = r(z), \quad z \in \Gamma,$$

$$\frac{1}{2}\frac{\partial u}{\partial l} = 2\text{Re}\left[\overline{\lambda(z)}u_{\bar{z}}\right] = r(z), \quad z \in L_3 = \sum_{j=0}^{[n/2]} A_{2j}A_{2j+1},$$

$$\frac{1}{2}\frac{\partial u}{\partial l} = 2\text{Re}\left[\overline{\lambda(z)}u_{\bar{z}}\right] = r(z), \quad z \in L_4 = \sum_{j=1}^{[(n+1)/2]} B_{2j-1}B_{2j}, \tag{3.2}$$

$$\text{Im}\left[\overline{\lambda(z)}u_{\bar{z}}\right]|_{z=A_{2j+1}} = c_{2j+1}, \quad j = 0, 1, \ldots, [n/2],$$

$$\text{Im}\left[\overline{\lambda(z)}u_{\bar{z}}\right]|_{z=B_{2j-1}} = c_{2j}, \quad j = 1, \ldots, [(n + 1)/2], \tag{3.3}$$

$$u(z_j) = b_j, j = 0, 1, \ldots, m, \ u(a_j) = b_{m+j}, \quad j = 1, \ldots, n,$$

where l is a vector at every point on $\Gamma \cup L_3 \cup L_4$, $b_j(j = 0, 1, \ldots, m + n)$, $c_j(j = 0, 1, \ldots, n + 1, c_0 = b_0)$ are real constants, $\lambda(z) = a(x) + ib(x) = \cos(l, x) - i\cos(l, y)$, $z \in \Gamma$, $\lambda(z) = a(x) + ib(x) = \cos(l, x) + i\cos(l, y)$, $z \in L_3 \cup L_4$, and $\lambda(z)$, $r(z)$, $c_j(j = 0, 1, \ldots, n + 1)$ satisfy the conditions

$$C_\alpha[\lambda(z), \Gamma] \leq k_0, \quad C_\alpha[r(z), \Gamma] \leq k_2, \quad |c_j| \leq k_2, \quad j = 0, 1, \ldots, n + 1,$$

$$C_\alpha[\lambda(z), L_j] \leq k_0, \quad C_\alpha[r(z), L_j] \leq k_2, \quad j = 3, 4, \quad |b_j| \leq k_2, j = 0, 1, \ldots, m+n, \tag{3.4}$$

$$\cos(l, n) \geq 0 \text{ on } \Gamma, \quad \max_{z \in L_3} \frac{1}{|a(x) - b(x)|} \leq k_0, \quad \max_{z \in L_4} \frac{1}{|a(x) + b(x)|} \leq k_0,$$

where n is the outward normal vector at every point on Γ, $\alpha\,(1/2 < \alpha < 1)$, k_0, k_2 are non-negative constants. The above discontinuous oblique derivative boundary value problem for (1.2) is called Problem P'. Problem P' for (1.2) with $A_3(z, u, u_z) = 0$, $z \in \bar{D}$, $r(z) = 0$, $z \in \Gamma \cup L_3 \cup L_4$, $b_j = 0 (j = 0, 1, \ldots, m + n)$ and $c_j = 0$ $(j = 0, 1, \ldots, n + 1)$ will be called Problem P_0'. Moreover we give the same definitions as in (5.10), (5.11), Chapter IV, but choose $K = (m + n - 1)/2$, or $K = (m + n)/2 - 1$ if $\cos(\nu, n) \equiv 0$ on Γ, and the condition $u(z_m) = b_m$ can be canceled. Besides we require that the solution $u(z)$ in D^+ satisfies the conditions

$$u_z = O(|z - z_j|^{-\delta}),\ \delta = \begin{cases} \beta_j + \tau, & \text{for } \gamma_j \geq 0, \quad \text{and } \gamma_j < 0, \quad \beta_j > |\gamma_j|, \\ |\gamma_j| + \tau, & \text{for } \gamma_j < 0, \quad \beta_j \leq |\gamma_j|, \quad j = 0, 1, \ldots, m + n, \end{cases}$$
(3.5)

in the neighborhood of $z_j\,(0 \leq j \leq m), a_j\,(1 \leq j \leq n)$ in D^+, where $\gamma_j' = \max(0, -\gamma_j)\,(j = 0, 1, \ldots, m + n), \gamma_0' = \max(0, -2\gamma_0), \gamma_m' = \max(0, -2\gamma_m)$ and $\gamma_j\,(j = 0, 1, \ldots, m + n)$ are real constants as stated in (5.10), Chapter IV, δ is a sufficiently small positive number. Now we explain that in the closed domain $\overline{D^-}$, the derivatives $u_x + u_y$, $u_x - u_y$ of the solution $u(z)$ in the neighborhoods of the $2n + 2$ characteristic lines $Z' = \{x + y = 0,\ x - y = 2,\ x \pm y = a_j\,(j = 1, \ldots, n), y \leq 0\}$ may be not bounded if $\gamma_j \leq 0 (j = 0, 1, \ldots, n + 1)$. Hence if we require that the derivative u_z of $u(z)$ in $\overline{D^-} \backslash Z'$ is bounded, then we need to choose $\gamma_j > 0$ $(j = 0, 1, \ldots, n + 1)$. If we only require that the solution $u(z)$ is continuous in $\overline{D}$, it suffices to choose $-2\gamma_0 < 1$, $-2\gamma_m < 1$, $-\gamma_j < 1\,(j = 1, \ldots, m - 1, m + 1, \ldots, m + n)$.

Furthermore, we need to introduce another oblique derivative boundary value problem.

Problem Q'   If $A_2(z) = 0$ in D, we find a continuously differentiable solution $u(z)$ of (1.2) in D_*, which is continuous in $\bar{D}$ and satisfies the boundary conditions (3.2),(3.3), but the point conditions in (3.3) are replaced by

$$u(2) = b_0 = d_0, \quad \text{Im}\,[\overline{\lambda(z)}u_z]|_{z = z_j'} = d_j, \quad j = 1, \ldots, m + n,$$
(3.6)

where $z_j'(\notin Z) \in \Gamma(j = 0, 1, \ldots, m + n)$ are distinct points, $d_j(j = 0, 1, \ldots, m + n)$ are real constants satisfying the conditions $|d_j| \leq k_2, j = 0, 1, \ldots, m + n$, but we do not assume $\cos(\nu, n) \geq 0$ on each $\Gamma_j(j = 1, \ldots, m)$.

3.2   Representations of solutions for the oblique derivative problem for (1.2)

First of all, we give the representation of solutions of Problem Q' for the equation

$$\begin{Bmatrix} u_{\bar{z}z} \\ u_{\bar{z}z^*} \end{Bmatrix} = 0 \text{ in } \begin{Bmatrix} D^+ \\ D^- \end{Bmatrix}.$$
(3.7)

It is clear that Problem Q' for (3.7) is equivalent to the following boundary value problem (Problem A') for the first order complex equation

$$Lw = \left\{ \begin{array}{c} w_{\bar{z}} \\ \bar{w}_{\bar{z}^*} \end{array} \right\} = 0 \ \text{in} \ \left\{ \begin{array}{c} D^+ \\ D^- \end{array} \right\}, \tag{3.8}$$

with the boundary conditions

$$\mathrm{Re}\,[\overline{\lambda(z)}w(z)] = r(z) \ \text{on} \ \Gamma,$$
$$\mathrm{Re}\,[\lambda(z)w(z)] = r(z) \ \text{on} \ L_3 \cup L_4,$$
$$\mathrm{Im}\,[\overline{\lambda(z)}w(z)]|_{z=z'_j} = b_j, \quad j = 1,\ldots,m+n, \tag{3.9}$$
$$\mathrm{Im}\,[\lambda(z)w(z)]|_{z=A_{2j+1}} = c_{2j+1}, \quad j = 0,1,\ldots,[n/2],$$
$$\mathrm{Im}\,[\lambda(z)w(z)]|_{z=B_{2j-1}} = c_{2j}, \quad j = 1,\ldots,[(n+1)/2],$$

and the relation

$$u(z) = 2\,\mathrm{Re} \int_2^z w(z)dz + b_0, \tag{3.10}$$

where the integral path is appropriately chosen. Thus from Theorem 5.2, Chapter IV, we can derive the following theorem.

Theorem 3.1 *The boundary value problem Q' for (3.7) in $\overline{D}$ has a unique continuous solution $u(z)$ as stated in (3.10), where the solution $w(z)$ of Problem A' for (3.8) in $\overline{D^-}$ possesses the form*

$$w(z) = \frac{1}{2}[(1-i)f(x+y) + (1+i)g(x-y)],$$
$$f(x+y) = \mathrm{Re}\,[(1+i)w(x+y)] \ \text{in} \ D^-\backslash\{\overline{D_2^-}\},$$
$$g(x-y) = k(x-y) \ \text{in} \ D_1^-, \tag{3.11}$$
$$f(x+y) = h(x+y) \ \text{in} \ D_2^-,$$
$$g(x-y) = \mathrm{Re}\,[(1-i)w(x-y)] \ \text{in} \ D^-\backslash\{\overline{D_1^-}\},$$

herein $w(x+y)(0 \le x+y \le 2), w(x-y)(0 \le x-y \le 2)$ are values of the solution $w(z)$ of Problem A' for (3.8) in $\overline{D^+}$ with the first boundary condition in (3.9) and the boundary condition

$$\mathrm{Re}\,[\overline{\lambda(x)}w(x)] = \left\{ \begin{array}{l} k(x) \ \text{on} \ L'_1 = \overline{D_1^-} \cap AB, \\ h(x) \ \text{on} \ L'_2 = \overline{D_2^-} \cap AB, \end{array} \right. \tag{3.12}$$

in which $k(x), h(x)$ can be expressed as

$$k(x) = \frac{2r((1-i)x/2) - [a((1-i)x/2) + b((1-i)x/2)]h_{2j}}{a((1-i)x/2) - b((1-i)x/2)}$$

$$\text{on } \tilde{L}_{2j+1} = \overline{\tilde{D}_{2j+1}^-} \cap AB,$$

$$h(x) = \frac{2r((1+i)x/2 + 1 - i)}{a((1+i)x/2 + 1 - i) + b((1+i)x/2 + 1 - i)} \tag{3.13}$$

$$- \frac{[a((1+i)x/2+1-i) - b((1+i)x/2+1-i)]k_{2j-1}}{a((1+i)x/2+1-i) + b((1+i)x/2+1-i)}$$

$$\text{on } \tilde{L}_{2j} = \overline{\tilde{D}_{2j}^-} \cap AB,$$

where $\tilde{D}_j^- (j = 1, 2, \ldots, 2n+1)$ are as stated before, and

$$h_{2j} = \text{Re}\,[\lambda(A_{2j+1})(r(A_{2j+1}) + ic_{2j+1})] + \text{Im}\,[\lambda(A_{2j+1})(r(A_{2j+1}) + ic_{2j+1})]$$

$$\text{on } \tilde{L}_{2j+1}, \quad j = 0, 1, \ldots, [n/2],$$

$$k_{2j-1} = \text{Re}\,[\lambda(B_{2j-1})(r(B_{2j-1}) + ic_{2j})] - \text{Im}\,[\lambda(B_{2j-1})(r(B_{2j-1}) + ic_{2j})]$$

$$\text{on } \tilde{L}_{2j}, \quad j = 1, \ldots, [(n+1)/2],$$

where $\tilde{L}_{2j+1} = \tilde{D}_{2j+1}^- \cap AB, \quad j = 0, 1, \ldots, [n/2], \quad \tilde{L}_{2j} = \tilde{D}_{2j}^- \cap AB, \quad j = 1, \ldots, [(n+1)/2]$.

Next we give the representation theorem of solutions of Problem Q' for equation (1.2).

Theorem 3.2 *Suppose that equation (1.2) satisfies Condition C. Then any solution of Problem Q' for (1.2) can be expressed as*

$$u(z) = 2\text{Re}\int_2^z w(z)dz + c_0, \quad w(z) = w_0(z) + W(z) \text{ in } D, \tag{3.14}$$

where $w_0(z)$ is a solution of Problem A' for the complex equation (3.8) with the boundary condition (3.2), (3.6) ($w_0(z) = u_{0z}$), and $w(z)$ possesses the form

$$w(z) = W(z) + w_0(z) \text{ in } D, \quad w(z) = \tilde{\Phi}(z)e^{\tilde{\phi}(z)} + \tilde{\psi}(z),$$

$$\tilde{\phi}(z) = \tilde{\phi}_0(z) + Tg = \tilde{\phi}_0(z) - \frac{1}{\pi}\int\int_{D^+} \frac{g(\zeta)}{\zeta - z}d\sigma_\zeta, \quad \tilde{\psi}(z) = Tf \text{ in } D^+,$$

$$\overline{W(z)} = \Phi(z) + \Psi(z) \text{ in } D^-, \tag{3.15}$$

$$\Psi(z) = \begin{cases} \displaystyle\int_{a_{2j+1}}^\nu g_1(z)d\nu e_1 + \int_0^\mu g_2(z)d\mu e_2 \text{ in } D_{2j+1}^-, \ j = 0, 1, \ldots, [n/2], \\[4mm] \displaystyle\int_2^\nu g_1(z)d\nu e_1 + \int_{a_{2j-1}}^\mu g_2(z)d\mu e_2 \text{ in } D_{2j}^-, \ j = 1, \ldots, [(n+1)/2], \end{cases}$$

in which $e_1 = (1+i)/2$, $e_2 = (1-i)/2$, $\mu = x+y$, $\nu = x-y$, $\tilde{\phi}_0(z)$ *is an analytic function in* D^+, $\mathrm{Im}\,[\tilde{\phi}(z)] = 0$ *on* $L' = (0,2)$, *and*

$$g(z)=\begin{cases} A_1/2+\overline{A_1}\bar{w}/(2w), & w(z)\neq 0,\\ 0, & w(z)=0, \quad z\in D^+, \end{cases} \qquad f(z)=\mathrm{Re}[A_1\tilde{\phi}_z]+A_2u+A_3,$$

$$g_1(z)=g_2(z)=A\xi+B\eta+Cu+D, \quad \xi=\mathrm{Re}w+\mathrm{Im}w, \quad \eta=\mathrm{Re}w-\mathrm{Im}w, \qquad (3.16)$$

$$A=\frac{\mathrm{Re}A_1+\mathrm{Im}A_1}{2}, \quad B=\frac{\mathrm{Re}A_1-\mathrm{Im}A_1}{2}, \quad C=A_2, D=A_3 \text{ in } D^+,$$

where $\tilde{\Phi}(z)$ *is an analytic function in* D^+ *and* $\Phi(z)$ *is a solution of the equation* (3.8) *in* D^- *satisfying the boundary conditions*

$$\mathrm{Re}\,[\overline{\lambda(z)}(\tilde{\Phi}(z)e^{\tilde{\phi}(z)}+\tilde{\psi}(z))] = r(z), \quad z\in\Gamma,$$

$$\mathrm{Re}\,[\overline{\lambda(x)}(\tilde{\Phi}(x)e^{\tilde{\phi}(x)}+\tilde{\psi}(x))] = s(x), \quad x\in L_0 = (0,2),$$

$$\mathrm{Im}\,[\overline{\lambda(z)}(\tilde{\Phi}(z)e^{\tilde{\phi}(z)}+\tilde{\psi}(z))]|_{z=z'_j} = b_j, \quad j=1,\dots,m+n,$$

$$\mathrm{Re}\,[\overline{\lambda(x)}\Phi(x)] = \mathrm{Re}\,[\overline{\lambda(x)}(\overline{W(x)}-\Psi(x))], \quad z\in L_0, \qquad (3.17)$$

$$\mathrm{Re}\,[\overline{\lambda(z)}\Phi(z)] = -\mathrm{Re}\,[\overline{\lambda(z)}\Psi(z)], \quad z\in L_3\cup L_4,$$

$$\mathrm{Im}\,[\overline{\lambda(z)}(\Phi(z)+\Psi(z))]|_{z=A_{2j+1}} = 0, \quad j=0,1,\dots,[n/2],$$

$$\mathrm{Im}\,[\overline{\lambda(z)}(\Phi(z)+\Psi(z))]|_{z=B_{2j-1}} = 0, \quad j=1,\dots,[(n+1)/2],$$

where $s(x)$ *can be written as in* (3.19) *below. Moreover, the solution* $u_0(z)$ *of Problem* Q' *for* (3.7) *the estimate*

$$\tilde{C}^1[u_0(z),\overline{D^-}] = C_\beta[u_0(z),\overline{D^-}] + C[w_0^\pm(\mu,\nu)Y^\pm(\mu,\nu),\overline{D^-}] \leq M_{19}(k_1+k_2) \quad (3.18)$$

in which $Y^\pm(z) = Y^\pm(\mu,\nu) = \prod_{j=0}^{n+1}|x\pm y-a_j|^{\gamma'_j+\delta}$, $w_0^\pm(\mu,\nu) = \mathrm{Re}\,w_0(z)\mp\mathrm{Im}\,w_0(z)$, $w_0(z) = w_0(\mu,\nu)$, $\mu = x+y$, $\nu = x-y$ *are as stated in* (5.24), *Chapter* IV *and* $u_0(z)$ *is as stated in the first formula of* (3.14), *where* $w(z) = w_0(z)$, $M_{19} = M_{19}(p_0,\beta,k_0,D)$ *is a non-negative constant.*

Proof Let $u(z)$ be a solution of Problem Q' for equation (1.2), and $w(z) = u_z$, $u(z)$ be substituted in the positions of w,u in (3.16). Thus the functions $g(z)$, $f(z)$, $\tilde{\psi}(z),\tilde{\phi}(z)$ in $\overline{D^+}$ and $g_1(z)$, $g_2(z)$, $\Psi(z)$ in $\overline{D^-}$ in (3.15),(3.16) can be determined. Moreover, we can find the solution $\tilde{\Phi}(z)$ in D^+ and $\Phi(z)$ in $\overline{D^-}$ of (3.8) with the boundary condition (3.17), where

$$s(x)=\frac{2r((1-i)x/2)-[a((1-i)x/2)+b((1-i)x/2)]h_{2j}}{a((1-i)x/2)-b((1-i)x/2)},$$

$$x\in(a_{2j},a_{2j+1}), \quad j=0,1,\dots,[n/2],$$

$$s(x)=\frac{2r((1+i)x/2+1-i)-[a((1+i)x/2+1-i)-b((1+i)x/2+1-i)]k_{2j-1}}{a((1+i)x/2+1-i)+b((1+i)x/2+1-i)}, \qquad (3.19)$$

$$x\in(a_{2j-1},a_{2j}), \quad j=1,\dots,[(n+1)/2],$$

in which the real constants $h_j (j = 0, 1, \ldots, n)$ are of the form

$$h_{2j} = \text{Re}\,[\lambda(A_{2j+1})(r(A_{2j+1}) + ic_{2j+1}) + \text{Im}\,[\lambda(A_{2j+1})(r(A_{2j+1}) + ic_{2j+1})]$$

$$\text{on } \tilde{L}_{2j+1}, \quad j = 0, 1, \ldots, [n/2],$$

$$k_{2j-1} = \text{Re}\,[\lambda(B_{2j-1})(r(B_{2j-1}) + ic_{2j}) - \text{Im}\,[\lambda(B_{2j-1})(r(B_{2j-1}) + ic_{2j})]$$

$$\text{on } \tilde{L}_{2j}, \quad j = 1, \ldots, [(n+1)/2],$$

where $\tilde{L}_{2j+1} = \tilde{D}_{2j+1}^{-} \cap AB$, $j = 0, 1, \ldots, [n/2]$, $\tilde{L}_{2j} = \tilde{D}_{2j}^{-} \cap AB$, $j = 1, \ldots, [(n+1)/2]$. Thus

$$w(z) = w_0(z) + W(z) = \begin{cases} \tilde{\Phi}(z)^{\tilde{\phi}(z)} + \tilde{\psi}(z) \text{ in } D^+, \\ w_0(z) + \overline{\Phi(z)} + \overline{\Psi(z)} \text{ in } D^-, \end{cases}$$

is the solution of Problem A' for the complex equation

$$\left\{ \begin{array}{c} w_{\bar{z}} \\ \overline{w_{\bar{z}^*}} \end{array} \right\} = \text{Re}\,[A_1 w] + A_2 u + A_3 \text{ in } \left\{ \begin{array}{c} D^+ \\ D^- \end{array} \right\}, \qquad (3.20)$$

and $u(z)$ is a solution of Problem Q' as stated in (3.14).

3.3   Unique solvability for the discontinuous oblique derivative problem for (1.2)

Theorem 3.3   *Suppose that equation (1.2) satisfies Condition C. Then Problem Q' for (1.2) has a unique solution in D.*

Proof   The proof is similar to the proof of Theorems 2.3 and 2.4, Chapter V, but the boundary condition on $L_j (j = 1 \text{ or } 2)$ and the point condition in which are modified. For instance the boundary condition on $L_j (j = 1 \text{ or } 2)$ and the point condition in (1.4), Chapter V are replaced by that on $L_3 \cup L_4$ and $\text{Im}\,[\lambda(z)w(z)]|_{z=A_{2j+1}} = 0$, $j = 0, 1, \ldots, [n/2]$, $\text{Im}\,[\lambda(z)w(z)]|_{z=B_{2j-1}} = 0$, $j = 1, \ldots, [(n+1)/2]$ respectively; the integral in (2.18), Chapter V is replaced by

$$\Psi_1(z) = \begin{cases} \displaystyle\int_{a_{2j+1}}^{\nu} g_0(z)e_1 d\nu + \int_0^{\mu} g_0(z)e_2 d\mu \text{ in } D_{2j+1}^-, \quad j=0, 1, \ldots, [n/2], \\ \displaystyle\int_2^{\nu} g_0(z)e_1 d\nu + \int_{a_{2j-1}}^{\mu} g_0(z)e_2 d\mu \text{ in } D_{2j}^-, \quad j=1, \ldots, [(n+1)/2], \end{cases} \qquad (3.21)$$

$$g_0(z) = A\xi_0 + B\eta_0 + Cu_0 + D \text{ in } D^-,$$

and so on; and the characteristic lines through the points $z_1 = 1 - i$ are replaced by the characteristic lines through the points $A_{2j+1}(j = 0, 1, \ldots, [n/2])$, B_{2j-1} $(j = 1, \ldots, [(n+1)/2]$.

Moreover we can obtain the estimates of solutions of Problem Q' for (1.2).

Theorem 3.4 *Suppose that equation* (1.2) *satisfies Condition C. Then any solution* $u(z)$ *of Problem Q′ for* (1.2) *satisfies the estimates*

$$\tilde{C}^1[u(z), \overline{D^-}] = C_\beta[u(z), \overline{D^-}] + C[w^\pm(\mu, \nu)Y^\pm(\mu, \nu), \overline{D^-}] \leq M_{20},$$

$$\tilde{C}^1_\beta[u(z), \overline{D^+}] = C_\beta[u(z), \overline{D^+}] + C_\beta[u_z X(z), \overline{D^+}] \leq M_{21}(k_1 + k_2), \qquad (3.22)$$

$$\tilde{C}^1[u(z), \overline{D^-}] \leq M_{21}(k_1 + k_2) = M_{22},$$

where

$$X(z) = \prod_{j=0}^{m+n} |z - a_j|^{\gamma'_j + \delta}, \quad Y^\pm(z) = Y^\pm(\mu, \nu) = \prod_{j=0}^{n+1} |x \pm y - a_j|^{\gamma'_j + \delta}, \qquad (3.23)$$

$$w_0^\pm(\mu, \nu) = \operatorname{Re} w_0(z) \mp \operatorname{Im} w_0(z), \quad w_0(z) = w_0(\mu, \nu), \quad \mu = x + y, \quad \nu = x - y,$$

in which $\gamma'_j = \max(0, -\gamma_j)$ $(j = 1, \ldots, m-1, m+1, m+n)$, $\gamma'_0 = \max(0, -2\gamma_0)$, $\gamma'_m = \max(0, -2\gamma_m)$ *and* γ_j $(j = 0, 1, \ldots, m+n)$ *are real constants as stated before,* β $(0 < \beta < \delta)$, δ *are sufficiently small positive numbers, and* $k = (k_0, k_1, k_2)$, $M_{20} = M_{20}(p_0, \beta, k, D)$, $M_{21} = M_{21}(p_0, \beta, \delta, k_0, D)$ *are two non-negative constants.*

From the estimate (3.22), we can see the regularity of solutions of Problem $Q′$ for (1.2).

Next, we consider the oblique derivative problem(Problem $P′$) for the equation (1.2).

Theorem 3.5 *Suppose that the mixed equation* (1.2) *satisfies Condition C. Then Problem P′ for* (1.2) *has a solution in D.*

Proof First of all, we prove the uniqueness of solutions of Problem $P′$ for (1.2). Suppose that there exist two solutions of Problem $P′$ for (1.2). By Condition C, it can be seen that $u(z) = u_1(z) - u_2(z)$ and $w(z) = u_z$ satisfy the homogeneous equation and boundary conditions

$$\begin{Bmatrix} w_{\bar{z}} \\ \bar{w}_{\bar{z}^*} \end{Bmatrix} = f, \quad f = \operatorname{Re}[A_1 w] + A_2 u \text{ in } \begin{Bmatrix} D^+ \\ D^- \end{Bmatrix},$$

$$u(z_j) = 0, \quad j = 0, 1, \ldots, m, \quad u(a_j) = 0, \quad j = 1, \ldots, n,$$

$$\operatorname{Re}[\overline{\lambda(z)}w(z)] = 0, \quad z \in \Gamma, \quad \operatorname{Re}[\lambda(z)w(z)] = 0, \quad z \in L_3 \cup L_4, \qquad (3.24)$$

$$\operatorname{Im}[\lambda(z)w(z)]|_{z = A_{2j+1}} = 0, \quad j = 0, 1, \ldots, [n/2],$$

$$\operatorname{Im}[\lambda(z)w(z)]|_{z = B_{2j}} = 0, \quad j = 1, \ldots, [(n+1)/2].$$

By using the method of proof in Theorem 3.3, $w(z) = u_z = 0$, $u(z) = 0$ in $\overline{D^-}$ can be verified. Thus we have

$$2\operatorname{Re}[\lambda(x)u_z] = \frac{\partial u}{\partial l} = 0 \text{ on } L' = (0, 2),$$

it is clear that $\lambda(x) = \cos(l, x) - i\cos(l, y) = \exp(-i\pi/4)$ on L_1' and $\lambda(x) = \cos(l, x) - i\cos(l, y) = \exp(i\pi/4)$ on L_2'. On the basis of the maximum principle of solutions for the equation

$$u_{z\bar{z}} = \text{Re}\,[A_1 u_z] + A_2 u, \quad z \in D^+, \tag{3.25}$$

if $\max_{\overline{D^+}} u(z) > 0$, then its maximum attains at a point $z^* \in \Gamma \cup L'$, obviously $z^* \neq z_j \, (j = 0, 1, \ldots, m), a_j \, (j = 1, \ldots, n)$, and we can prove $z^* \notin \Gamma$ by the method as stated in the proof of Theorem 3.4, Chapter III. Moreover, it is not difficult to prove that if $z^* \in L'$, then $\partial u/\partial l \neq 0$ at z^*. Hence $\max_{\overline{D^+}} u(z) = 0$. By the similar method, we can prove $\min_{\overline{D^+}} u(z) = 0$. Hence $u(z) = 0$, $u_1(z) = u_2(z)$ in $\overline{D}$.

Secondly, we first prove the existence of solutions of Problem P' for equation (1.2) in the linear case. From Theorem 3.3, it can be seen that Problem Q' for (1.2) has a solution $u^*(z)$ in D, if $u^*(z_j) = b_j, j = 0, 1, \ldots, m$, $u^*(a_j) = b_{m+j}, j = 1, \ldots, n$, then the solution $u^*(z)$ is just a solution of Problem P' for (1.2). Otherwise, $[u^*(a_1'), \ldots, u^*(a_{m+n}')] = [d_1^*, \ldots, d_{n+m}^*]$, in which $a_j' = z_j, j = 1, \ldots, m, a_j' = a_{j-m}, j = m+1, \ldots, m+n$, we find $m+n$ solutions $u_1(z), \ldots, u_{m+n}(z)$ of Problem Q' for (3.25) with the boundary conditions

$$\text{Re}\,[\overline{\lambda(z)}u_{kz}] = 0, \quad z \in \Gamma, \quad \text{Re}\,[\overline{\lambda(z)}u_{kz}] = 0, \quad z \in L_3 \cup L_4,$$

$$u_k(2) = 0, \quad \text{Im}\,[\overline{\lambda(z)}u_{kz}]|_{z=z_j'} = \delta_{jk}, \quad j, k = 1, \ldots, m+n,$$

$$\text{Im}\,[\overline{\lambda(z)}u_{kz}]|_{z=A_{2j+1}} = 0, \quad j = 0, 1, \ldots, [n/2], \tag{3.26}$$

$$\text{Im}\,[\overline{\lambda(z)}u_{kz}]|_{z=B_{2j-1}} = 0, \quad j = 1, \ldots, [(n+1)/2].$$

It is obvious that $U(z) = \sum_{k=1}^{m+n} u_k(z) \not\equiv 0$ in $\overline{D}$, moreover we can verify that

$$J = \begin{vmatrix} u_1(a_1') & \cdots & u_{m+n}(a_1') \\ \vdots & \ddots & \vdots \\ u_1(a_{m+n}') & \cdots & u_{m+n}(a_{m+n}') \end{vmatrix} \neq 0,$$

thus there exist $m+n$ real constants $c_1', c_2', \cdots, c_{m+n}'$, which are not equal to zero, such that

$$c_1' u_1(a_k') + c_2' u_2(a_k') + \cdots + c_{m+n}' u_{m+n}(a_k') = d_k^* - b_k, \quad k = 1, \cdots, m+n,$$

thus the function

$$u(z) = u^*(z) - \sum_{k=1}^{m+n} c_k' u_k(z) \quad \text{in } \overline{D}$$

is just a solution of Problem P' for the linear equation (1.2) with $A_2 = 0$ in D. In addition we can derive that the solution $u(z)$ of Problem P' for (1.2) satisfies the estimates similar to (3.22). Afterwards we consider the equation with the parameter $t \in [0, 1]$:

$$Lu_z = \begin{Bmatrix} u_{z\bar{z}} \\ u_{\bar{z}z^*} \end{Bmatrix} = \text{Re}\,[A_1 u_z] + t[A_2 u + A_3] + A(z) \text{ in } \begin{Bmatrix} D^+ \\ D^- \end{Bmatrix}, \tag{3.27}$$

where $A(z)$ is any function in D satisfying the condition $C[A(z)X(z), \overline{D^+}] + C[A^\pm(\mu, \nu)Y^\pm(\mu, \nu), \overline{D^-}] < \infty$. By using the method of parameter extension, namely when $t = 0$, we see that Problem P' for such equation has a unique solution by the above discussion. Moreover assuming that when $t = t_0 \in (0, 1]$, Problem P' for equation (3.27) has a solution, then we can prove that there exists a small positive constant ε, such that for any $t \in \{|t - t_0| \le \varepsilon, t \in [0, 1]\}$, Problem P' for such equation (3.27) has a solution. Thus we can derive that there exists a solution $u(z)$ of Problem P' for equation (3.27) with $t = 1$, especially when $A(z) = 0$ in D, i.e. Problem P' for equation (1.2) has a solution $u(z)$. This completes the proof.

4 Oblique Derivative Problems for Quasilinear Equations of Mixed Type in Multiply Connected Domains

In this section we discuss the oblique derivative boundary value problems for quasilinear second order equations of mixed (elliptic-hyperbolic) type in multiply connected domains. We first give a representation of solutions for the above boundary value problem, and then prove the uniqueness and existence of solutions of the above problem and give a priori estimates of solutions of the above problem. In the book [9]2), the author proposed the Dirichlet boundary value problem (Tricomi problem) for second order equations of mixed type in multiply connected domains. In [12] 1),3) the author only discussed the Dirichlet problem (Problem T_2) for the Lavrent'ev-Bitsadze equation of mixed (elliptic-hyperbolic) type: $u_{xx} + \mathrm{sgn}y\, u_{yy} = 0$ in a special doubly connected domain. Up to now we have not seen that other authors have solved it in multiply connected domains. In this section, we try to discuss the oblique derivative problem for quasilinear equations of mixed type in multiply connected domains, which includes the Dirichlet problem (Problem T_2) as a special case.

4.1 Formulation of the oblique derivative problem for second order equations of mixed type

Let D be an N-connected bounded domain D in the complex plane $\mathbb{C}$ with the boundary $\partial D = \Gamma \cup L$, where $\Gamma = \sum_{j=1}^N \Gamma_j \in C_\alpha^2 (0 < \alpha < 1)$ in $\{y > 0\}$ with the end points $z = a_1 = 0, b_1, a_2, b_2, \ldots, a_N, b_N = 2$, and $L = \cup_{j=1}^{2N} L_j$, $L_1 = \{x = -y, 0 \le x \le 1\}$, $L_2 = \{x = -y + b_1, b_1 \le x \le b_1 + (a_2 - b_1)/2\}$, $L_3 = \{x = y + a_2, b_1 + (a_2 - b_1)/2 \le x \le a_2\}$, $L_4 = \{x = -y + b_2, b_2 \le x \le b_2 + (a_3 - b_2)/2\}, \ldots$, $L_{2N-1} = \{x = y + a_N, b_{N-1} + (a_N - b_{N-1})/2 \le x \le a_N\}$, $L_{2N} = \{x = y + 2, 1 \le x \le ?\}$, in which $a_1 = 0 < b_1 < a_2 < b_2 < \cdots < a_N < b_N = 2$, and denote $D^+ = D \cap \{y > 0\}$, $D^- = D \cap \{y < 0\}$, $D_1^- = D^- \cap \{x + y < b_1\}$, $D_2 = D \cap \{b_1 < x + y < a_2\}$, $D_3^- = D^- \cap \{a_2 < x + y < b_2\}, \ldots, D_{2N-2}^- = D^- \cap \{b_{N-1} < x + y < a_N\}$, $D_{2N-1}^- = D^- \cap \{a_N < x + y\}$, and $z_1 = 1 - i, z_2 = b_1 + (a_2 - b_1)(1 - i)/2, \ldots, z_N = b_{N-1} + (a_N - b_{N-1})(1 - i)/2$. We assume that the inner angles $\pi/\alpha_{2j-1}, \pi/\alpha_{2j}$ of D^+ at the points $z = a_j, b_j (j = 1, \ldots, N)$ are greater than zero and less than π.

We consider the quasilinear second order equation of mixed type: (1.1) and its complex form (1.2) with Condition C.

The oblique derivative boundary value problem for equation (1.2) may be formulated as follows:

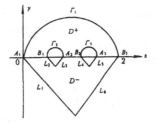

Problem P'' Find a continuous solution $u(z)$ of equation (1.2) in $\bar{D}$, which is continuously differentiable in $D_* = \bar{D}\backslash Z$ and satisfies the boundary conditions

Figure 4.1 $N = 3$

$$\frac{1}{2}\frac{\partial u}{\partial l} = \mathrm{Re}\,[\overline{\lambda(z)}u_z] = r(z), z \in \Gamma, \quad \frac{1}{2}\frac{\partial u}{\partial l} = \mathrm{Re}\,[\overline{\lambda(z)}u_{\bar z}] = r(z), \quad z \in L'. \quad (4.1)$$

$$\mathrm{Im}\,[\overline{\lambda(z)}u_{\bar z}]|_{z=z_j} = c_j, \quad j=1,\dots,N, \quad u(a_j)=d_j, \quad u(b_j)=d_{N+j}, \quad j=1,\dots,N, \quad (4.2)$$

where $Z = \{x \pm y = a_j, x \pm y = b_j, j = 1,\dots,N, y \le 0\}$, $L' = \cup_{j=1}^{N}L_{2j-1}$, l is a vector at every point on $\Gamma \cup L'$, $\lambda(z) = a(x) + ib(x) = \cos(l,x) \mp i\cos(l,y), z \in \Gamma \cup L'$, $\mp$ are determined by $z \in \Gamma$ and L' respectively, $c_j, d_j, d_{N+j}(j = 1,\dots,N)$ are real constants, and $\lambda(z), r(z), c_j, d_j, d_{N+j}(j = 1,\dots,N)$ satisfy the conditions

$$C_\alpha[\lambda(z),\Gamma]\le k_0, \quad C_\alpha[r(z),\Gamma]\le k_2, \quad C_\alpha[\lambda(z),L']\le k_0, \quad C_\alpha[r(z),L']\le k_2,$$

$$\cos(l,n) \ge 0 \text{ on } \Gamma, \quad |c_j|,|d_j|,|d_{N+j}| \le k_2, \quad j=1,\dots,N, \quad (4.3)$$

$$\max_{z\in L_1}\frac{1}{|a(x)-b(x)|}, \quad \max_{z\in L''}\frac{1}{|a(x)+b(x)|} \le k_0,$$

in which n is the outward normal vector on Γ, $L'' = \cup_{j=2}^{N}L_{2j-1}$, $\alpha\,(1/2 < \alpha < 1)$, k_0, k_2 are non-negative constants. Here we mention that if $A_2 = 0$ in D^+, then we can cancel the condition $\cos(l,n) \ge 0$ on Γ, and if the boundary condition $\mathrm{Re}\,[\overline{\lambda(z)}u_{\bar z}] = r(z), \ z \in L'$ is replaced by $\mathrm{Re}\,[\overline{\lambda(z)}u_z] = r(z), \ z \in L'$, then Problem P'' does not include the Dirichlet problem (Tricomi problem).

The boundary value problem for (1.2) with $A_3(z,u,u_z) = 0, \ z \in D, u \in \mathbb{R}, \ u_z \in \mathbb{C}$, $r(z) = 0, \ z \in \Gamma \cup L'$ and $c_j = 0\,(j = 0,1,\dots,N)$ and $d_j = 0\,(j = 1,\dots,2N)$ will be called Problem P_0''. The number

$$K = \frac{1}{2}(K_1 + K_2 + \dots + K_{2N}), \quad (4.4)$$

is called the index of Problem P'' and Problem P_0'' on the boundary ∂D^+ of D^+, where

$$K_j = \left[\frac{\phi_j}{\pi}\right] + J_j, \quad J_j=0 \text{ or } 1, \quad e^{j\phi_j} = \frac{\lambda(t_j-0)}{\lambda(t_j+0)}, \quad \gamma_j = \frac{\phi_j}{\pi} - K_j, \quad j=1,\dots,N, \quad (4.5)$$

in which $[a]$ is the largest integer not exceeding the real number a, and $t_1 = a_1 = 0$, $t_2 = b_1, t_3 = a_2, t_4 = b_2, \dots, t_{2N-1} = a_N, t_{2N} = b_N = 2$, and

$$\lambda(t) = e^{i\pi/4} \text{ on } l_j = (a_j,b_j), \quad \lambda(t_{2j-1}+0) = \lambda(t_{2j}-0) = e^{i\pi/4}, \quad j=1,\dots,N.$$

If $\cos(l, n) \not\equiv 0$ on each of Γ_j $(j = 1, \ldots, N)$, then we select the index $K = N - 1$ on ∂D^+. If $\cos(l, n) \equiv 0$ on Γ_j $(j = 1, \ldots, N)$, then we select the index $K = N/2 - 1$ on ∂D^+, and the last N point conditions in (4.2) can be eliminated. In this case, Problem P includes the Dirichlet problem (Tricomi problem) as a special case. Now we explain that in the closed domain $\overline{D^-}$, the derivative $u_x \pm u_y$ of the solution $u(z)$ in the neighborhoods of $4N$ characteristic lines $x \pm y = a_j$, $x \pm y = b_j (j = 1, \ldots, N)$ may not be bounded if $\gamma_j \alpha_j \leq 0 (j = 1, \ldots, 2N)$. Hence if we require that the solution $u(z)$ in $\overline{D^-} \backslash Z$ is bounded, where $Z = \{x + y = a_j, x + y = b_j, x - y = a_j, x - y = b_j, y \leq 0 (j = 1, \ldots, N)\}$, then it needs to choose $\gamma_j > 0 (j = 1, \ldots, 2N)$, herein γ_j $(j = 1, \ldots, 2N)$ are as stated in (4.5). If we require that solution $u(z)$ is only continuous in $\overline{D}$, it suffices to choose $-\gamma_j \alpha_j < 1 (j = 1, \ldots, 2N)$.

Moreover, we need to introduce another oblique derivative boundary value problem.

Problem Q'' If $A_2 = 0$ in D, one has to find a continuously differentiable solution $u(z)$ of (1.2) in D_*, which is continuous in $\overline{D}$ and satisfies the boundary conditions (4.1),(4.2), but the last N point conditions are replaced by

$$\text{Im} \, [\overline{\lambda(z)} u_{\overline{z}}]|_{z=z_j'} = d_j', \quad j = 1, \ldots, N, \tag{4.6}$$

where $z_j'(j = 1, \ldots, N - 1)$ are distinct points, such that $z_j' \in \Gamma$, $z_j' \notin Z (j = 1, \ldots, N - 1)$ and d_j' $(j = 1, \ldots, N)$ are real constants satisfying the conditions $|d_j'| \leq k_2$, $j = 1, \ldots, N$. In the case the condition $\cos(l, n) \geq 0$ on Γ can be canceled and we choose the index $K = N - 1$.

4.2 Representation and uniqueness of solutions for the oblique derivative problem for (1.2)

Now we give representation theorems of solutions for equation (1.2).

Theorem 4.1 *Suppose that equation (1.2) satisfies Condition C. Then any solution of Problem P'' for (1.2) can be expressed as*

$$u(z) = 2 \, \text{Re} \int_0^z w(z) dz + d_1, \; w(z) = w_0(z) + W(z), \tag{4.7}$$

where $w_0(z)$ is a solution of Problem A for the equation

$$\left\{ \begin{array}{c} w_{\overline{z}} \\ \overline{w_{\overline{z}^*}} \end{array} \right\} = 0 \; \text{in} \; \left\{ \begin{array}{c} D^+ \\ D^- \end{array} \right\}, \tag{4.8}$$

with the boundary conditions (4.1) and (4.2)($w_0(z) = u_{0z}$), and $W(z)$ possesses the form

$$W(z) = w(z) - w_0(z), \quad w(z) = \tilde{\Phi}(z)e^{\tilde{\phi}(z)} + \tilde{\psi}(z),$$

$$\tilde{\phi}(z) = \tilde{\phi}_0(z) + Tg = \tilde{\phi}_0(z) - \frac{1}{\pi} \int\int_{D^+} \frac{g(\zeta)}{\zeta - z} d\sigma_\zeta, \quad \tilde{\psi}(z) = Tf \text{ in } D^+,$$

$$W(z) = \overline{\Phi(z)} + \overline{\Psi(z)} \text{ in } \overline{D^-}, \tag{4.9}$$

$$\Psi(z) = \begin{cases} \displaystyle\int_0^\mu g_2(z)d\mu e_2 + \int_2^\nu g_1(z)d\nu e_1 \text{ in } D_{2j-1}^-, \quad j = 1, 2, \ldots, N, \\ \displaystyle\int_0^\mu g_2(z)d\mu e_2 + \int_{a_{j+1}}^\nu g_1(z)d\nu e_1 \text{ in } D_{2j}^-, \quad j = 1, 2, \ldots, N-1, \end{cases}$$

in which $\text{Im}\,[\tilde{\phi}(z)] = 0$ on $L_0 = \cup_{j=1}^N l_j$, $l_j = (a_j, b_j)$, $j = 1, \ldots, N$, $e_1 = (1+i)/2$, $e_2 = (1-i)/2$, $\mu = x+y$, $\nu = x-y$, $\tilde{\phi}_0(z)$ is an analytic function in D^+, and

$$g(z) = \begin{cases} A_1/2 + \overline{A_1}\bar{w}/(2w), & w(z) \neq 0, \\ 0, w(z) = 0, & z \in D^+, \end{cases} \quad f(z) = \text{Re}\,[A_1\tilde{\phi}_z] + A_2 u + A_3 \text{ in } D^+,$$

$$g_1(z) = g_2(z) = A\xi + B\eta + Cu + D, \quad \xi = \text{Re}\,w + \text{Im}\,w, \quad \eta = \text{Re}\,w - \text{Im}\,w,$$

$$A = \frac{\text{Re}\,A_1 + \text{Im}\,A_1}{2}, \quad B = \frac{\text{Re}\,A_1 - \text{Im}\,A_1}{2}, \quad C = A_2, \quad D = A_3 \text{ in } D^-,$$
$$\tag{4.10}$$

where $\tilde{\Phi}(z)$ is analytic in D^+ and $\Phi(z)$ is a solution of equation (4.8) in D^- satisfying the boundary conditions

$$\text{Re}\,[\overline{\lambda(z)}(\tilde{\Phi}(z)e^{\tilde{\phi}(z)} + \tilde{\psi}(z))] = r(z), \quad z \in \Gamma,$$

$$\text{Re}\,[\overline{\lambda(x)}(\tilde{\Phi}(x)e^{\tilde{\phi}(x)} + \tilde{\psi}(x))] = s(x), \quad x \in L_0,$$

$$\text{Re}\,[\overline{\lambda(x)}\Phi(x)] \doteq \text{Re}\,[\overline{\lambda(x)}(\overline{W(x)} - \Psi(x))], \quad z \in L_0, \tag{4.11}$$

$$\text{Re}\,[\overline{\lambda(z)}\Phi(z)] = -\text{Re}\,[\overline{\lambda(z)}\Psi(z)], \quad z \in L',$$

$$\text{Im}\,[\overline{\lambda(z_j)}\Phi(z_j)] = -\text{Im}\,[\overline{\lambda(z_j)}\Psi(z_j)], \quad j = 1, \ldots, N,$$

where $\lambda(x) = 1+i$ on L_0, and $s(x)$ is as stated in (4.14) below. Moreover, the solution $u_0(z)$ of Problem P'' for (4.8) in $\overline{D^-}$ satisfies the estimate in the form

$$C_\beta[u_0(z), \overline{D}] + C_\beta[X(z)w(z), \overline{D}] + C[w_0^\pm(\mu, \nu)Y^\pm(\mu, \nu), \overline{D^-}] \leq M_{23}(k_1 + k_2) \quad (4.12)$$

in which $X(z) = \Pi_{j=1}^{2N}[|z - t_j|^{|\gamma_j|\alpha_j + \delta}$, $Y^\pm(z) = Y^\pm(\mu, \nu) = \Pi_{j=1}^{2N}[|\mu - t_j||\nu - t_j|]^{|\gamma_j|\alpha_j + \delta}$, δ, β ($0 < \beta < \delta$), γ_j ($j = 1, \ldots, 2N$) are as stated in (4.5), $w_0^\pm(\mu, \nu) = \text{Re}\,w_0(z) \mp \text{Im}\,w_0(z)$, $w_0(z) = w_0(\mu, \nu)$, $\mu = x+y$, $\nu = x-y$, and

$$u_0(z) = 2\,\text{Re} \int_0^z w_0(z)dz + d_1 \tag{4.13}$$

and $M_{23} = M_{23}(p_0, \beta, k_0, D)$ is a non-negative constant.

Proof Let $u(z)$ be a solution of Problem P'' for equation (1.2), and $w(z) = u_z$, $u(z)$ be substituted in the positions of w, u in (4.10), thus the functions $g(z)$, $f(z)$, $\tilde{\psi}(z), \tilde{\phi}(z)$ in $\overline{D^+}$ and $g_1(z)$, $g_2(z)$, $\Psi(z)$ in $\overline{D^-}$ in (4.9),(4.10) can be determined. Moreover, we can find the solution $\tilde{\Phi}(z)$ in D^+ and $\Phi(z)$ in $\overline{D^-}$ of (4.8) with the boundary conditions (4.11), where

$$s(x) = \frac{2r((1-i)x/2) - 2\mathrm{Re}\,\overline{[\lambda((1-i)x/2)}\Psi((1-i)x/2)] - h(x)}{a((1-i)x/2) - b((1-i)x/2)} + \mathrm{Re}\,[\overline{\lambda(x)}\Psi(x)] \quad (4.14)$$

on L_0, in which

$$h(x) = \left[a\left(\frac{(1-i)x}{2}\right) + b\left(\frac{(1-i)x}{2}\right)\right]\,[\mathrm{Re}\,(\lambda(z_1)(r(z_1)+ic_1)) + \mathrm{Im}\,(\lambda(z_1)(r(z_1)+ic_1))],$$

and

$$\Psi(x) = \begin{cases} \displaystyle\int_2^\nu g_1(z)d\nu e_1 = \int_2^\nu g_1((1-i)\nu/2)d\nu e_1, \quad z = x+iy = (1-i)x \in L_1, \\[2mm] \displaystyle\int_0^\mu g_2(z)d\mu e_2 = \int_0^\mu g_2((1+i)\mu/2 + (1-i)a_j/2))d\mu e_2, \\[2mm] \quad z = x+iy = (1+i)x - a_j i \in L_{2j-1}, \quad j = 2,\ldots,N, \\[2mm] \displaystyle\int_0^\mu g_2(z)d\mu e_2 = \int_0^\mu g_2((1+i)\mu/2+1-i))d\mu e_2, \; z = (1+i)x - 2i \in L_{2N}. \end{cases}$$

Thus

$$w(z) = w_0(z) + W(z) = \begin{cases} \tilde{\Phi}(z)^{\tilde{\phi}(z)} + \tilde{\psi}(z) \text{ in } D^+, \\[2mm] w_0(z) + \overline{\Phi(z)} + \overline{\Psi(z)} \text{ in } D^-, \end{cases}$$

is the solution of Problem A for the complex equation

$$\begin{Bmatrix} w_{\bar{z}} \\ \bar{w}_{\bar{z}^*} \end{Bmatrix} = \mathrm{Re}\,[A_1 w] + A_2 u + A_3 \text{ in } \begin{Bmatrix} D^+ \\ D^- \end{Bmatrix}, \quad (4.15)$$

and $u(z)$ is a solution of Problem P for (1.2) as stated in the first formula in (4.7).

Theorem 4.2 *Suppose that equation (1.2) satisfies Condition C. Then Problem P'' for (1.2) has at most one solution in D.*

Proof Let $u_1(z), u_2(z)$ be any two solutions of Problem P'' for (1.2). It is clear that $u(z) = u_1(z) - u_2(z)$ and $w(z) = u_z$ satisfies the homogeneous equation and boundary conditions

$$\begin{Bmatrix} w_{\bar{z}} \\ \bar{w}_{\bar{z}^*} \end{Bmatrix} = \mathrm{Re}\,[A_1 w] + A_2 u \text{ in } \begin{Bmatrix} D^+ \\ D^- \end{Bmatrix},$$

$$\mathrm{Re}\,[\overline{\lambda(z)}w(z)] = 0, \quad z \in \Gamma, \quad \mathrm{Re}\,[\overline{\lambda(x)}w(x)] = s(x), \quad x \in L_0, \quad u(0) = 0,$$

$$\mathrm{Re}\,[\lambda(z)w(z)] = 0, \quad z \in L', \quad \mathrm{Im}\,[\lambda(z_j)w(z_j)] = 0, \quad j = 1,\ldots,N, \quad (4.16)$$

in which

$$s(x) = \frac{-2\text{Re}\left[\overline{\lambda((1-i)x/2)}\Psi((1-i)x/2)\right]}{a((1-i)x/2) - b((1-i)x/2)} + \text{Re}\left[\overline{\lambda(x)}\Psi(x)\right] \text{ on } L_0. \qquad (4.17)$$

From Theorem 4.1, the solution $w(z)$ can be expressed in the form

$$w(z) = \begin{cases} \tilde{\Phi}e^{\tilde{\phi}(z)} + \tilde{\psi}(z), \; \tilde{\psi}(z) = Tf, \tilde{\phi}(z) = \tilde{\phi}_0(z) + \tilde{T}g \text{ in } D^+, \\[2mm] \overline{\Phi(z)} + \overline{\Psi(z)} \text{ in } \overline{D^-}, \\[2mm] \Psi(z) = \begin{cases} \displaystyle\int_0^\mu [A\xi + B\eta + Cu]e d\mu e_2 + \int_2^\nu [A\xi + B\eta + Cu]d\nu e_1 \text{ in } D_{2j-1}^-, \\[2mm] \quad j = 1, 2, \dots, N, \\[2mm] \displaystyle\int_0^\mu [A\xi + B\eta + Cu]e d\mu e_2 + \int_{a_{j+1}}^\nu [A\xi + B\eta + Cu]d\nu e_1 \text{ in } D_{2j}^-, \\[2mm] \quad j = 1, 2, \dots, N-1, \end{cases} \end{cases} \qquad (4.18)$$

where $g(z)$ is as stated in (4.10), $\tilde{\Phi}(z)$ in D^+ is an analytic function and $\Phi(z)$ is a solution of (4.8) in $\overline{D^-}$ satisfying the boundary condition (4.11), in which $W(z) = w(z)$. If $A_2 = 0$ in D^+, then $\tilde{\psi}(z) = 0$, besides the functions $\tilde{\Phi}(z)$, $\Phi(z)$ satisfy the boundary conditions

$$\begin{cases} \text{Re}\left[\overline{\lambda(x)}\tilde{\Phi}(x)\right] = s(x), \\[2mm] \text{Re}\left[\overline{\lambda(x)}\Phi(x)\right] = \text{Re}\left[\overline{\lambda(x)}(\overline{W(x)} - \Psi(x))\right], \end{cases} \quad x \in L_0, \qquad (4.19)$$

where $s(x)$ is as stated before. Noting that

$$C[u(z), \overline{D^-}] \leq M_{24}\{C[X(z)w(z), \overline{D^+}] + C[w^\pm(z)Y^\pm(z), \overline{D^-}]\}$$

and applying the method of iteration, we can get

$$|w^\pm(z)Y^\pm(z)| \leq \frac{[2NM_{25}M((4 + M_{24})m + 1)R']^n}{n!} \text{ in } \overline{D^-},$$

where $M_{24} = M_{24}(D)$, $M_{25} = \max_{\overline{D^-}}[|A|, |B|, |C|]$, $M = 1 + 4k_0^2(1 + k_0^2)$, $m = C[w(z), \overline{D^-}]$, $R' = 2$. Let $n \to \infty$, we can derive $w^\pm(z) = 0$, i.e. $w(z) = w_1(z) - w_2(z) = 0, u(z) = 0, \Psi(z) = \Phi(z) = 0$ in $\overline{D^-}$, and $s(x) = 0$ on L_0. Besides, noting that the solution $u(z)$ of the equation

$$u_{z\bar{z}} = \text{Re}\left[A_1 u_z\right] + A_2 u \text{ in } D^+, \qquad (4.20)$$

$$\frac{1}{2}\frac{\partial u}{\partial l} = \text{Re}[\overline{\lambda(z)}u_z] = 0, \quad z \in \Gamma \cup L_0, \quad u(a_j) = 0, \quad u(b_j) = 0, \quad j = 1, \dots, N, \qquad (4.21)$$

and the index of the above boundary value problem is $K = N - 1$, on the basis of Theorem 3.7, Chapter III, we see that $u(z) = 0$ in D^+. This proves the uniqueness of solutions of Problem P'' for (1.2) in D. As for the general equation (1.2), we can prove the uniqueness of solutions of Problem P'' by the extremum principle for elliptic equations of second order by the method in the proof of Theorem 3.4, Chapter III.

4.3   The solvability for the oblique derivative problem for (1.2)

First of all, we prove the existence of solutions of Problem P'' for equation (3.7) in $\overline{D}$. It is obvious that Problem P'' for equation (3.7) is equivalent to the following boundary value problem (Problem A'') for (3.8) with the boundary conditions

$$\mathrm{Re}\,[\overline{\lambda(z)}w(z)] = r(z) \text{ on } \Gamma, \quad \mathrm{Re}\,[\lambda(z)w(z)] = r(z) \text{ on } L',$$

$$\mathrm{Im}\,[\lambda(z)w(z)]|_{z=z_j} = c_j, \quad j = 1,\ldots,N, \tag{4.22}$$

$$u(a_j) = d_j,\, u(b_j) = d_{N+j}, \quad j = 1,\ldots,N,$$

and the relation

$$u(z) = 2\,\mathrm{Re}\int_0^z w(z)dz + d_1. \tag{4.23}$$

Similarly to the method in the proof of Theorem 3.1, we can get the following theorem.

Theorem 4.3   *Problem P'' for (3.7) in D has a unique continuous solution $u(z)$.*

Proof   From the second and third boundary conditions in (4.22), we can obtain the following conditions:

$$\mathrm{Re}\,[\overline{\lambda(x)}w(x)] = k(x) \text{ on } L_0,$$

$$k(x) = \frac{2r((1-i)x/2) - [a((1-i)x/2)+b((1-i)x/2)]h}{a((1-i)x/2) - b((1-i)x/2)} \text{ on } L_0, \tag{4.24}$$

$$h = [\mathrm{Re}\,(\lambda(z_1)(r(z_1)+ic_1))+\mathrm{Im}\,(\lambda(z_1)(r(z_1)+ic_1))].$$

According to the method in the proof of Theorem 2.2, Chapter III, we can find a solution $w(z)$ of (3.8) in $\overline{D^+}$ with the first boundary condition in (4.22) and (4.24). Thus we can find the solution of Problem P'' for (3.7) in $\overline{D}$ as stated in (4.23), and the solution $w(z)$ of Problem A'' for (3.8) in $\overline{D_1^-}$ possesses the form

$$w(z) = \tilde{w}(z) + \overline{\lambda(z_1)}[r(z_1) - ic_1],$$

$$\tilde{w}(z) = \frac{1}{2}[(1-i)f(x+y) + (1+i)g(x-y)],$$

$$f(x+y) = \mathrm{Re}\,[(1+i)w(x+y)],$$

$$g(x-y) = \frac{2r((1-i)(x-y)/2)}{a((1-i)(x-y)/2) - b((1-i)(x-y)/2)} \text{ in } \overline{D_1^-}\backslash\{0,b_1\}.$$

Similarly we can write the solution of Problem P'' in $\overline{D_j^-}\,(j = 2, 3, \ldots, 2N - 1)$ as

$$w(z) = \tilde{w}(z) + \overline{\lambda(z_j)}[r(z_j) - ic_j],$$

$$\tilde{w}(z) = \frac{1}{2}[(1-i)f_j(x+y) + (1+i)g_j(x-y)] \text{ in } D_j^-, \quad j = 2,\ldots,N,$$

$$f_{2j}(x+y) = \frac{2r((1+i)(x+y)/2 + (1-i)a_{j+1}/2)}{a((1+i)(x+y) + (1-i)a_{j+1}/2) - b((1+i)(x+y)/2 + (1-i)a_{j+1}/2)},$$

$$g_{2j}(x-y) = \frac{2r((1-i)(x-y)/2)}{a((1-i)(x-y)/2) + b((1-i)(x-y)/2)} \text{ in } D_{2j}^-, \quad j = 1,2,\ldots,N-1,$$

$$f_{2j-1}(x+y) = \text{Re}[(1-i)w(x+y)],$$

$$g_{2j-1}(x-y) = \frac{2r((1-i)(x-y)/2)}{a((1-i)(x-y)/2) - b((1-i)(x-y)/2)} \text{ in } D_{2j-1}^-, \quad j = 2,\ldots,N, \tag{4.25}$$

in which $\overline{D_j^-}\,(j = 2, 3, \ldots, 2N - 1)$ are as stated before.

Theorem 4.4 *Suppose that the mixed equation (1.2) satisfies Condition C. Then Problem P'' for (1.2) in D has a solution.*

Proof It is clear that Problem P'' for (1.2) is equivalent to Problem A'' for the complex equation of first order and boundary conditions:

$$\left\{ \begin{matrix} w_{\bar{z}} \\ \bar{w}_{\bar{z}^*} \end{matrix} \right\} = \text{Re}\,[A_1 w] + A_2 u + A_3 \text{ in } \left\{ \begin{matrix} D^+ \\ D^- \end{matrix} \right\}, \tag{4.26}$$

$$\text{Re}\,[\overline{\lambda(z)}w(z)] = r(z), \quad z \in \Gamma, \quad \text{Re}\,[\lambda(z)w(z)] = r(z), \quad z \in L',$$

$$\text{Im}\,[\lambda(z_j)w(z_j)] = c_j, \quad u(a_j) = d_j, \quad u(b_j) = d_{N+j}, \quad j = 1,\ldots,N, \tag{4.27}$$

and the relation (4.23). In order to find a solution $w(z)$ of Problem A'' for (4.26) in D, we express $w(z)$ in the form (4.9)–(4.10) and use the successive iteration. First of all, denoting the solution $w_0(z)$ of Problem A'' for (4.26), and substituting $w_0(z)(= \xi_0 e_1 + \eta_0 e_2)$ and the corresponding function $u_0(z)$ into the positions of $w(z)$ $(= \xi e_1 + \eta e_2)$, $u(z)$ in the right hand side of (4.26),(4.9) and (4.10), thus the corresponding functions $g_0(z)$, $f_0(z)$ and the functions

$$W_1(z) = w_1(z) - w_0(z), \quad w_1(z) = \tilde{\Phi}_1(z)e^{\tilde{\phi}_1(z)} + \tilde{\psi}_1(z),$$

$$\tilde{\phi}_1(z) = \tilde{\phi}_0(z) - \frac{1}{\pi}\iint_{D^+}\frac{g_0(\zeta)}{\zeta - z}d\sigma_\zeta, \quad \tilde{\psi}_1(z) = Tf_0 \text{ in } D^+,$$

$$w_1(z) = w_0(z) + W_1(z), \quad \overline{W_1(z)} = \Phi_1(z) + \Psi_1(z), \tag{4.28}$$

$$g_0(z) = A\xi_0 + B\eta_0 + Cu_0 + D \text{ in } \overline{D^-},$$

$$\Psi_1(z) = \begin{cases} \displaystyle\int_0^\mu g_0(z)d\mu e_2 + \int_{2\nu}^\nu g_0(z)d\nu e_1 \text{ in } D_{2j-1}^-, \quad j = 1, 2, \ldots, N, \\ \displaystyle\int_0^\mu g_0(z)d\mu e_2 + \int_{a_{j+1}}^\nu g_0(z)d\nu e_1 \text{ in } D_{2j}^-, \quad j = 1, 2, \ldots, N - 1, \end{cases}$$

can be determined, where $\mu = x + y$, $\nu = x - y$, and the solution $w_0(z) = u_{0z}$, $u_0(z)$ satisfies the estimate (4.12). Moreover, we find a solution $\tilde{\Phi}_1(z), \Phi_1(z)$ of (4.8) satisfying the boundary conditions

$$\text{Re}\,[\overline{\lambda(z)}(\tilde{\Phi}_1(z)e^{\tilde{\phi}_1(z)} + \tilde{\psi}_1(z))] = r(z), \quad z \in \Gamma,$$

$$\text{Re}\,[\overline{\lambda(x)}(\tilde{\Phi}_1(x)e^{\tilde{\phi}_1(x)} + \tilde{\psi}_1(x))] = s_1(x), \quad x \in L_0,$$

$$\text{Re}\,[\overline{\lambda(x)}\Phi_1(x)] = \text{Re}\,[\overline{\lambda(x)}(W_1(x) - \Psi_1(x))], \quad z \in L_0, \qquad (4.29)$$

$$\text{Re}\,[\overline{\lambda(z)}\Phi_1(z)] = -\text{Re}\,[\overline{\lambda(z)}\Psi_1(z)], \quad z \in L',$$

$$\text{Im}\,[\overline{\lambda(z_j)}\Phi_1(z_j)] = -\text{Im}\,[\overline{\lambda(z_j)}\Psi_1(z_j)], \quad j = 1, \ldots, N,$$

where $\lambda(x) = 1 + i$ on L_0 and the function

$$s_1(x) = \frac{2r((1-i)x/2) - 2\text{Re}\,[\overline{\lambda((1-i)x/2)}\Psi_1((1-i)x/2)] - h(x)}{a((1-i)x/2) - b((1-i)x/2)}$$

$$+ \text{Re}\,[\overline{\lambda(x)}\Psi_1(x)],$$

$$h(x) = \left[a\left(\frac{(1-i)x}{2}\right) + b\left(\frac{(1-i)x}{2}\right)\right][\text{Re}\,(\lambda(z_1)(r(z_1) + ic_1))$$

$$+ \text{Im}\,(\lambda(z_1)(r(z_1) + ic_1))]\text{ on }L_0,$$

$$(4.30)$$

in which $w_1(z)$ satisfies the estimate

$$C_\beta[u_1(z), \overline{D}] + C_\beta[X(z)w_1(z), \overline{D^+}] + C[w_1^\pm(\mu, \nu)Y^\pm(\mu, \nu), \overline{D^-}] \le M_{26}, \qquad (4.31)$$

here $w_1^\pm(\mu, \nu) = \text{Re}\,w_1(\mu, \nu) \mp \text{Im}\,w_1(\mu, \nu)$, $X(z), Y^\pm(\mu, \nu)$ is as stated in (4.12), $M_{26} = M_{26}(p_0, \beta, k, D)$ is a non-negative constant. Thus we can obtain a sequence of functions: $\{w_n(z)\}$ and

$$w_n(z) = w_0(z) + W_n(z) = w_0(z) + \overline{\Phi_n(z)} + \overline{\Psi_n(z)} \text{ in } D^-,$$

$$\Psi_n(z) = \begin{cases} \displaystyle\int_0^\mu g_{n-1}(z)d\mu e_2 + \int_2^\nu g_{n-1}(z)d\nu e_1 \text{ in } D^-_{2j-1}, \quad j = 1, 2, \ldots, N, \\ \displaystyle\int_0^\mu g_{n-1}(z)d\mu e_2 + \int_{a_{j+1}}^\nu g_{n-1}(z)d\nu e_1 \text{ in } D^-_{2j}, \quad j = 1, 2, \ldots, N-1, \end{cases}$$

$$g_{n-1}(z) = B\xi_{n-1} + A\eta_{n-1} + Cu_{n-1} + D \text{ in } \overline{D^-},$$

$$(4.32)$$

and then

$$|[w_1^\pm(\mu, \nu) - w_0^\pm(\mu, \nu)]Y^\pm(\mu, \nu)|$$

$$\le |\Phi_1^\pm(\mu, \nu)Y^\pm(\mu, \nu)| + \sqrt{2}\{|Y^+(\mu, \nu)|\,|\int_0^\mu g_0(z)e_2 d\mu|$$

$$+ |Y^-(\mu, \nu)|[|\int_2^\nu g_0(z)e_1 d\nu| + \sum_{j=1}^{N-1}|\int_{a_{j+1}}^\mu g_0(z)e_1 d\nu|]$$

$$(4.33)$$

$$\le 2NM_{27}M((4 + M_{24})m + 1)R' \text{ in } \overline{D^-},$$

where $m = C[w_0^+(\mu,\nu)X^\pm(\mu,\nu), \overline{D^-}] + C[w_0^-(\mu,\nu)Y^\pm(\mu,\nu), \overline{D^-}]$, $M = 1 + 4k_0^2(1+k_0^2)$, $R' = 2$, $M_{27} = \max_{z \in \overline{D^-}}(|\tilde{A}|,|\tilde{B}|,|\tilde{C}|,|\tilde{D}|)$. It is clear that $w_n(z) - w_{n-1}(z)$ satisfies

$$\overline{w_n(z)} - \overline{w_{n-1}(z)} = \Phi_n(z) - \Phi_{n-1}(z) + \Psi_n(z) - \Psi_{n-1}(z) \text{ in } D^-,$$

$$\Psi_n(z) - \Psi_{n-1}(z) = \begin{cases} \displaystyle\int_0^\mu [g_n(z) - g_{n-1}(z)]d\mu e_2 + \int_2^\nu [g_n(z) - g_{n-1}(z)]d\nu e_1 \text{ in } D_{2j-1}^-, \\[2mm] j = 1,2,\ldots,N, \\[2mm] \displaystyle\int_0^\mu [g_n(z) - g_{n-1}(z)]d\mu e_2 + \int_{a_{j+1}}^\nu [g_n(z) - g_{n-1}(z)]d\nu e_1 \text{ in } D_{2j}^-, \\[2mm] j = 1,2,\ldots,N-1, \end{cases}$$

$$(4.34)$$

where $n = 1,2,\ldots$. From the above equality,

$$|[w_n^\pm(\mu,\nu) - w_{n-1}^\pm(\mu,\nu)]Y^\pm(\mu,\nu)|$$

$$\leq [2NM_{27}M((4+M_{24})m+1)]^n \int_0^{R'} \frac{R'^{n-1}}{(n-1)!}dR' \qquad (4.35)$$

$$\leq \frac{[2NM_{27}M((4+M_{24})m+1)R']^n}{n!} \text{ in } \overline{D^-}$$

can be obtained, and we can see that the sequences of functions $\{w_n^\pm(\mu,\nu)Y^\pm(\mu,\nu)\}$, i.e.

$$w_n^\pm(\mu,\nu)Y^\pm(\mu,\nu) = \{w_0^\pm + [w_1^\pm - w_0^\pm] + \cdots + [w_n^\pm - w_{n-1}^\pm]\}Y^\pm(\mu,\nu) \qquad (4.36)$$

$(n = 1,2,\ldots)$ in $\overline{D^-}$ uniformly converge to $w_*^\pm(\mu,\nu)X^\pm(\mu,\nu)$, and $w_*(z) = [w^+(\mu,\nu) + w^-(\mu,\nu) - i(w^+(\mu,\nu) - w^-(\mu,\nu))]/2$ satisfies the equality

$$\overline{w_*(z)} = \overline{w_0(z)} + \Phi_*(z) + \Psi_*(z) \text{ in } D^-,$$

$$\Psi_*(z) = \begin{cases} \displaystyle\int_0^\mu g_*(z)d\mu e_2 + \int_2^\nu g_*(z)d\nu e_1 \text{ in } D_{2j-1}^-, \quad j = 1,2,\ldots,N, \\[2mm] \displaystyle\int_0^\mu g_*(z)d\mu e_2 + \int_{a_{j+1}}^\nu g_*(z)d\nu e_1 \text{ in } D_{2j}^-, \quad j = 1,2,\ldots,N-1, \end{cases}$$

$$(4.37)$$

$$g_*(z) = B\xi_* + A\eta_* + Cu_* + D \text{ in } D^-,$$

and the corresponding function $u_*(z)$ is just a solution of Problem P'' for equation (1.2) in the domain D^- and $w_*(z)$ satisfies the estimate

$$C[w_*^\pm(\mu,\nu)Y^\pm(\mu,\nu), \overline{D^-}] \leq M_{28} = e^{2NM_{27}M((4+M_{24})m+1)R'}. \qquad (4.38)$$

Besides we see that the solution $w_*(z) = u_z$ of Problem A'' for (4.26) and the corresponding function $u_*(z)$ in $\bar{D}$ satisfy the estimate

$$C_\beta[u_*(z), \overline{D}] + C_\beta[w_*(z)X(z), \overline{D^+}] + C[w_*^\pm(\mu, \nu)Y^\pm(\mu, \nu), \overline{D^-}] \le M_{29}, \qquad (4.39)$$

where $M_{29} = M_{29}(p_0, \alpha, k, D)$ is a non-negative constant. Moreover the function $u(z)$ in (4.23) is a solution of Problem P'' for (1.2).

From the proof of Theorem 4.4, we can obtain the estimates of any solution $u(z)$ of Problem P'' and the corresponding function $w(z) = u_z$.

Theorem 4.5 *Suppose that equation (1.2) satisfies Condition C. Then any solution $u(z)$ of Problem P'' for (1.2) satisfies the estimates*

$$\tilde{C}^1[u(z), \overline{D}] = C_\beta[u(z), \overline{D}] + C_\beta[X(z)w(z), \overline{D^+}]$$

$$+ C[w^\pm(\mu, \nu)Y^\pm(\mu, \nu), \overline{D^-}] \le M_{30}, \quad \tilde{C}_\beta^1[u(z), \overline{D}] \le M_{31}(k_1 + k_2), \qquad (4.40)$$

where

$$X(z) = \prod_{j=1}^{2N} |z - t_j|^{|\gamma_j|\alpha_j + \delta}, \quad Y^\pm(z) = Y^\pm(\mu, \nu) = \prod_{j=1}^{2N} [|\mu - t_j||\nu - t_j|]^{|\gamma_j|\alpha_j + \delta}, \qquad (4.41)$$

herein $\gamma_j(j = 1, \ldots, 2N)$ are real constants in (4.5), $\beta(0 < \beta < \delta), \delta$ are sufficiently small positive constants, and $k = (k_0, k_1, k_2), M_{30} = M_{30}(p_0, \beta, k, D), M_{31} = M_{31}(p_0, \beta, \delta, k_0, D)$ are two non-negative constants.

Next, we consider the oblique derivative problem (Problem Q'') for the equation (1.2).

Theorem 4.6 *Suppose that the mixed equation (1.2) with $A_2 = 0$ satisfies Condition C. Then its Problem Q'' has a solution in D.*

Proof It is clear that Problem Q'' is equivalent to the following boundary value problem

$$\left\{ \begin{matrix} w_{\bar{z}} \\ \bar{w}_{\bar{z}^*} \end{matrix} \right\} = \text{Re}\,[A_1 w(z)] + A_3 \text{ in } \left\{ \begin{matrix} D^+ \\ D^- \end{matrix} \right\}, \qquad (4.42)$$

$$\text{Re}\,[\overline{\lambda(z)}w(z)] = r(z), \quad z \in \Gamma, \quad \text{Re}\,[\lambda(z)w(z)] = r(z), \quad z \in L',$$

$$\text{Im}\,[\overline{\lambda(z_j')}w(z_j')] = d_j', \quad j = 1, \ldots, N. \qquad (4.43)$$

Noting that $w(z)$ satisfies the second boundary condition in (4.11), namely

$$\text{Re}\,[\overline{\lambda(x)}w(z)] = s(x),$$

$$s(x) = \text{Re}\,[\overline{\lambda(x)}\Psi(x)]$$

$$+ \frac{2r((1-i)x/2) - 2\text{Re}\,[\overline{\lambda((1-i)x/2)}\Psi((1-i)x/2)] - h(x)}{a((1-i)x/2) - b((1-i)x/2)} \text{ on } L_0, \qquad (4.44)$$

and the index $K = N - 1$, similarly to the proof of Theorems 2.3 and 2.4, Chapter V, we can find a unique solution $w(z)$ of the boundary value problem (4.42)–(4.44), and the function $u(z)$ in (4.23) is just a solution of Problem Q'' in D.

Finally, we mention that the above result includes the Dirichlet problem (Tricomi problem) as a special case. In fact, if $\Gamma_1 = \{|z - 1| = 1\}$, $\Gamma_j = \{|z - \tilde{a}_j| = R_j\}$, $\tilde{a}_j = b_{j-1} + (a_j - b_{j-1})/2$, $R_j = (a_j - b_{j-1})/2$, $j = 2, \ldots, N$, $R_1 = 1$, the boundary condition of the Dirichlet problem is

$$u(z) = \phi(x) \text{ on } \Gamma \cup L', \tag{4.45}$$

which can be rewritten as

$$\mathrm{Re}\,[\overline{\lambda(z)}w(z)] = r(z) \text{ on } \Gamma, \quad \mathrm{Im}\,[\overline{\lambda(z_j)}w(z_j)] = c_j, \quad j = 1, \ldots, N,$$

$$\mathrm{Re}\,[\overline{\lambda(z)}w(z)] = r(z) \text{ on } L', \quad u(z) = 2\mathrm{Re}\int_0^z w(z)dz + d_0 \text{ in } D, \tag{4.46}$$

in which $d_0 = \phi(0)$,

$$\lambda(z) = a + ib = \begin{cases} \overline{i(z-1)}, & \theta = \arg(z-1) \text{ on } \Gamma_1, \\ \overline{i(z-\tilde{a}_j)}/R_j, & \theta = \arg(z-\tilde{a}_j) \text{ on } \Gamma_j, \quad j = 2, \ldots, N, \\ (1+i)/\sqrt{2} \text{ on } L_1, \\ (1-i)/\sqrt{2} \text{ on } L'' = \cup_{j=2}^N L_{2j-1}, \end{cases} \tag{4.47}$$

and

$$r(z) = \begin{cases} \phi_\theta/R_j \text{ on } \Gamma_j, \quad j = 1, \ldots, N, \\ \dfrac{\phi_x}{\sqrt{2}} \text{ on } L' = \cup_{j=1}^N L_{2j-1}, \end{cases}$$

$$c_1 = \mathrm{Im}\left[\frac{1+i}{\sqrt{2}}u_z(z_1 - 0)\right] = \frac{\phi_x - \phi_x}{\sqrt{2}}\Big|_{z=z_1-0} = 0,$$

$$c_j = \mathrm{Im}\left[\frac{1-i}{\sqrt{2}}u_z(z_j + 0)\right] = 0, \quad j = 2, \ldots, N, \tag{4.48}$$

in which $a = 1/\sqrt{2} \neq b = -1/\sqrt{2}$ on L_1 and $a = 1/\sqrt{2} \neq -b = -1/\sqrt{2}$ on L''.

If the index of Problem D on ∂D^+ is $K = N/2 - 1$, we can argue as follows: According to (4.46),(4.47), the boundary conditions of Problem D in D^+ possess the form

$$\mathrm{Re}\,[i(z - 1)w(z)] = r(z) = \phi_\theta, \quad z \in \Gamma,$$

$$\mathrm{Re}\left[\frac{1-i}{\sqrt{2}}w(x)\right] = s(x) = \frac{\phi'((1-i)x/2)}{\sqrt{2}},$$

$$x \in l_j = (a_j, b_j), \quad j = 1, \ldots, N.$$

It is clear that the possible points of discontinuity of $\lambda(z)$ on ∂D^+ are $t_1 = a_1 = 0$, $t_2 = b_1$, $t_3 = a_2$, $t_4 = b_2, \ldots, t_{2N-1} = a_N$, $t_{2N} = b_N = 2$, and

$$\lambda(a_1 - 0) = \lambda(b_j + 0) = e^{i\pi/2}, \quad j = 1, 2, \ldots, N - 1,$$

$$\lambda(b_N + 0) = \lambda(a_j - 0) = e^{3i\pi/2}, \quad j = 2, \ldots, N,$$

$$\lambda(a_j + 0) = \lambda(b_j - 0) = e^{i\pi/4}, \quad j = 1, \ldots, N,$$

$$\frac{\lambda(t_1 - 0)}{\lambda(t_1 + 0)} = e^{i\pi/4} = e^{i\phi_1}, \quad 0 < \gamma_1 = \frac{\phi_1}{\pi} - K_1 = \frac{1}{4} - 0 = \frac{1}{4} < 1,$$

$$\frac{\lambda(t_{2N} - 0)}{\lambda(t_{2N} + 0)} = e^{-5i\pi/4} = e^{i\phi_{2N}}, \quad -1 < \gamma_{2N} = \frac{\phi_{2N}}{\pi} - K_{2N} = -\frac{5}{4} - (-1) = -\frac{1}{4} < 0,$$

$$\frac{\lambda(t_j - 0)}{\lambda(t_j + 0)} = e^{5i\pi/4} = e^{i\phi_j}, \quad 0 < \gamma_j = \frac{\phi_j}{\pi} - K_j = \frac{5}{4} - 1 = \frac{1}{4} < 1, \quad j = 3, 5, \ldots, 2N-1,$$

$$\frac{\lambda(t_j - 0)}{\lambda(t_j + 0)} = e^{-i\pi/4} = e^{i\phi_j}, \quad -1 < \gamma_j = \frac{\phi_j}{\pi} - K_j = -\frac{1}{4} = -\frac{1}{4} < 0, \quad j = 2, 4, \ldots, 2N-2.$$

Thus $K_1 = K_2 = K_4 = \cdots = K_{2N-2} - 0$, $K_3 = K_5 - \cdots = K_{2N\ 1} = 1$, $K_{2N} = -1$, in the case, where the index of Problem D on ∂D^+ is

$$K = \frac{1}{2}(K_1 + K_2 + \cdots + K_{2N}) = \frac{N}{2} - 1.$$

Hence Problem D for (1.2) has a unique continuous solution $u(z)$ in $\overline{D}$. If we require that the derivative $w(z) = u_z$ of the solution $u(z)$ is bounded in $\overline{D} \backslash Z$, it suffices to replace $K_2 = K_4 = \cdots = K_{2N-2} = -1$, $K_{2N} = -2$ and then the index $K = -1$. In this case the problem has one solvability condition.

The oblique derivative problems for the Chaplygin equation of second order

$$K(y)u_{xx} + u_{yy} = 0, \quad K(0) = 0, \quad K'(y) > 0 \text{ in } D,$$

where D is a multiply connected domain, was proposed by L. Bers in [9]2), but the problem was still not solved. For more general second order quasilinear degenerate equations of mixed type and second order mixed equations in higher dimensional domains, many boundary value problems need to be investigated and solved.

The references for this chapter are [4],[6],[12],[14],[27],[31],[37],[39],[45],[56],[61], [64],[71],[72],[73],[85],[88],[90],[98].

References

[1] R. Adams. *Sobolev spaces.* Academic Press, New York, 1975.

[2] S. Agmon, L. Nirenberg and M. H. Protter. *A maximum principle for a class of hyperbolic equations and applications to equations of mixed elliptic–hyperbolic type.* Comm. Pure Appl. Math. 6 (1953), 455–470.

[3] H. Begehr. 1) *Boundary value problems for mixed kind systems of first order partial differential equations.* 3. Roumanian-Finnish Seminar on Complex Analysis, Bucharest 1976, Lecture Notes in Math. 743, Springer-Verlag, Berlin, 1979, 600–614.
2) *Complex analytic methods for partial differential equations.* World Scientific, Singapore, 1994.

[4] H. Begehr and A. Dzhureav. *An introduction to several complex variables and partial differential equations.* Addison Wesley Longman, Harlow, 1997.

[5] H. Begehr and R. P. Gilbert. 1) *Pseudohyperanalytic functions.* Complex Variables, Theory Appl. 9 (1988), 343–357.
2) *Transformations, transmutations, and kernel functions* I, II, Longman, Harlow, 1992, 1993.

[6] H. Begehr and A. Jeffrey. 1) *Partial differential equations with complex analysis.* Longman, Harlow, 1992.
2) *Partial differential equations with real analysis.* Longman, Harlow, 1992.

[7] H. Begehr and Wei Lin. *A mixed-contact boundary problem in orthotropic elasticity.* Partial Differential Equations with Real Analysis, Longman, Harlow, 1992, 219–239.

[8] H. Begehr and Guo-chun Wen. 1) *The discontinuous oblique derivative problem for nonlinear elliptic systems of first order.* Rev. Roumaine Math. Pures Appl. 33 (1988), 7–19.
2) *A priori estimates for the discontinuous oblique derivative problem for elliptic systems.* Math. Nachr. 142(1989), 307–336.
3) *Nonlinear elliptic boundary value problems and their applications.* Addison Wesley Longman, Harlow, 1996.

[9] L. Bers. 1) *Theory of pseudoanalytic functions.* New York University, New York, 1953.
2) *Mathematical aspects of subsonic and transonic gas dynamics.* Wiley, New York, 1958.

[10] L. Bers, F. John and M. Schechter. *Partial differential equations.* Interscience Publ., New York, 1964.

[11] L. Bers and L. Nirenberg. 1) *On a representation theorem for linear elliptic systems with discontinuous coefficients and its application.* Conv. Intern. Eq. Lin. Derivate Partiali, Trieste, Cremonense, Roma, 1954, 111–140.
2) *On linear and nonlinear elliptic boundary value problems in the plane.* Conv. Intern. Eq. Lin. Derivate Partiali, Trieste, Cremonense, Roma, 1954, 141–167.

[12] A. V. Bitsadze. 1) *Differential equations of Mixed type.* Mac Millan Co., New York, 1964.
2) *Boundary value problems for elliptic equations of second order.* Nauka, Moscow, 1966 (Russian); Engl. Transl. North Holland Publ. Co., Amsterdam, 1968.
3) *Some classes of partial differential equations.* Gordon and Breach, New York, 1988.
4) *Partial Differential Equations.* World Scientific, Singapore, 1994.

[13] A. V. Bitsadze, and A. N. Nakhushev. *Theory of degenerating hyperbolic equations.* Dokl. Akad. Nauk SSSR 204 (1972), 1289–1291 (Russian).

[14] B. Bojarski. 1) *Generalized solutions of a system of differential equations of first order and elliptic type with discontinuous coefficients.* Mat. Sb. N. S. 43(85) (1957), 451–563 (Russian).
2) *Subsonic flow of compressible fluid.* Arch. Mech. Stos. 18 (1966), 497–520; Mathematical Problems in Fluid Mechanics, Polon. Acad. Sci., Warsaw, 1967, 9–32.

[15] B. Bojarski and T. Iwaniec. *Quasiconformal mappings and non-linear elliptic equations in two variables* I, II. Bull. Acad. Polon. Sci., Sér. Sci. Math. Astr. Phys. 22 (1974), 473–478; 479–484.

[16] F. Brackx, R. Delanghe and F. Sommen. *Clifford analysis.* Pitman, London, 1982.

[17] S. A. Chaplygin. *Gas jets, Complete Works.* Moscow-Leningrad, Vol. 2, 1933.

[18] D. Colton. 1) *Partial differential equations in the complex domain.* Pitman, London, 1976.
2) *Analytic theory of partial differential equations.* Pitman, London, 1980.

[19] R. Courant and H. Hilbert. *Methods of mathematical physics* II. Interscience Publ., New York, 1962.

[20] I. I. Daniljuk. 1) *Nonregular boundary value problems in the plane.* Izdat. Nauka, Moscow, 1975 (Russian).
2) *Selected works.* Naukova Dumka, Kiev, 1995 (Russian etc.).

[21] Ju. V. Devingtal'. *Existence and uniqueness of the solution of the Frankl problem.* Uspehi Mat. Nauk 14 (1959), no.1 (85), 177–182 (Russian).

[22] A. Dzhuraev. 1) *Methods of singular integral equations.* Nauk SSSR, Moscow, 1987 (Russian); Engl. transl. Longman, Harlow, 1992.
2) *Degenerate and other problems.* Longman, Harlow, 1992.

[23] Zheng-zhong Ding. *The general boundary value problem for a class of semilinear second order degenerate elliptic equations.* Acta Math. Sinica 27 (1984), 177–191 (Chinese).

[24] Guang-chang Dong. 1) *Boundary value problems for degenerate elliptic partial differential equations.* Sci. Sinica 13 (1964), 697–708.
2) *Nonlinear second order partial differential equations.* Amer. Math. Soc., Providence, RI, 1991.

[25] Guang-chang Dong and Min-you Chi. *Influence of Tricomi's Mathematical work in China.* Mixed Type Equations, BSB Teubner, Leipzig, 90, 1986, 105–111.

[26] A. Douglis and L. Nirenberg. *Interior estimates for elliptic systems of partial differential equations.* Comm. Pure Appl. Math. 8 (1953), 503–538.

[27] Ai-nong Fang. *Quasiconformal mappings and the theory of functions for systems of nonlinear elliptic partial differential equations of first order.* Acta Math. Sinica 23 (1980), 280–292 (Chinese).

[28] F. I. Frankl. 1) *On the problems of Chaplygin for mixed subsonic and supersonic flows.* Izv. Akad. Nuak SSSR Ser. Mat. 9 (1945), 121–143.
2) *Two gas-dynamical applications of the Lavrent'ev-Bitsadze boundary value problem.* Vest. Leningrad Univ. Ser. Mat. Meh. Astronom. 6 (1951), 3–7 (Russian).
3) *Gas flows past profiles with a local supersonic zone ending in a direct shock wave.* Prikl. Mat. Meh. 20 (1956), 196–202 (Russian).
4) *Selected works on gas dynamics.* Nauka, Moscow, 1973 (Russian).

[29] A. Friedman. *Variational principles and free boundary problems.* Wiley, New York, 1982.

[30] F. D. Gakhov. *Boundary value problems.* Fizmatgiz, Moscow, 1963 (Russian); Pergamon, Oxford, 1966.

[31] D. Gilbarg and N. S. Trudinger. *Elliptic partial differential equations of second order.* Springer-Verlag, Berlin, 1977.

[32] R. P. Gilbert. 1) *Function theoretic methods in partial differential equations.* Academic Press, New York, 1969.
2) *Plane ellipticity and related problems.* Amer. Math. Soc., Providence, RI, 1981.

[33] R. P. Gilbert and J. L. Buchanan. *First order elliptic systems: A function theoretic approach.* Academic Press, New York, 1983.

[34] R. P. Gilbert and G. N. Hile. 1) *Generalized hypercomplex function theory.* Trans. Amer. Math. Soc. 195 (1974), 1–29.
2) *Degenerate elliptic systems whose coefficients matrix has a group inverse.* Complex Variables, Theory Appl. 1 (1982), 61–88.

[35] R. P. Gilbert and Wei Lin. 1) *Algorithms for generalized Cauchy kernels.* Complex Variables, Theory Appl. 2 (1983), 103–124.
2) *Function theoretic solutions to problems of orthotropic elasticity.* J. Elasticity 15 (1985), 143–154.

[36] R. P. Gilbert and Guo-chun Wen. 1) *Free boundary problems occurring in planar fluid dynamics.* Nonlinear Analysis 13 (1989), 285–303.
2) *Two free boundary problems occurring in planar filtrations.* Nonlinear Analysis 21 (1993), 859–868.

[37] Chao-hao Gu and Jia-xing Hong. *Some developments of the theory of partial differential equations of mixed type.* Mixed Type Equations, BSB Teubner, Leipzig, 90, 1986, 120–135.

[38] W. Haack and W. Wendland. *Lectures on Pfaffian and partial differential equations.* Pergamon Press, Oxford, 1972.

[39] G. N. Hile and M. H. Protter. 1) *Maximum principles for a class of first order elliptical systems.* J. Diff. Eq. 24(1) (1977), 136–151.
2) *Properties of overdetermined first order elliptic systems.* Arch. Rat. Mech. Anal. 66 (1977), 267–293.

[40] Zong-yi Hou. *Dirichlet problem for a class of linear elliptic second order equations with parabolic degeneracy on the boundary of the domain.* Sci. Record (N. S.) 2 (1958), 244–249 (Chinese).

[41] L. Hörmander. *Linear partial differential operators.* Springer-Verlag, Berlin, 1964.

[42] G. C. Hsiao and W. Wendland. *A finite element method for some integral equations of the first kind.* J. Math. Anal. Appl. 58 (1977), 449–481.

[43] Loo-Keng Hua. 1) *On Lavrent'ev's partial differential equation of the mixed type.* Sci. Sinica 13 (1964), 1755–1762 (Chinese).
2) *A talk starting from the unit circle.* Science Press, Beijing, 1977 (Chinese).

[44] Loo-keng Hua, Wei Lin and Tzu-chien Wu. *Second order systems of partial differential equations in the plane.* Pitman, London, 1985.

[45] T. Iwaniec. *Quasiconformal mapping problem for general nonlinear systems of partial differential equations.* Symposia Math. 18, Acad. Press, London, 1976, 501–517.

[46] T. Iwaniec and A. Mamourian. *On the first order nonlinear differential systems with degeneration of ellipticity.* Proc. Second Finnish-Polish Summer School in Complex Analysis, (Jyväskylä 1983), Univ. Jyväskylä, 28 (1984), 41–52.

[47] Xin-hua Ji and De-quan Chen. 1) *The equations of mixed type of elliptic and hyperbolic in the n-dimensional real projective space.* Acta Math. Sinica 23 (1980), 908–921.
2) *The non-homogeneous equations of mixed type in the real projective planer.* Mixed Type Equations, BSB Teubner, Leipzig, 90, 1986, 280–300.

[48] M. V. Keldych. *On certain cases of degeneration of equations of elliptic type on the boundary of a domain.* Dokl. Akad. Nauk SSSR(N.S.) 77 (1951), 181–183 (Russian).

[49] A. G. Kuz'min. *Equations of mixed type with non-classical behavior of characteristics.* Mixed Type Equations, BSB Teubner, Leipzig, 90, 1986, 180–194.

[50] O. A. Ladyshenskaja and N. N. Uraltseva. *Linear and quasilinear elliptic equations.* Academic Press, New York, 1968.

[51] E. Lanckau and W. Tutschke. *Complex analysis: Methods, trends, and applications.* Akademie-Verlag, Berlin, 1983.

[52] M. A. Lavrent'ev and A. V. Bitsadze. *The problem of equations of mixed type.* Dokl. AN SSSR 1950, Vol. 70, 373–376.

[53] M. A. Lavrent'ev and B. V. Shabat. *Methods of function theory of a complex variable.* GITTL Moscow, 1958 (Russian).

[54] J. Leray. *Hyperbolic differential equations.* Princeton Univ. Press, 1954.

[55] J. Leray and J. Schauder. *Topologie et équations fonczionelles.* Ann. Sci. École Norm. Sup. 51 (1934), 45–78; YMH 1 (1946), 71–95 (Russian).

[56] Ming-de Li and Yu-chun Qin. *On boundary value problems for singular elliptic equations.* Hangzhoudaxue Xuebao (Ziran Kexue), 1980., no.2, 1–8.

[57] Jian-bing Lin. *On some problems of Frankl.* Vestnik Leningrad Univ. 16 (1961), no.13, 28–34 (Russian).

[58] Chien-ke Lu. 1) *Boundary value problems for analytic functions.* World Scientific, Singapore, 1993.
2) *Complex variable methods in plane elasticity.* World Scientific, Singapore, 1995.

[59] L. G. Mikhailov. *A new class of singular integral equations and its applications to differential equations with singular coefficients.* Wolters-Noordhoff, Groningen, 1970.

[60] C. Miranda. *Partial differential equations of elliptic type.* Springer-Verlag, Berlin, 1970.

[61] V. N. Monakhov. 1) *Transformations of multiply connected domains by the solutions of nonlinear L-elliptic systems of equations.* Dokl. Akad. Nauk SSSR 220 (1975), 520–523 (Russian).
2) *Boundary value problems with free boundaries for elliptic systems.* Amer. Math. Soc., Providence, RI, 1983.

[62] N. I. Mushelishvili. 1) *Singular integral equations.* Noordhoff, Groningen, 1953.
2) *Some basic problems of the mathematical theory of elasticity.* Nauka, Moscow, 1946 (Russian); Noordhoff, Groningen, 1953.

[63] J. Naas and W. Tutschke. *Some probabilistic aspects in partial complex differential equations.* Complex Analysis and its Applications. Akad. Nauk SSSR, Izd. Nauka, Moscow, 1978, 409–412.

[64] L. Nirenberg. 1) *On nonlinear elliptic partial differential equations and Hölder continuity.* Comm. Pure Appl. Math. 6 (1953), 103–156.
2) *An application of generalized degree to a class of nonlinear problems.* Coll. Analyse Fonct. Liége, 1970, Vander, Louvain, 1971, 57–74.

[65] O. A. Oleinik. *On equations of elliptic type degenerating on the boundary of a region.* Dokl. Akad. Nauk SSSR (N.S.) 87 (1952), 885–888.

[66] M. H. Protter. 1) *The Cauchy problem for a hyperbolic second order equation.* Can. J. Math. 6 (1954), 542–553.
2) *An existence theorem for the generalized Tricomi problem.* Duke Math. J. 21 (1954), 1–7.

[67] M. H. Protter and H. F. Weinberger. *Maximum principles in differential equations.* Prentice-Hall, Englewood Cliffs, N. J., 1967.

[68] Deqian Pu. *Function-theoretic process of hyperbolic equations.* Integral Equations and Boundary Value Problems, World Scientific, Singapore, 1991,161–169.

[69] S. P. Pul'kin. *The Tricomi problem for the generalized Lavrent'ev-Bitsadze equation.* Dokl. Akad. Nuak SSSR 118 (1958), 38–41 (Russian).

[70] J. M. Rassias. 1) *Mixed type equations.* BSB Teubner, Leipzig, 90, 1986.
2) *Levture notes on mixed type partial differential equations.* World Scientific, Singapore, 1990.

[71] M. S. Salakhitdinov and B. Islomov. *The Tricomi problem for the general linear equation of mixed type with a nonsmooth line of degeneracy.* Soviet Math. Dokl 34 (1987), 133–136.

[72] M. Schneider. *The existence of generalized solutions for a class of quasilinear equations of mixed type.* J. Math. Anal. Appl. 107(1985), 425–445.

[73] M. M. Smirnov. *Equations of mixed type.* Amer. Math. Soc., Providence, RI, 1978.

[74] S. L. Sobolev. *Applications of functional analysis in mathematical physics.* Amer. Math. Soc., Providence, RI, 1963.

[75] He-sheng Sun. 1) *The problems of the rigidity of the surfaces with mixed Gauss curvature and boundary value problems for the equations of mixed type.* Proc. 1980 Beijing Sym. Diff. Geom. Diff. Eq., Beijing, 1982, 1441–1450.
2) *Tricomi problem for nonlinear equation of mixed type.* Sci. in China (Series A), 35 (1992), 14–20.

[76] Mao-ying Tian. *The general boundary value problem for nonlinear degenerate elliptic equations of second order in the plane.* Integral Equations and Boundary Value Problems, World Scientific, Singapore, 1991, 197–203.

[77] F. Tricomi. 1) *Sulle equazioni lineari alle derivate parziali di 2° ordine.* di tipo misto, atti Accad. Naz. Lincei Mem. Cl. Sci. Fis. Nat. (5) 14 (1923), 133–247 (1924).
2) *Repertorium der Theorie der Differentialgleichungen.* Springer–Verlag, Berlin, 1968.

[78] N. S. Trudinger. *Nonlinear oblique boundary value problems for nonlinear elliptic equations.* Trans. Amer. Math. Soc. 295 (1986), 509–546.

[79] W. Tutschke. 1) *The Riemann-Hilbert problem for nonlinear systems of differential equations in the plane.* Complex Analysis and its Applications. Akad. Nauk SSSR, Izd. Nauka, Moscow, 1978, 537–542.
2) *Boundary value problems for generalized analytic functions for several complex variables.* Ann. Pol. Math. 39 (1981), 227–238.

[80] I. N. Vekua. 1) *Generalized analytic functions.* Pergamon, Oxford, 1962.
2) *New methods for solving elliptic equations.* North-Holland Publ., Amsterdam, 1967.

[81] N. D. Vvedenskaya. *On a boundary problem for equations of elliptic type degenerating on the boundary of a region.* Dokl. Akad. Nauk SSSR(N.S.) 91 (1953). 711–714 (Russian).

[82] Chung-fang Wang. *Dirichlet problems for singular elliptic equations.* Hangzhoudaxue Xuebao (Ziran Kexue), 1978, no.2, 19–32 (Chinese).

[83] W. Warschawski. *On differentiability at the boundary in conformal mapping.* Proc. Amer. Math. Soc. 12 (1961), 614–620.

[84] E. Wegert. *Nonlinear boundary value problems for homeomorphic functions and singular integral equations.* Akademie-Verlag, Berlin, 1992.

[85] Guo-chun Wen. 1) *Modified Dirichlet problem and quasiconformal mappings for nonlinear elliptic systems of first order.* Kexue Tongbao (A monthly J. Sci.) 25 (1980), 449–453.
2) *The Riemann-Hilbert boundary value problem for nonlinear elliptic systems of first order in the plane.* Acta. Math. Sinica 23 (1980), 244–255 (Chinese).
3) *The singular case of the Riemann-Hilbert boundary value problem.* Beijing-daxue Xuebao (Acta Sci. Natur. Univ. Peki.), 1981, no.4, 1–14 (Chinese).
4) *The mixed boundary value problem for nonlinear elliptic equations of second order in the plane.* Proc. 1980 Beijing Sym. Diff. Geom. Diff. Eq., Beijing, 1982, 1543–1557.
5) *On a representation theorem of solutions and mixed boundary value problems for second order nonlinear elliptic equations with unbounded measurable coefficients.* Acta Math. Sinica 26 (1983), 533–537 (Chinese).
6) *Oblique derivative boundary value problems for nonlinear elliptic systems of second order.* Scientia Sinica, Ser. A, 26 (1983), 113–124.
7) *Some nonlinear boundary value problems for nonlinear elliptic equations of second order in the plane.* Complex Variables, Theory Appl. 4 (1985), 189–210.
8) *Applications of complex analysis to nonlinear elliptic systems of partial differential equations.* Analytic Functions of one Complex Variable, Amer. Math. Soc., Providence, RI, 1985, 217–234.
9) *Linear and nonlinear elliptic complex equations.* Shanghai Science Techn. Publ., Shanghai, 1986 (Chinese).
10) *Some boundary value problems for nonlinear degenerate elliptic complex equations.* Lectures on Complex Analysis, World Scientific, Singapore, 1988, 265–281.
11) *Conformal mappings and boundary value problems.* Amer. Math. Soc., Providence, RI, 1992.
12) *Function theoretic methods for partial differential equations and their applications and computations.* Advan. Math. 22 (1993), 391–402.
13) *Nonlinear irregular oblique derivative problems for fully nonlinear elliptic equations.* Acta Math. Sci. 15 (1995), 82–90.
14) *A free boundary problem in axisymmetric filtration with two fluids.* Acta Math. Appl. Sinica 1998, 139–143.
15) *Oblique derivative problems for linear second order equations of mixed type.* Science in China 41 (1998), 346–356.
16) *Approximate methods and numerical analysis for elliptic complex equations.* Gordon and Breach Science Publishers, Amsterdam, 1999.
17) *Linear and nonlinear parabolic complex equations.* World Scientific, Singapore, 1999.

18) *Nonlinear partial differential complex equations.* Science Press, Beijing, 1999 (Chinese).

[86] Guo-chun Wen and H. Begehr. 1) *Boundary value problems for elliptic equations and systems.* Longman, Harlow, 1990.
2) *Existence of solutions of Frankl problem for general Lavrent' ev-Bitsadze equations.* Revue Roumaine Math. Pure Appl. 45(2000), 141–160.

[87] Guo-chun Wen and Shi-xiang Kang. 1) *The Dirichlet boundary value problem for ultra-hyperbolic systems of first order.* J. Sichuan Normal Univ. (Natur. Sci.), 1992, no.1, 32–43 (Chinese).
2) *The Riemann-Hilbert boundary value problem of linear hyperbolic complex equations of first order.* J. Sichuan Normal Univ. (Natur. Sci.), 21 (1998),609–614.

[88] Guo-chun Wen and Ping-qian Li. *The weak solution of Riemann-Hilbert problems for elliptic complex equations of first order.* Appl. Anal. 45 (1992), 209–227.

[89] Guo-chun Wen and Zhao-fu Luo. *Hyperbolic complex functions and hyperbolic pseudoregular functions.* J. Ningxia Univ. (Natur. Sci.) 19 (1998), no.1, 12–18.

[90] Guo-chun Wen, Chung-wei Tai and Mao-ying Tain. *Function Theoretic methods of free boundary problems and their applications to mechanics.* Higher Education Press, Beijing, 1995 (Chinese).

[91] Guo-chun Wen and Mao-ying Tain. 1) *Solutions for elliptic equations of second order in the whole plane.* J. Math. 2 (1982), no.1, 23–36 (Chinese).
2) *Oblique derivative problems for quasilinear equations of mixed type in general domains* I. Progress in Natural Science 4 (1999), no.1, 85–95.
3) *Oblique derivative problems for nonlinear equations of mixed type in general domains.* Comm. in Nonlinear Sci. & Numer. Simu. 34 (1998), 148–151.

[92] Guo-chun Wen and Wen-sui Wu. *The complex form of hyperbolic systems of first order equations in the plane.* J. Sichuan Normal University (Natur. Sci.), 1994, no.2, 92–94 (Chinese).

[93] Guo-chun Wen and C. C. Yang. 1)*Some discontinuous boundary value problems for nonlinear elliptic systems of first order in the plane.* Complex Variables, Theory Appl. 25 (1994), 217–226.
2) *On general boundary value problems for nonlinear elliptic equations of second order in a multiply connected domain.* Acta Applicandae Mathematicae, 43 (1996), 169–189.

[94] Guo-chun Wen, Guang-wu Yang and Sha Huang. *Generalized analytic functions and their generalizations.* Hebei Education Press, Hebei, 1989 (Chinese).

[95] Guo-chun Wen and Zhen Zhao. *Integral equations and boundary value problems.* World Scientific, Singapore, 1991.

[96] W. Wendland. 1) *On the imbedding method for semilinear first order elliptic systems and related finite element methods.* Continuation methods, Academic Press, New York, 1978, 277-336.
2) *Elliptic systems in the plane.* Pitman, London, 1979.

[97] Xi-jin Xiang. *Pan-complex functions and its applications in mathematics and physics.* Northest Normal University Press, Changchun, 1988 (Chinese).

[98] C. C. Yang, G. C. Wen, K. Y. Li and Y. M. Chiang. *Complex analysis and its applications.* Longman, Harlow, 1993.

[99] Guo-chun Wen and Wei-ping Yin. *Applications of functions of one complex variable.* Capital Normal University Press, 2000 (Chinese).

Index

Printed in the United States
by Baker & Taylor Publisher Services